Microbial Ecosystems in Central Andes Extreme Environments

María Eugenia Farías
Editor

Microbial Ecosystems in Central Andes Extreme Environments

Biofilms, Microbial Mats, Microbialites and Endoevaporites

Springer

Editor
María Eugenia Farías
LIMLA-PROIMI-CONICET
Tucuman, Argentina

ISBN 978-3-030-36191-4 ISBN 978-3-030-36192-1 (eBook)
https://doi.org/10.1007/978-3-030-36192-1

This Springer imprint is published by the registered company Springer Nature Switzerland AG
The registered company address is: Gewerbestrasse 11, 6330 Cham, Switzerland

Preface

The Central Andean Altiplane, named Altiplane, (Bolivia and Peru), Atacama (Chile) or Puna (Argentine), represents a unique extreme environment since its locations are in high-altitude, closed basins that modulate salars and saline wetlands surrounded by deserts, largely influenced by volcanic activity. UV radiation, arsenic content, high salinity, alkalinity, and low dissolved oxygen content, together with extreme daily temperature fluctuations and oligotrophic conditions, shape an environment that recreates the early Earth and, even more, extraterrestrial conditions. The discovery of living microbialites and microbial mats in 2009 has increased the interest in this area as an early Earth counterpart.

Extreme microbial ecosystems thrive in different central Andean environments, such as salars, wetlands, lakes, volcanoes, hydrothermal vents, geysers, and desert soils. In this book we present the state of knowledge of these amazing ecosystems as a basis of their integral protection.

The book is divided into chapters grouped into parts per environments and their main microbial ecosystems.

The first part (Part I) includes a geographic and geological description of the area and the general considerations of microbial systems in Andean lakes. From there, the book is divided into environments and the most relevant microbiological ecosystems studied in each one. Part II starts with the lithobiontic microbial colonization found in the Atacama Desert, and its microbiology, biosignatures, and Vegas associated microbiology. Part III discusses the microbial life found at the highest Andean active volcanoes Socompa and Lluillaillaco (above 6000 m a.s.l.) and the Metagenomics approaches performed in Gaylussyte attached biofilms at Diamante, an extreme alkaline arsenic content lake placed in the Caldera of the Galán Supervolcano (4650 m a.s.l.). Part IV involves microbial ecosystems found in Salar and Salars wetlands. It includes studies of the rare biosphere at Salar de Huasco (3800 m a.s.l.), followed by Salar de Atacama where endoevaporite and microbial mat systems formed mainly of Archaea at Laguna Tebenquiche (2000 m a.s.l.). Updated knowledge of the endoevaporitic and mat systems and the metagenome analyses of Llamara (800 m a.s.l.) at Salar de Puquios is also included. The part will finish with a general overview of the halophile diversity of the main Argentinean and Chilean Andean Salars.

Part V contains updated knowledge of the amazing modern microbialites at Altiplano lakes, including new findings about Laguna Negra's Oncolites (4500 m a.s.l.) and Socompa and La Brava's microbialites.

Part VI includes an integral prospection of the Andean microbial ecosystem project implemented during the past 10 years in Argentine, Chile, and Bolivia. In addition, strategies that these microbial ecosystems developed to cope with extreme conditions, such as arsenic resistance and resistant genes dispersion by mobile elements, are discussed.

Tucumán, Argentina María Eugenia Farías

Contents

About the Editor

María Eugenia Farías Ph.D. (Argentina), Biological Sciences (1992), Ph.D. in Microbiology (1996), specialized in environmental microbiology. CONICET senior researcher CCT-PROIMI (Pilot Plant for Microbiological Industrial Processes). Head of the Laboratory of Microbiological Investigations of Lagunas Andinas (LIMLA). Her research is based on the study of the microbial ecosystems of Andean lagoons, prospecting studies, biodiversity, metagenomics as a basis for conservation and biotechnological application of strategic national ecosystems. Currently, she is doing a survey of the extremophile ecosystems associated with wetlands in Argentina, Chile and Bolivia. She is author of >100 manuscripts, and her h-index is 21.

About the Contributors

Virginia Helena Albarracín is a Biologist and Ph.D. in Biochemistry. Currently, she works as Head of Electron Microscopy Research Center and Core Facility (CIME), UNT-CONICET, Group Leader Lab of Molecular Microscopy and Ultrastructural Microbiology in CIME, UNT-CONICET and Independent Researcher, LIMLA-PROIMI-CONICET, Argentina, 2018. She also teaches as a Full Professor, Molecular Biology Department at the Health Institute, University of San Pablo-Tucumán and as an Assistant Professor, Molecular Biology Department at the Natural Sciences College, National University of Tucumán.

Ricardo N. Alonso Doctor in Geological Sciences. Professor at the University of Salta and Researcher (Conicet-Cega-Insugeo). Spent the past 40 years working on borates deposits, salars, lithium brines, evaporites and related subjects.

Octavio Artieda has a Ph.D. in Geology from the University of Zaragoza in 2004. Currently, he is a Lecturer and Researcher in the Department of Vegetal Biology, Ecology, and Earth Science at the University of Extremadura (Spain).

His main research topics are petrography, evaporitic minerals, soil micromorphology, soil genesis, geomorphology studies, and surficial processes in arid environments, and he is an expert in different study techniques: microscopy, X-ray, X-ray fluorescence, and geochemistry.

His current investigation lines are focused on arid soils and geomicrobiology of lithobiontic microorganisms in arid and hyperarid environments (currently in northern Chile), specifically characterization of lithic substrates in extremely arid environments and identification of biosignatures and microbial fossils. For the development of these works, he has participated in four international scientific expeditions to the Atacama Desert and Andean Mountains.

He is co-author of over 21 publications included in the Science Citation Index, with 300 total citations, and his h-index is 8.

Carmen Ascaso is a Research Professor at the National Museum of Natural Sciences, Spanish National Research Council (CSIC) Madrid and a head of the ECOGEO group. Her research interests have been focused on lichen symbiosis, and particularly, thallus-substrate studies. She has worked with the lichens samples exposed to outer space conditions at the Biopan—EXPOSE-E Program (ESA). In 1996 she was invited by NASA to study possible traces of life in the Martian meteorite ALH84001. Her team discovered the existence of life within the rocks of the Atacama Desert. Currently, C. Ascaso's research lines are focused on microbial ecology and geomicrobiology of lithobiontic microorganisms in hyperarid environments, lichenology, and biodeterioration of historical monuments. She is also an expert in electron microscopy, microanalysis, and micropaleontology. Some scientific achievements reached as results of her research projects are: (a) characterization of desertification processes in relation to climate change in the context of Earth history and its microscopic life, (b) developing a revolutionary research technique (SEM-BSE) for in situ visualization of lithobiontic microbiota, and (c) to set a new criterion of biogenecity for microbial fossils from Antarctic rocks. She is author of more than 150 JCR articles, and her h-index is 35.

Carolina Belfiore was born in Tucumán, Argentina, on July 11, 1977. She obtained a Degree in Biotechnology in 2004 and a Ph.D. in Biological Sciences in 2010. She is at present an Adjunt Researcher of the National Research Council (CONICET) in Argentina, working at the Laboratory of Microbial Research on Andean Lakes (LIMLA) of the Pilot Plant of Microbial Industrial Processes (PROIMI).

She is involved in an interdisciplinary team in the area of microbiology of salars and lagoons of Andean puna. She worked in biodiversity, microorganism metal resistance with a proteomics approach. Currently, she is in charge of the study of extreme microorganisms as potential plant growth promoting bacteria (biological inoculants). She published many papers in important peer-reviewed journals and is currently supervising two Ph.D. theses.

Flavia Jaquelina Boidi Ph.D. Candidate, Biological Sciences, B.S., Biology (2012), Universidad Nacional de Córdoba, Argentina. Fulbright-Bunge & Born scholar (2016), CONICET internal graduate scholar (2017). Her training and interests are oriented towards environmental microbiology. Currently, she is concluding her doctoral thesis from research performed at PROIMI-CONICET (Planta Piloto de Procesos Industriales Microbiológicos) and CICTERRA-CONICET-UNC (Centro de Investigaciones en Ciencias de la Tierra).

María Cristina Casero Ph.D., obtained her B.Sc. in Biology in 2013 and a M.Sc. in Environmental Microbiology in 2014 at the Universidad Autónoma de Madrid, Spain. She received her Ph.D. in Microbiology in 2019 by the Universidad Autónoma de Madrid, performed at the Biogeochemistry and Microbial Ecology Department at the National Museum of Natural Sciences, Spanish National Research Council (CSIC), Madrid, Spain. Her research is focused on the study of geomicrobiology of the Atacama Desert, with special interest in cyanobacteria, due to their peculiar adaptations to stressful environmental conditions. She uses a combination of different molecular tools together with microscopy and analytical chemistry in order to unmask their function within the microbial community as well as the biotechnological potential of the secondary metabolites produced by these group of organisms.

Marcela Cornejo She is the director of the Marine Biogeochemistry Laboratory. She obtained her doctorate in Oceanography at the Universidad de Concepción, Chile. She studies the nitrogen and carbon cycle in marine and freshwater ecosystems and the environmental drivers of the production and consumption of greenhouse gases in natural environments. She is part of the IMO (Instituto Milenio de Oceanografía, since 2013), the Oceanographic Time Series Station off Valparaiso (STOV, since 2015) and MOSE (Microbial Observatory of Salar Experiments, since 2014).

Escuela de Ciencias del Mar, Facultad de Ciencias del Mar y Geografía of the Pontificia Universidad Católica de Valparaíso, Valparaíso, Chile

John L. Darcy is a postdoctoral fellow in Computational Biology at the University of Colorado Anschutz Medical Campus. In his research, he uses mathematical and simulation models to understand microbial community assembly patterns. He received his PhD from the University of Colorado, Boulder, where his research focused on spatiotemporal patterns of microbial communities in the cryosphere. His postdoctoral work at the University of Hawaii, Mānoa, extended that work to foliar endophytic fungi within native Hawaiian plants. John's current research focuses on the human microbiome, but he is actively involved in projects spanning multiple systems.

Cristina Dorador She is the director of the Laboratory of Microbial Complexity and Functional Ecology at the Universidad de Antofagasta. She undertook her doctoral studies at the Max Planck Institute for Limnology in Ploen, Germany, and obtained her degree at the University of Kiel, Germany (2007). In the last 15 years she has focused her studies on understanding the microbiology of extreme environments, especially those located at the Chilean Altiplano and Atacama Desert. Her research includes the study of biogeochemical cycles, microbial diversity, mechanisms of adaptation to extreme conditions and different applications. She has led more than 40 research projects, and she is the author of more than 50 scientific articles.

Facultad de Ciencias del Mar y Recursos Biológicos, Departamento de Biotecnología, Instituto Antofagasta of the Universidad de Antofagasta, Antofagasta, Chile

Carlos Díaz-Gil specialist in biostatistics with a multidisciplinary education. Carlos obtained his Ph.D. in Marine ecology at the University of the Balearic Islands. He has collaborated with several different research fields varying from microbiology to marine ecology, aquaculture and individual behaviour plus the interactions among them and with the environmental effects.

Julián Rafael Dib is currently Adjunct Professor at the National University of Tucumán (UNT) and researcher at PROIMI-CONICET. He received his Bachelor in Biochemistry (2002) and Ph.D. (2009) at the UNT. After completing his Ph.D., Dr. Dib was granted a Humboldt Fellowship to work at the University of Münster, Germany. In 2013, Dr. Dib accepted an Adjunct Professor position at the UNT. His research interests are related to the biology of bacterial linear plasmids and the biological control in fruits by using killer yeasts. Much of his work has been on the biology of linear plasmids from rare actinobacteria. Dr. Dib has received numerous international awards and scholarships, including from the Fulbright Commission, DAAD, ERASMUS, and the Alexander von Humboldt Foundation. He was visiting professor at Technische Universität München, Germany (2015); Universidad de la República, Uruguay (2015); Universiti Teknologi Malaysia, Malaysia (2016-2019), and University of Innsbruck, Austria (2017).

Jocelyne DiRuggiero is a faculty member in the Departments of Biology and of Earth & Planetary Sciences at Johns Hopkins University. She obtained her Ph.D. at the University Claude Bernard Lyon I before leaving to the US on a Fulbright scholarship. She did her postdoctoral work at the University of Maryland first with Dr. Rita Colwell and then with Dr. Frank Robb. She was on faculty in the Department of Cell and Molecular Biology of the University of Maryland before moving to Johns Hopkins University. Her research focuses on the adaptive mechanisms of extremophiles to environmental stresses and in particular on the roles of small noncoding RNAs. She also investigates how those processes generate and maintain microbial diversity and affect ecosystem functioning at the dry limit for life. Her field sites include multiple deserts around the world and the Dry Valleys of Antarctica. She is a member of the Johns Hopkins University Center for Astrophysical Sciences and the director of the Institute for Planets and Life at Johns Hopkins University and the Space Telescope Science Institute.

Yoanna Eissler is a research director of the Laboratory of Biochemistry and Microbial Ecology housed in the Institute of Chemistry and Biochemistry at the University of Valparaíso. She obtained her Ph.D. in Oceanography at the University of Concepción, Chile, and has focused her research on viruses since then. She has studied how viruses could affect respiration rates of the microplanktonic community and algal populations. In addition, she isolated and characterized the infection cycle of a novel diatom virus and has investigated a virus that infect salmonids, affecting the aquaculture in Chile. Currently her research is focused on viral ecology, including topics such as new host–virus interactions in ecosystems that present extreme conditions for the development of life, such as Salar de Huasco and the oxygen minimum zone in the Humboldt current system in Chile.

Camila Fernández is a researcher at the laboratory of microbial ecology (CNRS) in France. She is also a visiting professor at the University of Concepcion. She coordinates the first French-Chilean International Associated Laboratory on Adaptive Strategies (LIA MAST) and coordinated for 10 years another collaborative structure, the International Laboratory of Marine Biogeochemistry and

Functional Ecology (LIA MORFUN). She obtained her Ph.D. in Marine Sciences at the University of the Mediterranean and has focused her research on marine biogeochemical cycles and microbial diversity since 2003. In 2017 she obtained the HDR degree at Sorbonne Université with her research on the re-evaluation of two key processes of the nitrogen marine cycle, nitrification and atmospheric nitrogen fixation. Extreme environments are natural laboratories for studying both processes, and Dr. Fernandez has done so in thermal hot springs, marine oxygen minimum zones and high-altitude environments such as Salar de Huasco. Recently she has been working on the impact and interactions between anthropogenic contaminants and microbial diversity and function. She has shown that pesticides used in aquaculture in Chile can alter carbon fluxes in microorganisms while also interfering with metabolic pathways in metazoans.

Ana Beatriz Fernández is a researcher specialized in microbial ecology and bioinformatics with experience analyzing microbial communities and bacterial genomes from high-throughput sequencing data. She obtained her Ph.D. in Molecular Biology and Biomedicine at the University of Sevilla in Spain where she deciphered the prokaryotic community present in saltern ponds from Santa Pola (Spain) isolating uncultured species in a salinity gradient using the metagenomic approach. Her postdoctoral experience was achieved through PROIMI-CONICET in Argentina, by studying microbial taxonomic diversity by pyrosequencing, genetic diversity by metagenomic and differences in physicochemical parameters that may influence changes on microbial diversity, in order to establish some biomarkers for the preservation of sedimentary ecosystems in areas of mining interest in the country. Currently, Dr. Fernández is developing her research career at the Public University of Navarra in Spain focused on the exploiting of novel genes for the production of new sustainable bioinsecticides from bacterial genomes.

Fátima Silvina Galván was born in Tucumán, Argentina, in 1989. She has a Bachelor in Biotechnology (2016) from the National University of Tucumán (UNT). Currently, she works as a Ph.D. student in the UNT as a Doctoral fellow of the National Research Council of Argentina (CONICET), working at the Ultrastructural and Molecular Microbiology Lab of the Electron Microscopy Research Center and Core Facility (CISME-CONICET-UNT).

Eli M.S. Gendron graduated with a BS in Biology from the University of New Mexico, USA, in 2012, and subsequently graduated with a PhD in Cellular, Molecular, and Developmental Biology from the University of Colorado, Boulder, USA, in 2019. Eli's doctoral research focused on identifying factors controlling the diversity of lotic microbial communities in alpine lakes along the Front Range of the Colorado Rockies using high-throughput sequencing techniques. He is currently working as a postdoctoral researcher in the Department of Nematology and Entomology at the University of Florida, USA. Eli's postdoctoral research focuses on utilizing mitochondrial metagenomics to improve the identification of individual nematode species from environmental samples in an effort to better inform crop pest management strategies.

Emmanuelle Gérard Ph.D. (France), Ph.D. in Microbiology (1999), Research Engineer at IPGP, specialized in environmental microbiology. Her research focuses on microorganism–mineral interactions in different contexts: geological sequestration of CO_2, formation of modern microbialites, and microbial consortia associated with hydrothermal vents. She is developing new methods for imaging and analyzing these interactions, such as confocal laser scanning microscopy coupled with Raman spectroscopy and laser microdissection coupled with single cell genomic. She is author of 29 manuscripts.

Fernando Javier Gomez CICTERRA-CONICET-UNC (Centro de Investigaciones en Ciencias de la Tierra, Universidad Nacional de Cordoba). Staff researcher at the CICTERRA. His research includes Carbonate Sedimentology and Basin Analysis, Carbonates Biogeochemistry and Geomicrobiology in ancient and modern environments, including hypersaline lakes and hot springs. He is also a Stratigraphy professor at the National University of Cordoba (Cordoba, Argentina) and currently a Visiting Professor in Sedimentology, Stratigraphy and Biogeochemistry at Washington University, Saint Louis (MO, USA).

Martha Hengst is a researcher at the Laboratory of Molecular Ecology and Applied Microbiology at Universidad Católica del Norte in Chile. She obtained her Ph.D. in Molecular Genetic and Microbiology at Pontificia Universidad Católica de Chile. She works in microbial ecology with a focus on microbial community processes and interactions in marine and extreme environments.

Her research is addressed to understand biodiversity patterns and molecular mechanisms evolved by microorganisms under biotic and abiotic environmental stresses.

For the last 10 years her group had worked in microbial communities inhabiting extreme environments of high-altitude ecosystems, from Chilean Altiplano at Atacama Desert, with a special emphasis on adaptive mechanisms and evolution processes to cope with environmental poly-stressor factors in evaporitic salt basins. Recently the scope of the study included microbial communities inhabiting in geothermal ecosystems dominated by deepest lineages of the tree of life, and thus the reason why they are proposed as a natural laboratory for understanding life at early Earth.

Daniel Kurth Ph.D. (Argentina). Master in Biotechnology (2002—Universidad Nacional de Rosario). Ph.D. in Biological Sciences (2010) obtained at the Instituto de Biología Celular y Molecular de Rosario (IBR)—Universidad Nacional de Rosario. Ph.D. subject was the genetic, molecular and structural characterization of acyl-CoA carboxylase complexes from *Mycobacterium tuberculosis*. His current position is as a CONICET researcher at CCT-PROIMI (Pilot Plant for Microbiological Industrial Processes). His research focuses on next generation sequencing data analysis, applied to microbial genomes and environmental microbiology. Research lines include plasmid prediction in microbial genomic and metagenomic data and bioprospection of enzymes of biological interest.

Manuel Contreras Leiva is an advisor and expert in ecological evaluation of wetlands and in the restoration of aquatic ecosystems, with an M.Sc. and Ph.D. in Biological Sciences and has vast experience in evaluating the ecological status of complex and chained ecosystems. From 1993 to date, Manuel has directed and developed more than 500 environmental studies related to the evaluation of continental and marine aquatic systems, monitoring of aquatic systems and conservation and management.

Agustina I. Lencina is a geologist from Catamarca, Argentina. She dedicated her years of youth to mining and environmental work and research. Although, she has always been enthusiastic about the natural beauty of the Puna. At present, she is passionately working on her Ph.D. degree focusing on modern microbialites as regards this altiplanic region in Catarmarca. Accompanied by an amazing research team, she is learning about the beginning of life and discovering new microbialite deposits in the most recondite places of the Puna.

Victoria Meslier received her Ph.D. in Microbiology in 2013 at INRA Jouy-en-Josas, France, on the interplay between complex dairy starters and industrial cheese processes. She then joined Dr. Jocelyne DiRuggiero's lab at the Johns Hopkins University, USA, to explore endolithic microbial communities in hyperarid deserts. She focused particularly on the factors driving the assembly of endolithic microorganisms at the dry limit of life. Since 2018, she joined MetaGenoPolis at INRA to elucidate the effect of diet on the gut microbiome composition and its impact on human health.

Estela Cecilia Mlewski Ph.D. (Argentina), Biological Sciences (2005), Ph.D. in Biology (2011). CONICET assistant researcher from CICTERRA (Centro de Investigaciones en Ciencias de la Tierra). Her research focuses on the study of microbial life in modern microbialites and extreme environments. Currently, she is studying the microbial and mineral interaction related to the carbonatation process, including the removal of metal within the carbonate matrix.

Verónica Molina She is the director of the Laboratorio Observatorio de Ecología Microbiana and currently coordinates the Doctorado Interdisciplinario en Ciencias Ambientales. She obtained her master's and Ph.D. in Oceanography at the Universidad de Concepción, Chile. Her main research studies since 2002 are related with the role of microbial communities to biogeochemical cycles mainly associated with nitrogen recycling in contrasting aquatic ecosystems including marine, oxygen minimum zones, coastal upwelling areas, fjord and estuarine and continental, including high-altitude wetlands from extreme environments. One of her main interests is focused on the response of microbial communities to environmental changes including temporal and spatial and its interaction with other organisms using molecular and biogeochemical methods. She is part of the following networking initiatives: French-Chilean International Associated Laboratory on Adaptive Strategies (LIA MAST, since 2019, previously in LIA MORFUN, since 2010),

MOSE (Microbial Observatory of Salar Experiments, since 2011) and the Oceanographic Time Series Station off Valparaiso (STOV, since 2015).

Facultad de Ciencias Naturales y Exactas, Departamento de Biología, Universidad de Playa Ancha, Valparaíso, Chile

Merit del Rocío Mora-Ruiz has a background in Microbiology. She has worked on different topics ranging from malaria to aquaculture. Merit obtained her Ph.D. in Environmental Microbiology and Biotechnology at the University of the Balearic Islands of Mallorca. Her main interest is the study of microbial diversity using culture-dependent and culture-independent approaches. Merit is currently a researcher for an aquaculture company.

María Florencia Pérez received her Bachelor in Biotechnology (2015) at the National University of Tucumán (UNT, Argentina). Currently, she is a Ph.D. student at PROIMI-CONICET and works on the analysis of the plasmidome of microbial communities in high-altitude lakes of the Andean Puna. In 2017, she performed a research internship at the Department of Genomic and Applied Microbiology (Georg-August-University, Göttingen, Germany).

Vilma Pérez is a researcher at the Laboratory of Molecular Ecology and Applied Microbiology, at Universidad Católica del Norte in Chile, and currently she is a researcher at the Australian Centre for Ancient DNA, University of Adelaide, Australia. She obtained her Ph.D. in Biological Sciences at the Universidad de Antofagasta, Chile. Her main interest is the study of adaptation strategies to biotic and abiotic factors in microbial communities of polyextreme environments. She has developed this area through the use of different approaches, such as proteomics, NGS and genome sequencing.

Dorota L. Porazinska is an Assistant Professor at the Department of Entomology and Nematology, University of Florida. She is interested in patterns of terrestrial and freshwater microbial diversity, their interactions with plants, and overall role of biodiversity in ecosystem functioning under different scenarios of environmental change. In addition, her interests extend into areas of biogeography, community assembly, and molecular diagnostics. Most of her work has been conducted in a wide variety of natural ecosystems ranging from tropical rainforests to high alpine to glacial cryoconite holes.

María Cecilia Rasuk obtained her Ph.D. in Biological Sciences from the National University of Tucuman (2017). Her Ph.D. was based on the study of the microbial ecosystems of Andean lakes, prospecting studies, biodiversity, metagenomics as a basis for conservation and biotechnological application of strategic national ecosystems. This work was carried out at PROIMI-CONICET. Currently, she is doing a postdoctorate awarded from the Alexander Von Humboldt Foundation at the Technical University of Munich, Department of Microbiology. She is working on Functional Metagenomics using a new thermophilic host to provide necessary support to the idea that the use of alternative hosts for functional metagenomic screening can reveal genes for enzymes that would not be detected by using classical expression hosts such as *E. coli*.

Walter Rojas has a background in regional geology in northern Argentina and Chile. Researched the origin of lithium brines at the Potsdam University (Germany), Bremen University (Germany) and BGRM institute (France). Specialist in 3-D geology modelling and resource estimation for the mining industry. Author of several publications about metallogeny and lithium, oxygen and hydrogen isotopes. Member of the Geological Society of France since 2014.

Ana Paula Santos was born in Tucumán, Argentina in 1993. She obtained a Degree in Biotechnology in 2017 from the National University of Tucumán (UNT). She is a Ph.D. student, and received a fellowship from the National Research Council of Argentina (CONICET).

Luis Alberto Saona Acuña was born in Ecuador, he moved to Santiago de Chile when he was 5 years old and now lives in San Miguel de Tucumán in Argentina where he is working on his Ph.D. at the Laboratory of Microbiological Investigations of Lagunas Andinas (LIMLA) of CCT-PROIMI (Pilot Plant for Microbiological Industrial Processes). Luis did his M.Sc. at the Universidad de Chile and he graduated with high distinction. Now Luis is a Ph.D. student and works with microbial ecosystems of the Andean Puna Region. There, he studied both the metabolic and genomic relationship between arsenate and phosphate. He uses an interdisciplinary approach to complete their goals, using tools from bioinformatics to molecular biology and classic microbiology. He is the author of six scientific manuscripts and has two patents.

Steven K. Schmidt is a professor in the Department of Ecology and Evolutionary Biology at the University of Colorado, Boulder, USA. His research interests include the biogeochemistry and microbial diversity of high-elevation and high-latitude ecosystems throughout the world. He has led or participated in expeditions to many of the most extreme terrestrial landscapes on Earth including the Himalayas, Andes, Alaska Range, Atacama Desert, and the Dry Valleys of Antarctica. Besides a long-term love affair with soil microbes, he is also fascinated with microbial life in and under high-elevation snow packs, glaciers, lakes, and nieves penitentes. Basically, he is happiest when he in a harsh and beautiful landscape, where he can pretend to be doing research.

Zachary R. Schubert graduated with a BA degree (with high honors) from the University of Colorado, Boulder, in 2014. His honors thesis was entitled "Dew Formation and Water Availability at High Elevation in the Atacama Desert, Chile." He is currently working as a data analyst in private industry, but hopes someday to return to science.

Adam J. Solon is a PhD student at the University of Colorado, Boulder, and member of the Alpine Microbial Observatory (AMO). He investigates the processes and mechanisms governing microbial community assembly in soils of cold, hyper-arid ecosystems. His research involves work in the Dry Valleys of Antarctica and in the Puna de Atacama of South America. Additionally, he studies the resistance and resilience of soil microbial communities after human-mediated disturbances in the semi-arid grasslands of Colorado, USA.

Pacifica Sommers earned her PhD in Ecology and Evolutionary Biology from the University of Arizona in 2016, and currently is completing postdoctoral research at the University of Colorado in Boulder. Her research focuses on the patterns and processes maintaining biological diversity, particularly of microbial ecosystems in extreme environments.

Mariana N. Soria is a biotechnologist from the National University of Tucumán. Currently, she is a Ph.D. student in Biology Science at the National University of Tucumán, working on the molecular biology and microbiology of Archaea from Diamante Lake to ascertain the biological role in the energy metabolism of arsenic in this microorganism. This work is carried out at the Laboratory of Microbiological Research of Lagunas Andinas (LIMLA—Tucumán, Argentina) PROIMI—CONICET.

Preston Sowell is an environmental scientist and consultant, naturalist, photographer, and explorer residing in Boulder, Colorado. He founded the environmental consulting firm Geotic Solutions in 2015. When not consulting, Preston uses his scientific and mountaineering expertise to support research expeditions studying high altitude ecosystems and environmental responses to climate change in the South American Andes. Preston's work in the Sibinacocha watershed has led to the discovery of new animal and plant species, and established altitude records for several species and populations. In 2002, Preston collected amphibian samples that documented the most remote and the highest infections of chytridiomycosis (a pathogenic fungus) in frogs. He has led, supported, and photographed expeditions to 18 countries around the world, including 17 scientific expeditions to remote areas of South America.

Tatiana Stepanenko is a specialist in molecular microbiology, she obtained her Ph.D. in Biology Science from the National University of Rosario (2016), working on the molecular biology of opportunistic pathogens. During her postdoctoral fellowship (2019), she studied the molecular mechanisms involved in the arsenic metabolism in extremophilic organisms of the Central Andes. Dr. Stepanenko is currently part of the research group of the Laboratory of Microbiological Research of Lagunas Andinas (LIMLA—Tucumán, Argentina), where she continues to work on the study of the adaptations of Andean Microbial Ecosystems with potential biotechnological applications.

Alejandra Leonor Valdez was born in Tucumán, Argentina in 1987. She obtained a Degree in Biotechnology in 2013 from the National University of Tucumán

(UNT). She has been a Ph.D. student since 2015, and received a fellowship from the National Research Council of Argentina (CONICET).

Patricio G. Villafañe was born in Tucumán, Argentina, and is 28 years old. He is a Geologist from National University of Tucumán and has a M.Sc. in Applied Paleontology from the University of Valencia. Currently, Patricio is a Ph.D. student in Geology at the National University of Tucumán, working on the geological and paleoenvironmental implications of the modern microbialites of La Puna and its comparison with fossil analogs deposits. This work is carried out at the Laboratory of Microbiological Research of Lagunas Andinas (LIMLA—Tucumán, Argentina) PROIMI–CONICET, and in the Higher Institute of Geological Correlation (INSUGEO—Tucumán, Argentina)—CONICET.

Lara Vimercati received her PhD from the University of Colorado in 2019. Her research focuses primarily on how microbes function and survive in cold and dry, high-altitude and high-latitude environments, with an emphasis on applications to the study of Astrobiology. Her work and her love for wild places has led her to conduct research on the highest volcanoes of the Atacama, the highest mountains of Africa, and in the Dry Valleys and Palmer Station in Antarctica, among other remote and extreme environments.

Pieter T. Visscher has been investigating microbial mats and microbialites for over 35 years at study sites on six continents, including the High Andean Lakes in Chile and Argentina. His main interest is in biosignatures, both gases and minerals, produced by the microbial communities in these ancient ecosystems. Visscher combines field measurements (including in situ microelectrode deployments) with laboratory experimentation to reveal geochemical and microbial mechanisms of carbonate precipitation and sulfur gas production as well as cycling of arsenic, sulfur, oxygen, and carbon. The ultimate goal of his work is to link the modern systems to the fossil record and to provide information that could help in the search for life beyond Earth. Visscher is professor at the University of Connecticut in the Departments of Marine Sciences and Geosciences and professeur de classe exceptionelle at the Université de Bourgogne Franche-Comté in Dijon, France.

Petr Vítek is a junior research scientist at the Laboratory of Ecological Plant Physiology at the Global Change Research Institute of the Czech Academy of Sciences, Brno, Czech Republic. Petr obtained his undergraduate degree, followed by Ph.D. at Charles University in Prague in 2010 (awarded the Bolzano Prize), where he continued as a post-doc until 2014. He is an expert in Raman spectroscopy and its application to biological materials and geo-biological systems. He is interested in analysis of biomolecules from microorganisms to higher plants, especially in the context of various environmental stress factors. One of his main research topics is focused on extreme desert environments, responses of desert microorganisms (especially phototrophic) to an excess of solar radiation (including UV), extreme dryness or salinity. He has participated in extremophile research in the Atacama Desert since his Ph.D. in 2009, when the collaboration with Jacek Wierzchos began. Researcher ID of the author: R-8022-2016.

Martin Wagenknecht studied Biology with majors in Microbiology, Genetics, Virology, and Biochemistry at the Philipps-University Marburg, Germany. He did his diploma thesis on phytopathogenic fungi at the Max Planck Institute for Terrestrial Microbiology, Marburg, Germany, and received his Diploma in Biology in 2005. After an interim stay as visiting scientist at Aalborg University, Aalborg, Denmark, in 2006, Dr. Wagenknecht moved on to the University of Münster, Germany, for his graduation in the area of actinobacterial linear plasmids. After receiving his Dr. rer. nat. (Ph.D.) in 2010, he worked as a postdoctoral research associate in PolyModE, a project funded by the European Union, focusing on the development of enzymes for industrial applications. In 2015, Dr. Wagenknecht worked as Group Leader Industrial Biotechnology at Fraunhofer Institute for Machine Tools and Forming Technology (project group RMV), Augsburg, Germany. Since 2015, he has been active as assistant lecturer for the Wilhelm Büchner University of Applied Sciences, Darmstadt, Germany. Since 2016, Dr. Wagenknecht has held the position of Associate Director Molecular Biology and Expression Systems at Boehringer Ingelheim RCV GmbH & Co KG, Vienna, Austria.

Jacek Wierzchos is a Research Scientist at the National Museum of Natural Sciences (Spanish National Research Council). He is an expert in geomicrobiology and microbial ecology of extremely arid environments. He is co-author of a pioneer method to study extant or extinct microorganisms inside rocks. In 1996 he was invited by NASA to study possible traces of life in the Martian meteorite ALH84001. J. Wierzchos is a lead of the group that discovered several novel microbial endolithic ecosystems in the core of the Atacama Desert. His current investigation lines are focused on microbial ecology and geomicrobiology of lithobiontic microorganisms in arid and hyperarid environments, specifically: (a) characterization of endolithic microbial communities in relation to desertification processes, (b) study of bioreceptivity of lithic substrates in extremely arid environments, (c) promoting knowledge on the adaptation strategies used by lithobiontic microbial communities within a range of lithic substrates and climate regimes at the Atacama Desert, (d) determination of the limits of lithobiontic microbial life in hyperarid environments and identification of biosignatures and microbial fossils—witnesses of extinct life, and (e) identification of microbial secondary metabolites with possible biotechnological applicability. He is author of more than 100 JCR articles, and his h-index is 33.

Part I
General Considerations

Chapter 1
Origin and Evolution of the Central Andes: Deserts, Salars, Lakes, and Volcanoes

Ricardo N. Alonso and Walter Rojas

1.1 Introduction

The Central Andes region has received great attention since the plate tectonics paradigm was introduced. Many international researchers have been attracted to investigate this young and active orogen, which is still evolving and where exogenous and endogenous phenomena interact and can be analyzed in situ. This is a place of tectonics and volcanism versus the climate. It is a natural laboratory in which to investigate oceanic plate subduction underneath a continental plate. This provides an analogous comparison with old range systems such as the Laramide in the USA or with modern collisional orogens such as the Himalaya and the great Himalayan Plateau in Tibet. French universities have preferentially worked in Bolivia and Peru, whereas Germans and Americans have worked generally in Chile and Argentina. Many congresses, symposia, and doctoral thesis have been published, along with thousands of articles in scientific journals. In the Andes in the north of Argentina and Puna, there were two main stages. The first one was run by Cornell University (Ithaca, NY, USA), through the Institute for the Study of the Continents (INSTOC), and the second one was run by Potsdam University (Potsdam, Germany). The turning point in the understanding of the origin and evolution of the Andes came in the classic paper by Isacks (1988), followed by a paper by Allmendinger et al. (1997). The German research is well represented by Oncken et al. (2006) and Strecker et al. (2007). Such publications and their references represent a synthesis of all that has been achieved.

R. N. Alonso
Universidad Nacional de Salta–Consejo Nacional de Investigaciones Científicas y Técnicas (CONICET) (Centro Argentino (CEGA)–Instituto Superior de Correlación Geológica (INSUGEO)), Salta, Argentina

W. Rojas (✉)
Eramine Sudamérica SA, Salta, Argentina

M. E. Farías (ed.), *Microbial Ecosystems in Central Andes Extreme Environments*, https://doi.org/10.1007/978-3-030-36192-1_1

1.2 Noncollisional Orogen

The Central Andes is the result of the interaction between the Nazca oceanic plate and the South American continental plate. This interaction has produced subduction, which means the sinking of the oceanic plate underneath the continental plate. It is very interesting to note that the Central Andes conformation is due to the plate dipping angle, which is around 30°E. That angle drives the plate to reach deeper regions with a temperature of 1200 °C, where it is smelted, and the magma reaches the surface via volcanoes. Moreover, the crust is heated, so it can be easily deformed. This heating produces crust shortening, generating an orogenic formation (Allmendinger et al. 1983; Jordan et al. 1983; Jordan and Alonso 1987; Isacks 1988).

It is very different from other orogenic formations, such as the Himalaya, which is the result of a collision of two continents: India and Asia. Although it is true that today they are merged, in the past, India was a drift continent, which moved fast toward the north (by 16–19 cm/year) until it collided with Asia. From this tremendous collision emerged the Himalaya, including Mount Everest and the extraordinary high plateau of Tibet. So, whereas the Himalaya is the result of collisional orogenic building, the Central Andes is the result of noncollisional orogenic building.

The Himalayas were born from this amazing collision, and some of the peaks there reach a height of approximately 9 km above sea level (asl). Likewise, the Tibetan plateau was formed with large basins such as Qaidam, with an axis of 1000 km. On the basis of its size, the Andes take second place in terms of both the height of its mountains and the extension of the Altiplano–Puna, which is about 2000 km long. The Andean structure reaches 900 km at the latitude of Arica (Chile), where it has a deep inflection (an orocline or Arica elbow); this has led to all sorts of interpretations in search of an explanation for the peculiar tectonics (Fig. 1.1).

In the Pacific Ocean, there is an oceanic trench along the whole border of South America. The Peruvian–Chilean Ocean Trench reaches a depth of 8 km at the latitude of Antofagasta, but approximately 200–300 km away, the Andean chain reaches around 7 km asl; thus, this is the biggest disruption of relief known on the Earth.

The continental crust reaches its maximum thickness (80 km) below Cerro Zapaleri, which is located between Argentina, Bolivia, and Chile. This is the thickest continental crust known on Earth. The depth at which earthquakes occur allows us to know the geometry of the subducted slab. The shallower hypocenters (at a depth of 60 km) are near the Pacific coast, while toward the center of the continent they are at a depth of above 400 km, giving the form of the sinking shape of the oceanic trench.

At the same time that the oceanic plate sinks, it heats up, and then it begins to evaporate and eventually melts. The formation of mineral deposits is linked to this process. Native sulfur appears through the sublimation of vapors in the volcanic crests, with belts of antimony, bismuth, and other metals from the internal circulation of hydrothermal fluids. After erosion, many volcanoes show their old roots, which can contain copper deposits, such as Chuquicamata (the largest open-pit

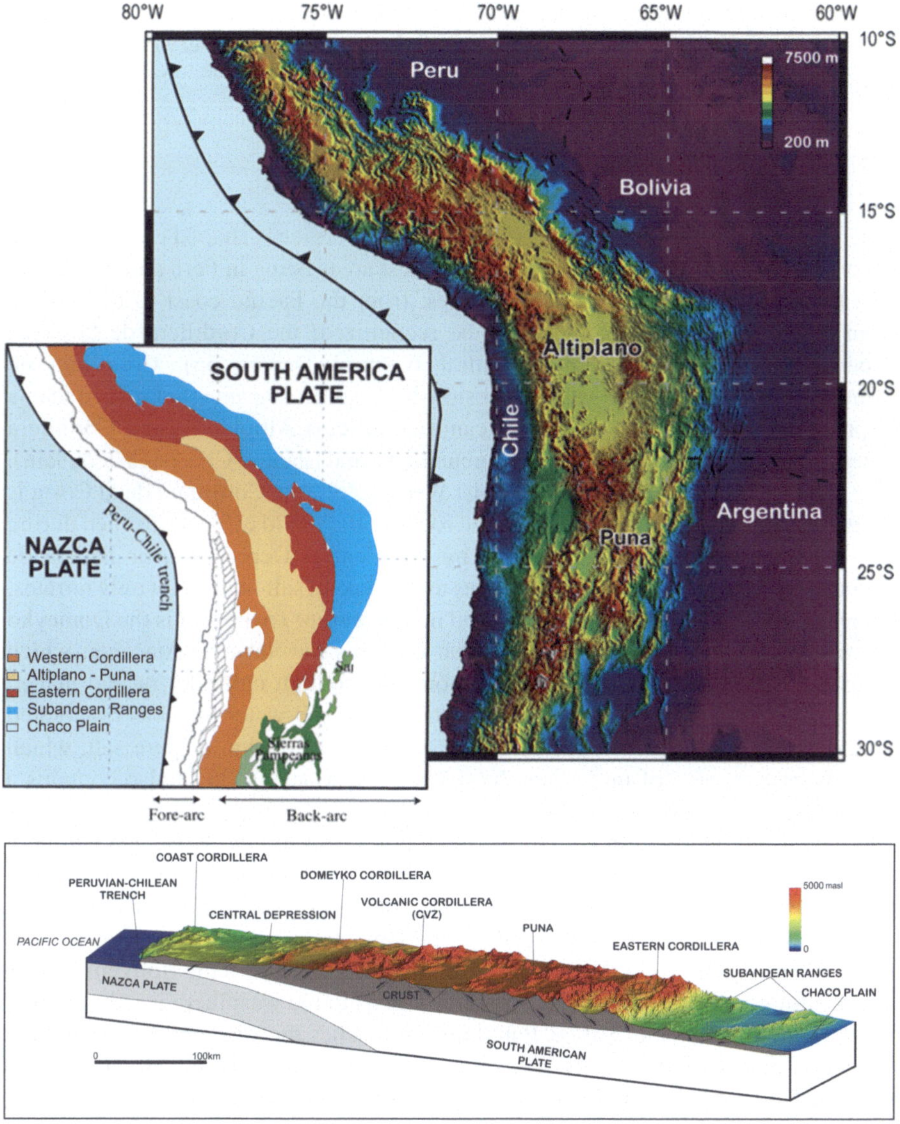

Fig. 1.1 The main morphotectonic units in the Andes and a cross-section of the Central Andes at 25°S

mine in the world) or Cerro Rico de Potosí (a mountain that contains the largest geochemical concentration of silver in the world).

The oceanic plate sinking angle defines all of the genesis of the Andean formation. In the case of a lower angle, there is no plate melting, as occurs north and south of the Central Andes, where the subduction angle is almost horizontal, so volcanoes tend to disappear. There are shallow subduction zones or flat slab regions. For instance, if the angle were to steep, then the volcanoes would be very close to the Pacific coast and the rest of the Andean configuration would be very different.

1.3 Morphostructural Units

When observed from space, the Central Andes is a crescent-shaped chain, with the wider portion being in Bolivia and the narrowest areas being in Peru and Argentina. In a transversal cross-section of the Andes, from the Pacific coast to the eastern Gran Chaco, the following units can be recognized: the Cordillera de la Costa (Coastal Cordillera), Depresión Central (Central Depression), Cordillera de Domeyko (Domeyko Cordillera), Depresión del Salar de Atacama (Salar de Atacama Depression), Cordillera Volcánica (Volcanic Cordillera), Altiplano–Puna, Cordillera Oriental (Eastern Cordillera), Sierras Subandinas, and Llanura Chaco (Chaco Plain).

As described by the German geologist Werner Zeil, the Cordillera de la Costa is the greatest thickness of basaltic volcanic rocks in the world (15 km), formed during the Jurassic period. This is followed by the Central Depression, or Pampa de Tamarugal, containing deposits of nitrates and iodine in salt fields. Not only nitrates, but also large deposits of copper are found in this region. To the east is the Domeyko Cordillera and then the curious basin that contains the Salar de Atacama, which holds the largest concentration of common salt (sodium chloride) on the planet. Over millions of years, the natural concentration of salt water that was evaporated by the dry atmosphere has formed a thick layer of over 1 km of pure salt, which extends over an area of more than 3000 km^2. The Atacama Basin also contains a lithium-rich deposit in the salt beds.

In the north of Argentina, there are some units between the Volcanic Cordillera and the Chaco Plain, which continues on to Paraguay. Another point to consider is that the Andean growth has blocked the access of eastern wet winds generated in the Amazon region, Mato Grosso, and the Gran Pantanal. This effect generates a wet face, with rainfall rates of thousands of millimeters per year (in the Bolivian Yungas and the Tucumano–Oranense rain forest), whereas to the west there is a diminution in the rainfall (Calchaqui Valley, Puna) until it reaches the Atacama Desert and the hyperarid desert of the Pacific coast, where exotic minerals are precipitated as nitrates. This conjunction of eastern wet regions and the aridity on the western side overlaps the orographic units and determines the relief and its evolution. To be more precise, the Central Andes region is a result of inner (endogenous) forces, which generate the growth and building of landforms (ranges, volcanoes), whereas exogenous forces (such as the climate) are responsible for its destruction. Because Andean building is modern on a geological scale, the relief is well conserved, making it very attractive for scientific research (Lamb and Davis 2003; Alonso et al. 2006; Strecker et al. 2007).

The north of Argentina shows a particular geographic framework. That is a consequence of several factors, including polychrome rocks covering almost all of the geological periods; the enormous diversity of its lithologies; the variety of morphostructural environments varying from high cordilleran volcanoes to shallow eastern plains through salars, mountains, valleys, and ranges; a sharp relief with summits passing through the ceiling of clouds and great canyons with deep rivers reaching deeper portions of the crust; and a wonderful variety of climate and microclimates, from tropical rain forests to dry and cold deserts. Also, a wide range of natural environments feature in the physiography of northern Argentina.

Only some regions of the world are characterized by enormous elevation differences with mountains higher than 6 km asl, from which there are several steps occupied by valleys and creeks until a shallow plain is reached. This situation occurs between high volcanic summits such as the Llullaillaco volcano (with a height of ±7 km asl) and the Chaco Plain, with deep and shallow valleys in the middle.

1.4 Geological Provinces

The main cross-section of the Central Andes shows the following morphotectonic units from west to east: an oceanic trench, a coastal range, the Central Depression (Pampa de Tamarugal), the Domeyko Range, the Atacama–Punta Negra Depression, the main Volcanic Cordillera or Central Volcanic Zone, the Altiplano–Puna High Plateau, the Eastern Cordillera, the Subandean Belts, and the Chaco Plain.

1.4.1 The Peruvian–Chilean Trench

The Peruvian–Chilean Ocean Trench is 8–9 km deep (at the latitude of Antofagasta) and is one of the world's deepest ocean trenches. There are two main reasons for this. The first one is related to the 30°E dip of the Wadati–Benioff plane on the Nazca plate. The second one is associated with a lack of sediments, as the coast is located in the most arid regions of the world. As a consequence, this holds detrital sediments back from the ocean trench as there are no transport elements such as fluvial transport. The shallow contact between the Nazca plate and the South American continental plate generates tensions that have produced the most powerful earthquakes and tsunamis in the Earth's historical record.

1.4.2 The Coastal Cordillera

The Coastal Cordillera is a low-elevation relief that runs parallel to the Pacific coast. This range is composed of granitic rocks and a complex of Jurassic lava flows and marine sedimentary rocks. The sediments contain a great quantity of ammonites, and sometimes fossil bones from marine reptiles are also found there, as are copper, gold, and mainly iron ore in the form of magnetite. On the flanks, some dunes are well developed. Along with the rest of the region, it is hyperarid, with scarce xerophyte vegetation. The Camanchaca is a very common phenomenon, characterized by a marine fog that covers the western slope of the Coastal Cordillera. Also, the famous *guaneras* and *covaderas* (guano deposits) are located on the coast and on the islands, where several meters of bird guano have accumulated over thousands of years.

1.4.3 The Central Depression

The Central Depression, or Pampa de Tamarugal, is a low zone at 500 m asl, bounded by the coastal range in the west and the Domeyko Range in the east. The hyperaridity of this depressed belt has resulted in accumulation of great nitrate ores. These exotic salts, which have formed in hyperarid conditions, are interlayered with iodate and chromate mineralization. It is known also as the Pampa de Tamarugal because of the presence of *Prosopis*, which is commonly known as *tamarugo*. Sodium and potassium nitrate (nitratine and nitro) were intensively exploited from the middle of the nineteenth century until the first decades of the twentieth century. There are still some important vestiges of those old mining works, such as Humberstone. The possession of *nitrateras* (nitrates) and *guaneras* triggered the Pacific War in 1879, in which Chile defeated the Peruvian–Bolivian coalition, annexing the Antofagasta, Iquique, and Arica regions. At the end of the nineteenth century, the natural fertilizer triangle (phosphorus (P), nitrogen (N), and potassium (K)) was considered strategic. The *guaneras* were the sources of P and the *nitrateras* provided N and K. Water resources are very scarce, and there is only one river (the Loa River) that flows from the Andean Ranges through the *salitre* (saltpeter) plain. The brackish and confined underground water resources were pumped from wells to supply the nitrate mines. Also of note are the natural hot springs that form oases in the desert, such as the Pica and Mamiña Springs. One of the main archeological vestiges is geoglyph figures, drawn by the ancient inhabitants, which have been preserved for several centuries by the extreme aridity of the region. Similar preservation has happened with the old carriage and animal tracks that ran to the *salitreras* for use by the miners. The herd of bulls were provided from the northern region of Argentina, being carried on foot to the mines (Fig. 1.2).

Fig. 1.2 The extreme aridity in the Salar de Atacama

1.4.4 The Domeyko Cordillera

The Domeyko Cordillera is a physiographic feature marked mainly by Cretaceous rocks, which contain Paleogene porphyries. These are the source of the main copper–molybdenum ores in Chile, such as at Chuquicamata. Most of those porphyries are interlayered in the West Fault—a megastructure that borders and raises the Domeyko Range in the west. This range and other important physiographic features have received the names of famous natural scientists such as Domeyko, Philippi, Gay, and Pissis.

The Atacama–Punta Negra basin is situated at 2600 m asl, being one of the tectonic steps that gradually rise from the Pacific border to the highest mountains in the Volcanic Cordillera. The depression region is bounded by the Domeyko Range to the west and the Central Volcanic Zone to the east. Also, it contains the Salar de Atacama, one of the world's biggest salars, which contains the highest punctual concentration of halite, with a 3000 km^3 volume. It represents a mature salar, with a halite core 1.5 km thick. As this salar has a high concentration of lithium brine, it is considered the first world brine producer of this ore type. On the eastern side, there are some freshwater springs where the original *atacameños* (Atacama people) lived. The water comes from the snowmelt that occurs higher up in the main volcanic range. Many archeological objects have been unearthed and are now exhibited in the Le Paige Museum in the Atacama town of San Pedro.

The slopes that extend up to the Salar de Atacama are composed by volcanic tuff and ignimbrites, which come from the high Volcanic Cordillera. Some young volcanic cones, such as Lincancabur and Lascar, are visible, and they sometimes have fumaroles showing their recent volcanic activity. The cordillera volcanoes belong to the stratovolcano typology and are usually above 6000 m asl. The slopes, coming down from 4 km high, cover a megastructure of the Andean crust known as the Monocline Occidental, which is considered a magma pathway.

1.4.5 The Central Volcanic Zone

A relief analysis from the west shows firstly the impressive Volcanic Cordillera—high ranges forming a natural boundary with Chile. It includes some of the highest world-class volcanic formations such as Ojos del Salado (6891 m asl), Piscis (6882 m asl), Llullaillaco (6739 m asl), and Socompa (6051 m asl). The stratovolcano structure was formed by piling-up of lava and ashes. There are more than 1000 volcanoes in the Central Andes region, with an interesting portion of them being distributed in Salta and Jujuy Provinces. There are many volcano types (stratovolcanoes, basaltic cones, domes, etc.), showing their different stages (active, inactive, extinct, etc.). Some of them have collapsed in the past, generating enormous calderas. An emblematic example is El Galán Caldera in the Puna of Catamarca, with a 40 km diameter and ignimbrite flows extending to the Calchaqui Valley around 100 km away. A small portion of that ignimbrite is represented in the Fuerte de Tacuil. In the Jujuy Puna region there is La Pacana Caldera, close to Jama, which is broken and deformed, with a diameter of approximately 60 km. In the same region can be found other calderas such as Pairique, Coranzuli, and Rachaite, among others. All of these are associated with steep walls and white and pink flows, resulting from a high-energy eruption, known as pyroclastic flow, originating on the edge of the caldera ring after the collapse and sinking of the volcano superstructure. The collapse of these structures produced volcanic boilers and ignimbrites. The Central Andes region contains the biggest ignimbrite "province" in the world, covering more than 500 km^2. The volcanic ash from this great volcanic eruption of the cordillera was moved by the wind toward the east and was deposited for millions of years in inland South America. Besides the main chain of volcanoes, there are some transverse branches that cross through the Puna to the east. In Jujuy, there is the Coyaguaima volcanic complex; in Salta, the Rincon–Tul-Tul–del Medio–Pocitos chain and El Quevar complex (Cerro Verde, Aguas Calientes, Azufre, Mamaturi, and Quironcolo); and in Catamarca, the Archibarca and Galán volcanic chains. Many of these volcanoes are higher than 6000 m asl and were formed during the Miocene epoch between 10 million and 5 million years ago. Also, some magnificent examples of isolated volcanoes can be found, such as Tuzgle (5560 m asl) in Jujuy Province, an old stratovolcano with an active magmatic chamber, which generated a young lava flow spread around 10,000 years ago because it was not affected by the last glaciation. This volcanic formation is surrounded by hot springs, providing very clear evidence of the still-active steam heating system. Many volcanoes have given way to glass trails, which, in the obsidian state, has been valued by native groups to produce projectiles and, as a hydrated material, has resulted in perlite, a rock that expands in heat, increasing its volume by up to ten times, and is widely used in the construction industry. In the last 1.3 million years, at least four catastrophic volcanic eruptions have buried northwestern region of Argentina (NOA) in ashes (1,300,000, 300,000, 100,000, and 5000 years ago). The remains of these ashes can still be found; they are known as *puloil* and are used by people as an abrasive to clean kitchen utensils. The last two eruptions were produced by Lascar (Fig. 1.3) (Coira et al. 1993; Kay and Kay 1993; Kay et al. 1994, 1999).

Fig. 1.3 The Socompa and Antofalla volcanoes in the Puna region

1.4.6 The Altiplano–Puna High Plateau

The Puna (in Argentina) and the Altiplano (in Bolivia) are an elevated region rising to an average of 3700 m asl, characterized by an arid-to-semiarid climate and interior drainage that allows the formation of lacustrine environments of evaporation, resulting in salt lakes. The Uyuni Salt Flats in Bolivia (the biggest in the world) and the Arizaro in Argentina are two examples of this type of environment. Today, they are mostly dry salt lakes, but in the Quaternary period, during glaciation and deglaciation, they were filled with water, forming enormous lakes; the only things left of them are the ancient high banks (Placzek et al. 2006). Deposits of diatomaceous earth, made up of siliceous microfossils, are common in these banks.

The Altiplano and Puna regions, which form a single unit limited by the Volcanic Cordillera and the Eastern Cordillera, are a high-elevation region, separated internally by blocks and crossed by isolated and chained volcanoes spreading as transverse branches of the main volcanic chain. Among those volcanoes and ranges, some close and large depressions are formed, where water comes up to the surface and evaporates, creating the salars. More precisely, most of the salt content is due to leaching from the volcanic formation and the surrounding hot springs, and there is no relationship with the sea, as is usually thought. The Puna has a great variety of salars, saline lakes, playa muds, swamps, and other similar environments. The dryness increases steadily to the west and also to the south. In accordance with this subdivision, in north Puna there are some lakes (Pozuelos, Guayatayoc, Vilama) and saline lakes (Olaroz, Salinas Grandes), whereas in south Puna there are salars such as Rincon, Pocitos, Arizaro, Antofalla, and Hombre Muerto West, among others. The salars are surrounded by tectonic blocks that are meridionally oriented, and by volcanic formations on the north and south edges. All of the surface water goes into the depressions, discharging all of its saline content, which, because of the dryness, can be converted into crystals. Also, the hot springs are considered direct sources of ions, represented by some enormous geysers, which are extinct today. On the basis of solubility, there is a salt precipitation order, called evaporites: calcium carbonate (travertine), calcium sulfate (gypsum), sodium–calcium borate (borax–ulexite), and sodium chloride (halite) near the center of the depression. Also, sodium carbonate (at the Santa Maria Lagoon) and sodium sulfate (at the Rio Grande and Pocitos Salars) can be found.

Associated with the volcanoes are hundreds of active and inactive hot springs. The active hot springs can be used for medicinal purposes, and the Incas used them well. A good example is the hot springs of Incachule (a Quechua word, which means "Inca bath"), located near San Antonio de los Cobres. The Inca road system (Caminos del Inca), besides joining strategic points of the empire, passed the thermal springs and the copper, gold, salt, and obsidian mines, along with the sources of other products the Incas exploited. Many thermal fountains are related to formation of salts, which are found in the salt beds, borates being among them. The borates are formed by boron-rich thermal water, spilled into salt lakes. These lakes existed in the Puna about 14 million years ago, but they gave way to great concentrations of borates between 7 million and 5 million years ago during the Late Miocene. The beds of Tincalayu and Sijes in Salta and Loma Blanca in Jujuy rank third among the reserves of borates worldwide, after those in Turkey and the USA. Actually, 50 salt beds of the Altiplano, the Puna, and the Andean volcanic arc contain borates, ulexite being the most important.

The brine of the salt flats is rich in chemical elements such as lithium, potassium, magnesium, boron, cesium, and rubidium. Lithium is a light metal and is found as an ion in underground saline waters, exploited mainly in the Salar de Atacama in

Chile and Hombre Muerto (Salta–Catamarca) and Olaroz (Jujuy) in Argentina. The old salars formed in the Miocene and were completely deformed by the Andean structure, and they contain enormous quantities of rock salt and gypsum.

Climatically, the Puna is a cold region, with a mean annual temperature of 8 °C, a harsh winter season during which the temperature can reach −30 °C, great thermic amplitude showing sometimes 50 °C differences in a single day, cold and dry winds, high heliophany, intense solar radiation due to the absence of clouds and high elevation, low oxygen partial pressure, strong reflectivity of the salars, scarce rainfalls during the summer, and some snowfalls in winter (these can vary during El Niño–La Niña years), among other parameters, which make us consider the Puna as an inhospitable region with a rigorous climate and one of the smallest numbers of inhabitants per square kilometer (Turner 1972; Alonso et al. 1984, 1991; Marrett et al. 1994; Vandervoort et al. 1995; Allmendinger et al. 1997; Strecker et al. 2007; Gibert et al. 2009).

1.4.7 The Eastern Cordillera

The Eastern Cordillera is a morphotectonic region formed by a system of faulted mountain blocks cut by deep valleys and canyons spreading parallel to the Central Andes from Peru to the north of Argentina. The valleys lead to the Quebrada de Humahuaca, Calchaqui Valley, Escoipe, El Toro, and others. These valleys have been routes for communication and transport on both sides of the Andes since ancient times. Paleo-Indians, Incas, and Spanish were mobilized by these natural access ways, which were used for transport, trading, or warring for control.

The Calchaqui Valley has a western flank composed mostly of igneous and metamorphic rocks, including the Sierra de Cajon o Quilmes, the Cachi–Palermo Range, and Acay Hill. The snowy hills of Chuscha, Cachi, and El Chani in Jujuy are part of the eastern border of the Puna, being formed by great granite bodies (Ramos 2008). To the east of Calchaqui Valley are the Cumbres Calchaquies, composed mostly of red Cretaceous sedimentary rocks.

Another Andean feature, Lerma Valley, situated around 1200 m asl, includes Salta City and is flanked by Cordón de Lesser and the Mojotoro Range to the west and east, respectively. Both mountain blocks are composed of marine sedimentary rocks dating from the Precambrian, Cambrian, and Ordovician periods. To the east and at the same latitude as Salta City, there is the Siancas Valley, where the town of General Guemes is located. This is a wide valley with a lower elevation of around 700 m asl, flanked by the Mojotoro Range to the west and the del Gallo Range and its southern extension (the San Antonio Range) to the east (Hain et al. 2011).

The Eastern Cordillera reaches its greatest expression in the mountains located just on the boundary with the Puna region in the form of the Cachi and Palermo, Acay, Chai, and El Aguilar Ranges, where some peaks reach 5000 or 6000 m asl. All of these mountains and their respective valleys have resulted from strong Andean pressure, which produced deep faulting that uplifted the rocks, with consequent thick rock overlapping. Purmamarca Village and its Cerro de los Siete Colores (Seven-Color Hill) is a very good example of these different ages of rock overlapping. The fragmentation feature predominates over folding. Among the great valleys and canyons of this unit can be outlined Humahuaca, Calchaqui, Toro, Escoipe, Lerma, Siancas, Jujuy, etc. The Eastern Cordillera is an extense unit that borders the Altiplano–Puna region on the eastern side. The Andes are a single unit until they reach the northwest of Argentina, where they bifurcate into two as the Volcanic Cordillera and the Eastern Cordillera to the east, leaving the Puna between both. So it is not correct to use the term "precordillera" (as in San Juan and Mendoza) when referring to the mountains located on the west flank of the cities of Salta and Jujuy, because they really belong to the Prepuna region, taking into account the fact that the Cordillera Principal location is situated much further west. The Eastern Cordillera ends around Tucumán City where the Pampean Ranges begin. Among the main ranges that contain the Eastern Cordillera can be mentioned Leon Muerto, Carahuasi, Cebilar, Metan, Zamaca, Lesser, Castillo, Mojotoro, Castillejo, Alta, Tilcara, Mal Paso, Aparzo, and Santa Victoria. With some exceptions, these ranges have elevations varying from 3000 to 4000 m asl (Jordan and Alonso 1987; Mon and Salfity 1995; Trauth et al. 2000; Hain et al. 2011).

1.4.8 The Subandean Ranges

The Subandean Ranges were originally named by Count Guido de Bonarelli, an Italian geologist who studied them in the early twentieth century. There, oil-bearing rocks dating back to the Devonian, Carboniferous, and Cretaceous periods lie under a thick bed of Neogene sediments. Gas and oil accumulated in the folds of the Earth's crust were deformed by the thrust of the Andes. The giant gas fields discovered in Bolivia include the fields of San Alberto and Sábalo.

The Subandean Ranges is a group of small ranges, with a continuous decrease in elevation until they disappear completely on the surface. This unit is bordered on the west by the Eastern Cordillera and by the extended Chaco Plain on the eastern side, and it is characterized by large folds crossed by faults. Folding predominates over faulting, so the deformed structures such as synclines and anticlines are becoming smoothed until they disappear completely under the eastern plain. This region has dense vegetation and is crossed by many rivers. In the north of Argentina, it is on the eastern side of the Eastern Cordillera until the latitude of Tucumán, but the names "the Santa Barbara System" and "the Southern Subandean Ranges" have also been proposed for the area immediately to the south of the Juramento River. This system includes many ranges such as Medina, La Candelaria, Cerro Colorado, Cuchuma, San Antonio, Lumbrera, del Gallo, Centinela, and Maiz Gordo. The northern Subandean System includes ranges such as Puesto Viejo, Zapla, Ledesma, Calilegua, Zenta, Caimancito, Porongal, Las Pavas, Candado, Ramos, San Antonio, Aguarague, Madrejones, Macueta, and others, which continue on to Bolivia, where they reach their greatest development. In general, they are no higher than 2500 m asl on the eastern side and 3500 m asl on the western side (Jordan and Alonso 1987; Decelles et al. 2011).

1.4.9 The Chaco Plain

The Chaco Plain is an extended floodplain composed of materials derived from the volcanic formation that were deposited there afterward. This region is crossed by large rivers such as the Pilcomayo, Bermejo, and Juramento. As part of the foreland domain of the Central Andes, it is located in a wide region known as the Chaco–Pampeana Plain, the Chaco–Saltena Plain, and the Chaco–Paranense Plain, among other names, which emphasizes its low and plain relief covering a big eastern portion of the subtropical region. This plain environment is extended to the east from the last Andean features known as the Subandean Ranges. From a geodynamic point of view, this plain region constitutes an enormous basin of sedimentation, which is subsiding by sinking as a consequence of isostatic compensation by the Andean uplift. Its elevation decreases to the east, averaging 300 m asl. The Chaco–Saltena Plain is limited on the north side by the Pilcomayo River, the Formosa and Chaco Provinces to the east, and Santiago del Estero to the south. The semiarid Chaco is an extense plain located in the central portion of the South American Chaco, with rainfalls varying between 450 and 650 mm per year and maximum temperatures that are considered extreme (48.9 °C in Rivadavia is the highest temperature registered in the zone). This region is also well known as the "warm South American Pole." The Chaco Plain is a relatively uniform relief where two main environments can be differentiated: the stabilized Chaco Plain and the spillage Chaco Plain. The fine materials on the surface, such as silt and clays, are the basis of the well-developed soils, which have good agricultural potential. If, someday, the excess of water that is discharged by large rivers into the Atlantic Ocean were to be channeled through the plain, this region could be transformed into a new Pampa Húmeda (Humid Pampas).

1.5 Climatic Extremes: Jungle and Deserts

The displacement of the Andes lifted the mountains and pushed them toward the east, to the point where today the deformed Andean front is found on the edge of the Chaco Plain. During this process, some valleys opened and others closed. The Amazonian moist winds find high mountains as they advance toward the west. The discharge of their moisture on the eastern slopes of the Andes results in significant biodiversity. Places such as Chapare in Bolivia are found, where the annual rainfall is almost 7000 mm, or the Sierra de Aconquija, with 3000 mm a year. This rainy area of Bolivia, on the eastern side of the Andes and in the deep valleys, gives rise to the hot and humid Yungas, where coca and other tropical plants grow.

The winds gradually lose their moisture and become dry when they reach the Puna and the Altiplano. For this reason, the regions become deserts in the shadow of the rains. The aridity progresses more and more toward the west until it reaches extreme values on the Chilean–Peruvian coast. There, the Atacama Desert is known for its hyperdryness, which is a product of the eastern front of the Amazonian rainfall on one hand and the absence of evaporation of the cold oceanic waters of the Humboldt oceanic current to the west on the other. This extreme aridity has given way to the formation of nitrate and phosphate beds, which brought riches to Chile at the end of the nineteenth century and the beginning of the twentieth century. The phosphates were formed by the dry climate, which allowed the accumulation of marine bird droppings over hundreds of thousands of years.

The desert strip that runs along the Chilean–Peruvian coast explains the great depth of the oceanic trench, which is free of sediments. It is interesting to observe how one side of the Andes has 7000 mm of annual rains and the other side has almost 0 mm, although they are at the same latitude. The same thing occurs with the temperature. To the east of the Andes Rivadavia (Salta) is the warm South American Pole, with a maximum temperature of 50 °C, while at the same latitude during winter in the Puna, there are temperatures of around −30 °C, with an annual average of 8 °C (Fig. 1.4).

Fig. 1.4 Two faces of the Central Andes: rain forest to the east and deserts to the west

1.6 Discussion and Conclusions

To give an idea of the magnitude of the deformation of the Andes, there is a good example of a well-known rock in northern Argentina. This is the Yacoraite Formation, a unit of limestone from the late Cretaceous period, with abundant dinosaur tracks. That rock was formed before the Andes in a large inland basin of South America, extending from Peru to Tucumán and from the north of Chile in the west to the base of the basement of Paraguay in the east (Marquillas et al. 2005). This calcareous plank, formed at sea level 66 million years ago, was split with the rupture of the Andes, and today it is 5000 m high in the mountains of the highlands and 6000 m deep in the oil fields of the Subandean Ranges. This is a result of stacking and deformation measuring 11 km vertically.

The north of Argentina is a vast geography, modeled during the last millions of years by exogenous agents that have carved the rocks that form the South American foundation. The Andean crust has suffered several glaciations periods, whose contents remain in some snow peaks. Also, it has been heated by magma rising from deeper zones, generating great volcanic formations loaded with metals. It has been torn into deep valleys and canyons, and has been folded into the low Subandean Ranges covered by tropical vegetation, whose cores hold precious reservoirs of hydrocarbons. This natural distribution means we can find salars at the same latitude in the west and abundant vegetation of tree ferns on the eastern side. In the north of Argentina, some places are named after icons of this architectonic brochure of landscape elements: the Llullaillaco volcano, Socompa volcano, Arizaro Salar, Calchaqui Valley, El Toro Canyon, Escoipe Canyon, Las Conchas Canyon, Cachi, Humahuaca Canyon, and Baritu National Park, among others.

The Central Andes and the Andes of northern Argentina are a single entity resulting from tectonic interaction between an oceanic plate and a continental plate. The Central Andes are a noncollisional orogene of a tectonovolcanic nature, formed by the subduction of the Nazca oceanic plate beneath the South American continental plate. Their evolution is the product of dynamic interaction between endogenous and exogenous forces, including the internal and external energies of the Earth.

The origin and evolution of the Central Andes and the high Andean plateau called Altiplano–Puna are studied by scientists who study the Earth in many countries. All of this is due to the fact that they have distinct elements related to tectonics, climate, volcanoes, physiography, mineral deposits, salars, and other phenomena that occur and interact dynamically. The Andes and the highlands are a mountainous part of a structure currently in formation. The Andes are growing vertically and pushing horizontally. This has led to the development of a particular relief in the region of Argentina, Bolivia, and Chile, where volcanic peaks reach a height of around 7 km asl to the west, while a plain lies to the east. On this plain, there is a huge basin in subsidence, where hundreds of meters of debris have been deposited, torn from the Andean structure. This chapter demonstrates the uniqueness and singularity of the Central Andean orogenic formation.

Acknowledgements We wish to express our gratitude to Dr. María Eugenia Farías for inviting us to write this text, Dr. Manfred Strecker for provision of geological information, Prof. Maria Angelica Aramayo for revision of the manuscript in English, the Consejo Nacional de Investigaciones Científicas y Técnicas (CONICET)–Centro Argentino (CEGA)–Instituto Superior de Correlación Geológica (INSUGEO) investigation program, and Consejo de Investigación de la Universidad Nacional de Salta (CIUNSa).

References

Allmendinger RW, Ramos VA, Jordan TE, Palma M, Isacks BL (1983) Paleogeography and Andean structural geometry, northwest Argentina. Tectonics 2:1–16

Allmendinger RW, Jordan TE, Kay SM, Isacks BL (1997) Evolution of the Puna--Altiplano Plateau of the Central Andes. Annu Rev Earth Planet Sci 25:139–174

Alonso RN, Viramonte J, Gutierrez R (1984) Puna Austral – Bases para el subprovincialismo Geológico de la Puna Argentina. Noveno Congreso Geológico Argentino S.C. Bariloche Actas, pp 43–63

Alonso RN, Jordan TE, Tabbutt KT, Vandervoort DS (1991) Giant evaporite belts of the Neogene Central Andes. Geology 19:401–404

Alonso RN, Bookhagen B, Carrapa B, Coutand I, Haschke M, Hilley GE, Schoenbohm L, Sobel ER, Strecker MR, Trauth MH, Villanueva A (2006) Chapter 12: Tectonics, climates, and landscape evolution of the southern Central Andes: the Argentine Puna Plateau and adjacent regions between 22 and 30° lat. In: Oncken et al (eds) The Andes—active subduction orogeny. Springer, Berlin, pp 265–283

Coira B, Kay SM, Viramonte J (1993) Upper Cenozoic magmatic evolution of the Argentina Puna—a model for changing subduction geometry. Int Geol Rev 35:677–720

Decelles PG, Carrapa B, Horton BK, Gehrels GE (2011) Cenozoic foreland basin system in the Central Andes of northwestern Argentina: implications for Andean geodynamics and modes of deformation. Tectonics 30:TC6013. https://doi.org/10.1029/2011TC002948

Gibert RO, Taberner C, Saez A, Giralt S, Alonso RN, Lawrence Edwards R, Pueyo JJ (2009) Igneous origin of CO_2 in ancient and recent hot springs waters and travertines from the Northern Argentinean Andes. J Sediment Res 79(8):554–567

Hain MP, Strecker MR, Bookhagen B, Alonso RN, Pingel H, Schmitt AK (2011) Neogene to Quaternary broken foreland formation and sedimentation dynamics in the Andes of NW Argentina (25°S). Tectonics 30:TC2006. https://doi.org/10.1029/2010TC002703

Isacks BL (1988) Uplift of the Central Andean Plateau and bending of the Bolivian orocline. J Geophys Res 93:3211–3231

Jordan TE, Alonso RN (1987) Cenozoic stratigraphy and basin tectonics of the Andes Mountains 20°–28° south latitude. Am Assoc Pet Geol Bull 71:49–64

Jordan TE, Isacks BL, Allmendinger RW, Brewer JA, Ramos VA, Ando CJ (1983) Andean tectonics related to the geometry of the subducted Nazca plate. Geol Soc Am Bull 94:341–361

Kay RW, Kay SM (1993) Delamination and delamination magmatism. Tectonophysics 219:177–189

Kay S, Coira B, Viramonte J (1994) Young mafic back arc volcanic rocks as indicators of continental lithospheric delamination beneath the Argentine Puna Plateau, Central Andes. J Geophys Res 99:24323–24339

Kay SM, Mpodozis C, Coira B (1999) Magmatism, tectonism, and mineral deposits of the Central Andes (22°–33°S latitude). In: Skinner BJ (ed) Geology and ore deposits of the Central Andes: Society Economic Geology Special Publication no. 7, pp 27–59

Lamb S, Davis P (2003) Cenozoic climate change as a possible cause for the rise of the Andes. Nature 425:792–797

Marquillas RA, del Papa C, Sabino IF (2005) Sedimentary aspects and paleoenvironmental evolution of a rift basin: Salta Group (Cretaceous--Paleogene), northwestern Argentina. Int J Earth Sci 94:94–113

Marrett RA, Allmendinger RW, Alonso RN, Drake R (1994) Late Cenozoic tectonic evolution of the Puna Plateau and adjacent foreland, northwestern Argentine Andes. J S Am Earth Sci 7:179–207

Mon R, Salfity JA (1995) Tectonic evolution of the Andes of northern Argentina. In: Tankard AJ, Suárez Soruco R, Welsink HJ (eds) Petroleum basins of South America. American Association of Petroleum Geology. Memoir 62:269–283

Oncken O, Chong G, Franz G, Giese P, Götze H-J, Ramos V, Strecker M, Wigger P (eds) (2006) The Andes: active subduction orogeny. Frontiers in earth sciences, XXII. Springer, Berlin

Placzek CJ, Quade J, Patchett PJ (2006) Geochronology and stratigraphy of late Pleistocene lake cycles on the southern Bolivian Altiplano: implications for causes of tropical climate change. Geol Soc Am Bull 118:515–532

Ramos VA (2008) The basement of the Central Andes: the Arequipa and related terranes. Annu Rev Earth Planet Sci 36:289–324

Strecker MR, Alonso RN, Bookhagen B, Carrapa B, Hilley GE, Sobel ER, Trauth MH (2007) Tectonics and climate of the southern Central Andes. Annu Rev Earth Planet Sci 35:747–787

Strecker MR, Alonso RN, Bookhagen B, Carrapa B, Coutand I, Hain MP, Hilley G, Mortimer E, Schoenbohm L, Sobel R (2009) Does the topographic distribution of the Central Andean Puna Plateau result from climatic or geodynamic processes? Geology 37(7):643–646

Trauth MH, Alonso RN, Haselton KR, Hermanns RL, Strecker MR (2000) Climate change and mass movements in the northwest Argentine Andes. Earth Planet Sci Lett 179:243–256

Turner JCM (1972) Puna. In: Leanza AF (ed) Geología regional Argentina. Academia Nacional de Ciencias, Córdoba. Publicación Especial, pp 91–116

Vandervoort DS, Jordan TE, Zeitler PK, Alonso RN (1995) Chronology of internal drainage development and uplift, southern Puna Plateau, Argentine Central Andes. Geology 23:145–148

Chapter 2
Modern Microbial Mats and Endoevaporite Systems in Andean Lakes: A General Approach

María Eugenia Farías and Luis Alberto Saona Acuña

2.1 Background

The area extending through the southwest of Bolivia and the north of Argentina and Chile (the Central Andes of South America) displays a geography of closed basins at high altitudes, where salt flats are very abundant (at Uyuni, Atacama, Arizaro, and Antofalla, among others). These high-altitude environments—also known as the Puna (in Argentina), the Altiplano (in Bolivia), and the Atacama Desert (in Chile)—present not only salt deserts but also a large reservoir of high-altitude Andean lakes (HAALs) and active volcanoes, where inhabitants of both evaporitic deposits and different polyextremophilic microbial ecosystems flourish (Albarracín et al. 2015) (Fig. 2.1 [image 1]). Microbes inhabiting extreme environments face these conditions with different strategies, including formation of complex extremophile microbial communities driven by light and O_2 availability. In that way, biofilms, mats, endoevaporitic mats, domes, and microbialites have been found to exist in association with salars, lagoons, and even volcanic fumaroles in Central Andean extreme environments.

The interactions of microbes and minerals—together with the main concept of the microbial–mineral ecosystems that result from that interaction, and the importance of the Andean Puna as an early Earth or extraterrestrial environment counterpart—are discussed in this chapter.

M. E. Farías (✉)
LIMLA-PROIMI-CONICET, Tucuman, Argentina

L. A. Saona Acuña
Laboratorio de Investigaciones Microbiológicas de Lagunas Andinas (LIMLA), Planta Piloto de Procesos Industriales Microbiológicos (PROIMI), Centros Científico Tecnológicos (CCT), Consejo Nacional de Investigaciones Científicas y Técnicas (CONICET), San Miguel de Tucumán, Tucumán, Argentina

M. E. Farías (ed.), *Microbial Ecosystems in Central Andes Extreme Environments*, https://doi.org/10.1007/978-3-030-36192-1_2

Fig. 2.1 Layered structure of Andean microbial ecosystems. (*1*) Endoevaporitic ecosystems associated with gypsum domes in Llamara, Chile. (*2*) Microbial mat in Tebenquiche, chile. (*3*) Thrombolites in Laguna La Brava, Chile. (*4*) Oncolites in Trs Quebradas Argentina

2.2 Importance of Microbe–Mineral–Earth Changes

To reconstruct the sedimentary geological history of the Earth, it is essential to study the role played by microorganisms in organomineralization (Reid et al. 2000; Dupraz et al. 2004, 2009; Dupraz and Visscher 2005; Glunk et al. 2011). This process has been key since the Precambrian era (Knoll and Golubic 1992; Gerdes 2007; Dupraz et al. 2009) because of microorganism metabolic activity interventions in (1) stabilization of terrestrial sediments (Paterson 1994), (2) changes in both the morphology and the mineral content in sedimentary macrostructures (Braissant et al. 2004), and, consequently (3) changes in the forms of sedimentary macrostructures (Burne and Moore 1987; Reid et al. 2000). In other words, through the organomineralization process, microorganisms can determine the geochemistry of their microenvironment (Visscher and Stolz 2005), impacting on mineral precipitation (Beveridge 1981; Cody and Cody 1989; Gerdes et al. 1994; Ludwig 2004; Douglas 2005; Ali-Bik et al. 2011). To perform these functions, microorganisms have different characteristics. The microbial surface has an anionic charge; through this, microorganisms act as nucleating agents for mineral precipitation (Beveridge 1981; Douglas 2005; Dupraz and Visscher 2005). The production of exopolymeric substances (EPSs) is equally relevant. EPSs form an organic matrix with anionic functional groups that are able to bind metal ions such as Ca^{2+} (Braissant et al. 2007). Thereby, when EPSs are themselves degraded by microorganisms, carbonate minerals precipitate through a process referred to as organomineralization (Dupraz and Visscher 2005; Dupraz et al. 2009). On the basis of evidence from silicified fossil microbialites, it has been suggested that microorganisms have engaged in organomineralization for most of the Earth's history (Awramik and Riding 1988). In this sense, investigations of modern microbialites, including endoevaporitic deposits, may aid the interpretation of ancient analogs and even provide a unique model for understanding of the origin of saline basins on Mars.

2.3 Extreme Conditions in the Andean Altiplano

The area extending through the southwest of Bolivia (the Western Andean flank) and the north of Argentina and Chile (the Central Andes of South America) has a geography of closed basins at high altitude, where salt flats are very abundant (e.g., at Uyuni, Atacama, and Antofalla). These high-altitude deserts—also known as the Puna (in Argentine), the Altiplano (in Bolivia), and the Atacama Desert (in Chile)—present not only salt deserts but also a large reservoir of HAALs and active volcanoes where evaporitic deposits and different polyextremophilic microbial ecosystems flourish (Albarracín et al. 2015) (Fig. 2.1 [image 1]). Because of the climatic conditions and their geographic characteristics, these ecosystems are exposed to a series of multiple extreme conditions such as high ultraviolet (UV) radiation, high dryness, low oxygen pressure, large thermal fluctuations, and volcanic eruptions, among others (Vuille et al. 2000; Vuille and Keimig 2004; Garreaud et al. 2009).

In terms of UV radiation, the Central Andes, and especially the Puna–Altiplano, have recorded the highest solar radiation levels in the world (Duffie and Beckman 2013; Albarracín et al. 2015). To survive these conditions, microbial communities have developed and acquired strategies such as (1) expression of photolyases to minimize damage from direct exposure to UV radiation (Sancar 2003; Albarracín et al. 2013), (2) repair of accumulated damage in cyclobutane pyrimidine dimers (CPDs) (Fernández Zenoff et al. 2006; Albarracín et al. 2012), (3) synthesis of antioxidant enzymes to avoid oxidative stress damage (Albarracín et al. 2011; Di Capua et al. 2011), (4) pigment production (Weber 2005; Kurth et al. 2015; Albarracín et al. 2016b), and (5) biofilm associations with minerals and rocks to avoid direct exposure to UV (Albarracín et al. 2016b; Rascovan et al. 2016).

The microbial communities inhabiting the HAALs are also influenced by the aquatic conditions, which are characterized by a high pH and salinity. In most lagoons studied in Argentina, Chile, and Bolivia, the pH remains alkaline; this is because of the alkalinity of the salt flats (called soda or carbonate salars) (Oremland et al. 2009; Albarracín et al. 2015), which coincides with the presence of carbonates and calcium.

In addition to UV radiation, a high pH, and salinity, these lagoons have high concentrations of heavy metals and metalloids; among them, arsenic is the one most studied (Demergasso et al. 2007; Escudero et al. 2013; Kurth et al. 2017; Rascovan et al. 2016). The presence of arsenic in HAALs is due to geological activity, mainly of volcanic origin (Smedley and Kinniburgh 2002), and its average concentration varies between 0.8 and 11.8 mg L^{-1} (Albarracín et al. 2015). The highest recorded As concentration is in Laguna Diamante, located in the crater of the Galán volcano, which has an arsenic concentration of 230 mg L^{-1} (Rascovan et al. 2016)—almost ten times greater than the concentration present in Mono Lake, a lagoon that has also been studied for its high concentration of arsenic (Oremland et al. 2009; Wolfe-Simon et al. 2011). The implications of the arsenic in Andean microbial ecosystems (AMEs) are discussed in Chap. 19 of this book.

2.4 The Andean Altiplano: A Counterpart of the Early Earth, Where the Oldest Forms of Life on the Planet are Still Alive

To compare these environments as being analogous to the early Earth and, more specifically, analogous to the conditions in which life originated, it is necessary to take into account the physical, geological, chemical, and mineralogical conditions, in addition to the energy sources existing at that time (Russell et al. 2013) (Table 2.1). It is widely assumed that life originated in the oceans at least 3.8 billion years ago (Mojzsis et al. 1996; Rosing 1999; Watanabe et al. 2000; Kempe and Kazmierczak 2002), while the oldest fossil records (stromatolites) date back 2.7 billion years (Fischer 1965). The early oceans had 1.5–2 times higher salt concentrations than the oceans do today (Knauth 1998; Knauth et al. 2005; Cullum et al. 2016). This high salinity lasted for billions of years, during which no biological evolution toward more complex life-forms was evident (Knauth 1998). It was not until the Cambrian explosion that the salt started to come out of the oceans through evaporation processes (Fischer 1965; Knauth 1998; Cullum et al. 2016); in that moment, large evaporitic deposits were formed, accumulating about 50% of the salt present (Knauth 1998). At present, the Atacama region contains the largest and most extensive salt flats in the world, where closed basins with large evaporitic deposits are abundant, at Uyuni, Salinas Grandes Atacama, Arizaro, Antofalla, etc. (Díaz del Río et al. 1972; Rosacher 2002; Fricker et al. 2005). It is commonly assumed that the early Earth's alkaline oceans favored biogenesis (Kempe and Kazmierczak 2002), whereas an acid composition would have favored hydrolysis of proteins (Kempe and Kazmierczak 2002), making accumulation of larger molecules—and thereby also the origin of membranes and other organelles that allow life—impossible. In this sense, the current hypersaline lakes of the Altiplano region, the great majority of which present alkaline conditions (soda lakes) (Albarracín et al. 2016a), acquire greater interest as models analogous to the primitive Earth (Table 2.1).

Another key issue is obtaining energy from free sources in the highly reducing environment where life originated. One of the engines for obtaining energy that is assumed to have been present even before the last unique common ancestor (LUCA) is the arsenite oxidase (Aio) system for oxidizing As (Duval et al. 2008). Thus, what

Table 2.1 Environmental conditions of modern Central Andes Puna vs early Earh

Condition	Primitive Earth conditions (Archean eon, 3.2–3.4 billion years ago)	Extant Atacama conditions	References
Saline water	Continents had not yet formed, and all of the currently accumulated salt was diluted in the sea, which was 1.5–2 times more saline than it is now	Closed basins accumulate salt, forming extensive salt flats and hypersaline lagoons; they are 3–8 times more saline than the sea	Benedetto (2010); Knauth (1998); Knauth et al. (2005)
Highs arsenic and sulfur concentrations	Because of volcanic activity, the primitive Earth was 4 times hotter than Earth is today; sulfur and arsenic are compounds in the Earth's mantle and were released in massive amounts during Archean volcanic activity; at present, the Andes have huge volcanic activity, which influences the chemical composition of underground waters, filling them with arsenic, sulfur, and other compounds		Olsen et al. (1999); Hamaguchi et al. (1969); Oremland et al. (2009); Bundschuh et al. (2008)
High pH (alkalinity)	pH was high (alkaline) because of the close contact of ocean waters with mantle rocks (peridotite) and the influence of hydrothermal springs	pH is high because of the composition of the water, which is full of compounds such as carbon and sulfur, coming from underground waters	Bundschuh et al. (2008); Albarracín et al. (2015); Oremland et al. (2009)
Life ruled by microorganisms	Last unique common ancestor (LUCA)	Bacteria, archaea, cyanobacteria, and diatoms	Rascovan et al. (2016); Delaye (2011)
High ultraviolet radiation	Due to the absence of the ozone layer	Due to the high altitude and desert environment	Holland (1994); Pla-Garcia and Menor-Salván (2017); Albarracín et al. (2013, 2015)
	Due to volcanoes		
Low O_2 pressure	Early earth atmosphere had no free O_2; photosynthesis, the primary process that generates O_2, had not started happening yet	O_2 pressure is lower at high altitude than at sea level; saline waters contain less O_2 than freshwater	Holland (1994); Pla-Garcia and Menor-Salván (2017); Wetzel et al. (2001)

initially started as an independent electrochemical transformation between As(III) and As(V) today is defined as bioenergetic metabolism through which bacteria can obtain energy, using arsenic molecules (Newman et al. 1998; Oremland and Stolz 2003; Kulp et al. 2008; Kulp 2014). This bioenergetic mechanism has been evidenced in lakes in the Altiplano region, which are characterized by their high arsenic concentrations, such as Mono Lake and Searles Lake, among others. They are inhabited by resistant microorganisms that are able to breathe through redox reactions of arsenic molecules, using Aio and Arr enzymes (see Chaps. 8, 15, and 19 of this book) (Albarracín et al. 2016a; Kurth et al. 2017; Rascovan et al. 2016).

In this sense, the evaporitic basins of the Puna region—formed in an environment with a high volcanic incidence; lagoons that present high salinity, an alkaline pH, and very high concentrations of arsenic (as discussed below); and high UV radiation exposure (Albarracín et al. 2011, 2016a, b)—are the best current analogs of the early Earth, where the first forms of life arose. Without going any further, the presence of modern living stromatolites—the same kind of life that inhabited the Earth 3.6 billion years ago—supports this concept (Farías et al. 2013).

2.5 The Andean Altiplano and the Red Planet

Today, the conditions on Mars are extremely hostile to life as we know it (Lederberg and Sagan 1962; Sagan and Lederberg 1976). Nevertheless, discoveries of liquid water in brine from Mars (Fischer et al. 2014) have increased interest in the study of Andean lakes, not only because they are a window to the origin of life but also because they represent an analog of the planet Mars on Earth. They are a unique thalassic environment over 4000 m above sea level (asl) with high UV incidence and temperatures under 40° C, among other relevant characteristics (Whittaker et al. 1997; Cabrol et al. 2007; Wierzchos et al. 2011). It is estimated that Mars—before it lost its atmosphere—had a geological history similar to that of the primitive Earth (Rothschild 1990); thus, if the conditions of that time allowed the origin of life on Earth, it can be hypothesized that life could also have originated on the early Mars (Sagan 1971, 1973; Sagan and Lederberg 1976).

As was discussed earlier, the Andean Altiplano region contains the driest desert in the world (Davila et al. 2008). Despite the scarcity of water, many endolithic microorganisms are able to colonize different soil types (Friedmann et al. 1967; Warren-Rhodes et al. 2006; Wierzchos et al. 2006), including those rich in sulfate (gypsum soils) and halite—minerals that are also present on Mars (Rothschild 1990; Parnell et al. 2004; Dong et al. 2007). Examination of gypsum soils of the Atacama region has shown that they are inhabited mainly by cyanobacteria, which have managed to settle in the maximum-humidity area (Dong et al. 2007). The crusts of salt (halite) can be considered the last available niche for life in hyperarid environments on Earth (Davila et al. 2008).

Evaporitic minerals such as sulfate and halite (which are present in the Atacama and on Mars) are the best candidates to use in the study of the hypothetical habitability of Mars (Farmer and Des Marais 1999; Szynkiewicz et al. 2010; Summons et al. 2011; Glamoclija et al. 2012). In this sense, endolithic microbial communities inhabiting evaporitic rocks in extreme environments, such as the Andean Altiplano region, have been described as the main counterparts of life on Mars (Davila and Schulze-Makuch 2016).

These life-forms have become established in conditions similar to those on Mars (dryness, UV radiation, and salinity). Study of these mechanisms of adaptation and evolution is key to determining what kind of life could exist or could have existed on Mars. Finally, a new concept has been proposed recently: "follow the salt" (Jones 2018). This emphasizes that life on Mars could occur in shallow water brines; therefore, the search for life on Earth—analogous to Mars—would have to be focused on that which subsists under hypersaline conditions (Jones 2018).

2.6 Biofilms, Mats, Evaporitic Mats, and Microbialites

Biofilms, mats, endoevaporitic mats and domes, and microbialites are microbial ecosystems, formed by microbial communities, where the availability of light and O_2 creates a decrease in stratification that promotes functional group diversity (Fig. 2.1). This microbial diversity, together with the geochemistry, may favor the precipitation of minerals. Below, we review the characteristics of different types of ecosystems. Biofilms have been described as a collection of microorganisms and their extracellular products, bound to a solid surface (Marshall 1984) such as rock, plastic, wood, etc. The bacterial colonization of the surface is dependent on the formation of EPSs, and the EPSs can trap inorganic and abiotic compounds, as well as immobilizing water. In that respect, biofilms are similar to microbial mats. Biofilms may also be involved in the formation of mineral precipitates. The species composition may vary from a single species to complex communities. When these microbial communities are distributed in layers in terms of their light and O_2 requirements, then we have a microbial mat. Microbial mats are fibrillar, slime-supported, coherent coatings on sedimentary and rocky surfaces, which—by their morphology, physiology, and arrangement in space and time—interact with the physical and chemical environment to produce a laminated pattern (Krom and Berner 1983). The term "stromatolite" was coined after the discovery of fossil records of these laminated buildups (Kalkowsky 1908). Cyanobacteria constitute the predominant organisms in most microbial mats, although diatoms and fungi may also be present. Microbial mats form on bedding surfaces. Most have a characteristic filamentous morphology; the filaments have rigid cell walls and polymeric sheaths. Laminations develop from interaction of the microbes with the environment in which they are deposited. Of main importance is the capacity of the mat-constructing biota to migrate vertically to escape burial by sediments and to recolonize the newly deposited surface. Such biolaminated slicks can reach several centimeters and even decimeters in thickness. Microbial mats develop in a wide variety of shallow aquatic environments, ranging from freshwater to hypersaline water, and even within a halite or gypsum crust (Stivaletta and Barbieri 2009). The microbial mats associated with halite and gypsum crusts have been referred to as endoevaporitic mats (Rothschild et al. 1994). Precipitation and evaporation are proposed as the main mechanisms in the formation of endoevaporitic mats, including gypsum crusts, some forming "stromatolite-like" domal structures (Babel 2004; Stivaletta et al. 2010). The exact roles of microbial metabolic activity and growth in the formation of these gypsum domes are not well established. In fact, the domes have also been described as evaporites with an associated laminated endolithic microbial community, also referred to as endoevaporites (Rothschild et al. 1994; Spear et al. 2003; Canfield et al. 2004; Sahl et al. 2008). These photosynthetic microbial ecosystems should therefore be present in evaporitic environments where the precipitating salts, such as gypsum, have favorable physical properties, such as translucency and hygroscopy (Rothschild et al. 1994; Oren et al. 1995; Stivaletta and Barbieri 2009).

Extant continental evaporites are found on all continents and originate from the evaporation of saline water in the shallow areas of hypersaline lakes. Evaporites typically consist of chloride-, sulfate-, and potassium-containing minerals.

The most widely studied types of microbial-induced organomineralization are those of carbonate with calcium (as calcite and/or aragonite; see Reid et al. (2000)), calcium and magnesium (as magnesium calcite, or dolomite; see Glunk et al. (2011) and Vasconcelos and McKenzie (1997)), and magnesium (magnesite; see Sanz-Montero and Rodríguez-Aranda (2008) and Thompson and Ferris (1990)). Lithifying microbial mats can form microbialites, which are preserved in the rock record.

Microbialites are organosedimentary deposits accreted by sediment trapping, binding, and in situ precipitation due to the growth and metabolic activities of microorganisms (Walter 1976; Burne and Moore 1987). Geological records indicate that microbialites first appeared 3.5 billion years ago and were the main evidence of life on Earth for the next 2 billion years (Allwood et al. 2006; Schopf 2006). Stromatolites are layered forms of microbialites. As the first communities performing significant oxygenic photosynthesis, they are thought to have played a major role in oxygenation of the Earth's atmosphere (Kasting 1991; Holland 1994).

Most microbialites can be classified into one of three main categories based on their macroscopic features (Dupraz and Strasser 1999; Riding 1991): stromatolites, showing a laminated macrofabric (Monty 1977; Semikhatov et al. 1979); thrombolites, displaying a clotted (mesoclot) macrofabric (Aitken 1967; Kennard and James 1986; Shapiro 2000; Turner et al. 2000); and leiolite, without a well-defined macrofabric (structureless) (Braga et al. 1995). These three types of microbialites can display a wide range of microstructures, including micropeloidal, densely micritic, or agglutinated microfabrics (Riding 1991; Dupraz and Strasser 1999).

References

Aitken JD (1967) Classification and environmental significance of cryptalgal limestones and dolomites, with illustrations from the Cambrian and Ordovician of Southwestern Alberta. J Sediment Petrol 37:1163–1178. https://doi.org/10.1306/74D7185C-2B21-11D7-8648000102C1865D

Albarracín VH, Dib JR, Ordoñez OF, Farías M (2011) A harsh life to indigenous proteobacteria at the Andean mountains: microbial diversity and resistance mechanisms towards extreme conditions. In: Sezenna ML (ed) Proteobacteria: phylogeny, metabolic diversity and ecological effects. Nova Science, Hauppauge, pp 91–131

Albarracín VH, Gärtner W, Farias ME (2016a) Forged under the sun: life and art of extremophiles from Andean lakes. Photochem Photobiol 92:14–28. https://doi.org/10.1111/php.12555

Albarracín VH, Gärtner W, Farías ME (2013) UV resistance and photoreactivation of extremophiles from high-altitude Andean Lakes. J Photochem Photobiol B Biol:1–4
Albarracín VH, Kraiselburd I, Bamann C et al (2016b) Functional green-tuned proteorhodopsin from modern stromatolites. PLoS One 11:1–18. https://doi.org/10.1371/journal.pone.0154962
Albarracín VH, Kurth D, Ordoñez OF et al (2015) High-up: a remote reservoir of microbial extremophiles in Central Andean wetlands. Front Microbiol 6:1404. https://doi.org/10.3389/fmicb.2015.01404
Albarracín VH, Pathak GP, Douki T et al (2012) Extremophilic acinetobacter strains from high-altitude lakes in Argentinean Puna: remarkable UV-B resistance and efficient DNA damage repair. Orig Life Evol Biosph 42:201. https://doi.org/10.1007/s11084-012-9276-3
Ali-Bik MW, Metwally HIM, Kamel MG, Wali AMA (2011) Gypsum and dolomite biomineralization in endoevaporitic microbial niche, EMISAL, Fayium, Egypt. Environ Earth Sci 62:151–159. https://doi.org/10.1007/s12665-010-0510-x
Allwood AC, Walter MR, Kamber BS et al (2006) Stromatolite reef from the Early Archaean era of Australia. Nature 441:714–718. https://doi.org/10.1038/nature04764
Awramik SM, Riding R (1988) Role of algal eukaryotes in subtidal columnar stromatolite formation. Proc Natl Acad Sci U S A 85:1327–1329. https://doi.org/10.1073/pnas.85.5.1327
Babel M (2004) Models for evaporite, selenite and gypsum microbialite deposition in ancient saline basins. Acta Geol Pol 54:219–249
Benedetto JL (2010) El continente de Gondwana a través del tiempo. Una Introducción a la Geología Histórica. Academia Nacional de Ciencias, Córdoba
Beveridge TJ (1981) Ultrastructure, chemistry, and function of the bacterial wall. Int Rev Cytol 72:229–317. https://doi.org/10.1016/S0074-7696(08)61198-5
Braga JC, Martin JM, Riding R (1995) Controls on microbial dome fabric development along a carbonate-siliciclastic shelf-basin transect, Miocene, SE Spain. PALAIOS 10:347. https://doi.org/10.2307/3515160
Braissant O, Cailleau G, Aragno M, Verrecchia EP (2004) Biologically induced mineralization in the tree *Milicia excelsa* (Moraceae): its causes and consequences to the environment. Geobiology 2:59. https://doi.org/10.1111/j.1472-4677.2004.00019.x
Braissant O, Decho AW, Dupraz C et al (2007) Exopolymeric substances of sulfate-reducing bacteria: interactions with calcium at alkaline pH and implication for formation of carbonate minerals. Geobiology 5:401. https://doi.org/10.1111/j.1472-4669.2007.00117.x
Bundschuh J, Giménez-Forcada E, Guerequiz R, Pérez-Carrera A, Garcia ME, Mello J, Deschamps E (2008) Fuentes geogénicas de arsénico y su liberación al medio ambiente. In: Bundschuh J, Pérez-Carrera A, Litter MI (eds) Distribución del arsénico en las regiones Ibérica e Iberoamericana. Editorial Programa Iberoamericano de Ciencia y Tecnologia para el Desarrollo, Buenos Aires, pp 311–321
Burne RV, Moore LS (1987) Microbialites: organosedimentary deposits of benthic microbial communities. PALAIOS 2:241. https://doi.org/10.2307/3514674
Cabrol NA, Grin EA, Hock AN (2007) Mitigation of environmental extremes as a possible indicator of extended habitat sustainability for lakes on early Mars. In: Hoover RB, Levin GV, Rozanov AY, Davies PCW (eds) Proceedings Vol 6694, Instruments, methods, and missions for astrobiology X. Proceedings of Optical Engineering + Applications, San Diego, 26–30 Aug 2007. https://doi.org/10.1117/12.731506
Canfield DE, Sorensen KB, Oren A (2004) Biogeochemistry of a gypsum-encrusted microbial ecosystem. Geobiology 2:133–150. https://doi.org/10.1111/j.1472-4677.2004.00029.x
Cody AM, Cody RD (1989) Evidence for micro-biological induction of {101} montmartre twinning of gypsum ($CaSO_4 \cdots 2H_2O$). J Cryst Growth 98:721. https://doi.org/10.1016/0022-0248(89)90310-2
Cullum J, Stevens DP, Joshi MM (2016) Importance of ocean salinity for climate and habitability. Proc Natl Acad Sci U S A 113:4278–4283. https://doi.org/10.1073/pnas.1522034113
Davila AF, Gomez-Silva B, De los Ríos A, et al. (2008) Facilitation of endolithic microbial survival in the hyperarid core of the Atacama Desert by mineral deliquescence. J Geophys Res Biogeosci 113(G1). doi:10.1029/2007JG000561

Davila AF, Schulze-Makuch D (2016) The last possible outposts for life on Mars. Astrobiology 16:159–168. https://doi.org/10.1089/ast.2015.1380

Delaye L (2011) Last universal common ancestor. In: Encyclopedia of astrobiology. Springer, Berlin, pp 906–909

Demergasso CS, Guillermo CD, Lorena EG et al (2007) Microbial precipitation of arsenic sulfides in Andean salt flats. Geomicrobiol J 24:111–123. https://doi.org/10.1080/01490450701266605

Di Capua C, Bortolotti A, Farías ME, Cortez N (2011) UV-resistant *Acinetobacter* sp. isolates from Andean wetlands display high catalase activity. FEMS Microbiol Lett 317:181–189

Díaz del Río G, Bonilla Parra R, Peralta Toro F (1972) Geología de superficie, sub-superficie y geoquímica del Salar de Atacama. CORFO, Departamento de Recursos Hidráulicos, Chile

Dong H, Rech JA, Jiang H et al (2007) Endolithic cyanobacteria in soil gypsum: occurrences in Atacama (Chile), Mojave (United States), and Al-Jafr Basin (Jordan) deserts. J Geophys Res Biogeosci 112:1–11. https://doi.org/10.1029/2006JG000385

Douglas S (2005) Mineralogical footprints of microbial life. Am J Sci 305:503–525

Duffie JA, Beckman WA (2013) Solar engineering of thermal processes. 4th edn. Hoboken: Wiley

Dupraz C, Reid RP, Braissant O et al (2009) Processes of carbonate precipitation in modern microbial mats. Earth Sci Rev 96:141–162

Dupraz C, Strasser A (1999) Microbialites and micro-encrusters in shallow coral bioherms (Middle to Late Oxfordian, Swiss Jura mountains). Facies 40:101. https://doi.org/10.1007/BF02537471

Dupraz C, Visscher PT (2005) Microbial lithification in marine stromatolites and hypersaline mats. Trends Microbiol 13:429

Dupraz C, Visscher PT, Baumgartner LK, Reid RP (2004) Microbe–mineral interactions: early carbonate precipitation in a hypersaline lake (Eleuthera Island, Bahamas). Sedimentology 51:745–765. https://doi.org/10.1111/j.1365-3091.2004.00649.x

Duval S, Ducluzeau AL, Nitschke W, Schoepp-Cothenet B (2008) Enzyme phylogenies as markers for the oxidation state of the environment: the case of respiratory arsenate reductase and related enzymes. BMC Evol Biol 8:206

Escudero LV, Casamayor EO, Chong G et al (2013) Distribution of microbial arsenic reduction, oxidation and extrusion genes along a wide range of environmental arsenic concentrations. PLoS One 8:e78890. https://doi.org/10.1371/journal.pone.0078890

Farías ME, Rascovan N, Toneatti DM et al (2013) The discovery of stromatolites developing at 3570 m above sea level in a high-altitude volcanic lake Socompa, Argentinean Andes. PLoS One 8:e53497. https://doi.org/10.1371/journal.pone.0053497

Farmer JD, Des Marais DJ (1999) Exploring for a record of ancient Martian life. J Geophys Res Planets 104:26977–26995. https://doi.org/10.1029/1998JE000540

Fernández Zenoff V, Siñeriz F, Farías ME (2006) Diverse responses to UV-B radiation and repair mechanisms of bacteria isolated from high-altitude aquatic environments. Appl Environ Microbiol 72:7857. https://doi.org/10.1128/AEM.01333-06

Fischer AG (1965) Fossils, early life, and atmospheric history. Proc Natl Acad Sci U S A 53:1205–1215

Fischer E, Martínez GM, Elliott HM, Rennó NO (2014) Experimental evidence for the formation of liquid saline water on Mars. Geophys Res Lett 41:4456–4462. https://doi.org/10.1002/2014GL060302

Fricker HA, Borsa A, Minster B et al (2005) Assessment of ICESat performance at the Salar de Uyuni, Bolivia. Geophys Res Lett 32:1–5. https://doi.org/10.1029/2005GL023423

Friedmann EI, Lipkin Y, Ocampo-Paus R (1967) Desert algae of the Negev (Israel). Phycologia 6:185–196

Garreaud RD, Vuille M, Compagnucci R, Marengo J (2009) Present-day South American climate. Palaeogeogr Palaeoclimatol Palaeoecol 281:180–195. https://doi.org/10.1016/J.PALAEO.2007.10.032

Gerdes G (2007) Structures left by modern microbial mats in their host sediments. In: Atlas of microbial mat features preserved within the siliciclastic rock record. Elsevier, Amsterdam, pp 5–38

Gerdes G, Dunajtschik-Piewak K, Riege H et al (1994) Structural diversity of biogenic carbonate particles in microbial mats. Sedimentology 41:1273–1294. https://doi.org/10.1111/j.1365-3091.1994.tb01453.x

Glamoclija M, Fogel ML, Steele A, Kish A (2012) Microbial nitrogen and sulfur cycles at the gypsum dunes of White Sands National Monument. Geomicrobiol J 29:733. https://doi.org/10.1080/01490451.2011.608111

Glunk C, Dupraz C, Braissant O et al (2011) Microbially mediated carbonate precipitation in a hypersaline lake, Big Pond (Eleuthera, Bahamas). Sedimentology 58:720–736. https://doi.org/10.1111/j.1365-3091.2010.01180.x

Hamaguchi H, Onuma N, Hirao Y, Yokoyama H, Bando S, Furukawa M (1969) The abundances of arsenic, tin and antimony in chonditic meteorites. Geochim Cosmochim Acta 33:507–518

Holland HD (1994) Early Proterozoic atmospheric change. In: Bengston S (ed) Early life on Earth. Columbia University Press, New York, pp 237–244

Jones EG (2018) Shallow transient liquid water environments on present-day Mars, and their implications for life. Acta Astronaut 146:144–150. https://doi.org/10.1016/j.actaastro.2018.02.027

Kalkowsky E (1908) Oolith und Stromatolith im nord-deutschen Buntsandstein. Z D Geol Gesell 60:68–125

Kasting JF (1991) Box models for the evolution of atmospheric oxygen: an update. Glob Planet Change 97:125–131

Kempe S, Kazmierczak J (2002) Biogenesis and early life on Earth and Europa: favored by an alkaline ocean? Astrobiology 2:123. https://doi.org/10.1089/153110702753621394

Kennard JM, James NP (1986) Thrombolites and types of stromatolites: two microbial structures distinct. PALAIOS 1:492–503. https://doi.org/10.2307/3514631

Knauth LP (1998) Salinity history of the Earth's early ocean. Nature 395:554–555

Knauth LP, Burt DM, Wohletz KH (2005) Impact origin of sediments at the opportunity landing site on Mars. Nature 438:1123–1128. https://doi.org/10.1038/nature04383

Knoll AH, Golubic S (1992) Proterozoic and living cyanobacteria. In: Schidlowski M, Golubic S, Kimberley M et al (eds) Early organic evolution. Springer, Berlin, pp 450–462

Krom MD, Berner RA (1983) A rapid method for the determination of organic and carbonate carbon in geological samples. J Sediment Petrol 53:660–663

Kulp TR (2014) Early Earth: arsenic and primordial life. Nat Geosci 7:785–786. https://doi.org/10.1038/ngeO2275

Kulp TR, Hoeft SE, Asao M et al (2008) Arsenic(III) fuels anoxygenic photosynthesis in hot spring biofilms from Mono Lake, California. Science 321:967–970

Kurth D, Amadio A, Ordoñez OF et al (2017) Arsenic metabolism in high altitude modern stromatolites revealed by metagenomic analysis. Sci Rep 7:1024. https://doi.org/10.1038/s41598-017-00896-0

Kurth D, Belfiore C, Gorriti MF, et al. (2015) Genomic and proteomic evidences unravel the UV-resistome of the poly-extremophile *Acinetobacter* sp. Ver3. Front Microbiol 6:328. https://doi.org/10.3389/fmicb.2015.00328

Lederberg J, Sagan C (1962) Microenvironments for life on Mars. Proc Natl Acad Sci U S A 48:1473–1475

Ludwig R (2004) Carbon cycling and calcification in hypersaline microbial mats. Thesis, Universität Bremen

Marshall KC (1984) Microbial adhesion and aggregation. In: Dahlem Konferenzen. Springer, Berlin

Mojzsis SJ, Arrehenius G, McKeegan KD et al (1996) Evidence for life on Earth before 3,800 million years ago. Nature 384:55–59

Monty C (1977) Evolving concepts on the nature and the ecological significance of stromatolites. In: Flügel E (ed) Fossil algae: recent results and developments. Springer, Berlin, pp 15–35

Newman DK, Ahmann D, Morel FMM (1998) A brief review of microbial arsenate respiration. Geomicrobiology 15:255–268

Olsen EJ, Kracher A, Davis AM, Steele IM, Hutcheon ID, Bunch TE (1999) The phosphates of IIIAB iron meteorites. Meteorit Planet Sci 34:285–300

Oremland RS, Saltikov CW, Wolfe-Simon F, Stolz JF (2009) Arsenic in the evolution of Earth and extraterrestrial ecosystems. Geomicrobiol J 26:522. https://doi.org/10.1080/01490450903102525

Oremland RS, Stolz JF (2003) The ecology of arsenic. Science 300:939–944. https://doi.org/10.1126/science.1081903

Oren A, Kuhl M, Karsten U (1995) An endoevaporitic microbial mat within a gypsum crust: zonation of phototrophs, photopigments, and light penetration. Mar Ecol Prog Ser 128:151. https://doi.org/10.3354/meps128151

Parnell J, Lee P, Cockell CS, Osinski GR (2004) Microbial colonization in impact-generated hydrothermal sulphate deposits, Haughton impact structure, and implications for sulphates on Mars. Int J Astrobiol 3:247–256

Paterson DM (1994) Microbiological mediation of sediment structure and behavior. In: Stal LJ, Caumette P (e) (eds) Microbial mats. Springer, Berlin, pp 97–109

Pla-Garcia J, Menor-Salván C. (2017) La composición química de la atmósfera primitiva del planeta Tierra. An Quím 113(1):16-26

Rascovan N, Maldonado J, Vazquez MP, Eugenia Farías M (2016) Metagenomic study of red biofilms from Diamante Lake reveals ancient arsenic bioenergetics in haloarchaea. ISME J 10:299–309. https://doi.org/10.1038/ismej.2015.109

Reid RP, Visscher PT, Decho AW et al (2000) The role of microbes in accretion, lamination and early lithification of modern marine stromatolites. Nature 406:989–992. https://doi.org/10.1038/35023158

Riding R (1991) Classification of microbial carbonates. In: Riding R (ed) Calcareous algae and stromatolites. Springer, New York, pp 21–51

Rosing MT (1999) ^{13}C-depleted carbon microparticles in ≥3700-Ma sea-floor sedimentary rocks from west Greenland. Science 283:674–676. https://doi.org/10.1126/science.283.5402.674

Rothschild L (1990) Earth analogs for Martian life: Microbes in evaporites, a new model system for life on Mars. Icarus 88:246–260

Rothschild LJ, Giver LJ, White MR, Manceinelli RL (1994) Metabolic activity of miroorganisms in evaporites. J Phycol 30:431–438

Russell MJ, Nitschke W, Branscomb E (2013) The inevitable journey to being. Philos Trans R Soc B Biol Sci 368:20120254. https://doi.org/10.1098/rstb.2012.0254

Sagan C (1971) The long winter model of Martian biology: a speculation. Icarus 15:511–514

Sagan C (1973) Ultraviolet selection pressure on the earliest organisms. J Theor Biol 39:195–200

Sagan C, Lederberg J (1976) The prospects for life on Mars: a pre-Viking assessment. Icarus 28:291–300. https://doi.org/10.1016/0019-1035(76)90039-7

Sahl JW, Pace NR, Spear JR (2008) Comparative molecular analysis of endoevaporitic microbial communities. Appl Environ Microbiol, https://doi.org/10.1128/AEM.00879-08 74:6444

Sancar A (2003) Photolyase and cryptochrome blue-light photoreceptors. Chem Rev 103:2203–2237

Sanz-Montero ME, Rodríguez-Aranda JP (2008) Participación microbiana en la formación de magnesita dentro de un ambiente lacustre evaporítico: Mioceno de la Cuenca de Madrid. Rev Soc Esp Mineral 9:231–232

Schopf JW (2006) Fossil evidence of Archaean life. Philos Trans R Soc B Biol Sci 361:869–885. https://doi.org/10.1098/rstb.2006.1834

Semikhatov MA, Gebelein CD, Cloud P, Awkamik SM, Benmore WC (1979) Stromatolite morphogenesis—progress and problems. Can J Earth Sci 16:992–1015. https://doi.org/10.1139/e79-088

Shapiro R (2000) A comment on the systematic confusion of thrombolites. PALAIOS 15:166–169

Smedley PL, Kinniburgh DG (2002) A review of the source, behaviour and distribution of arsenic in natural waters. Appl Geochem 17:517–568

Spear JR, Ley RE, Berger AB, Pace NR (2003) Complexity in natural microbial ecosystems: the Guerrero Negro experience. Biol Bull 204:168–173

Stivaletta N, Barbieri R (2009) Endolithic microorganisms from spring mound evaporite deposits (southern Tunisia). J Arid Environ 73:33–39. https://doi.org/10.1016/j.jaridenv.2008.09.024

Stivaletta N, Ĺopez-Garćia P, Boihem L et al (2010) Biomarkers of endolithic communities within gypsum crusts (southern Tunisia). Geomicrobiol J 27:101–110. https://doi.org/10.1080/01490450903410431

Summons RE, Amend JP, Bish D et al (2011) Preservation of Martian organic and environmental records: final report of the Mars Biosignature Working Group. Astrobiology 11:157. https://doi.org/10.1089/ast.2010.0506

Szynkiewicz A, Ewing RC, Moore CH et al (2010) Origin of terrestrial gypsum dunes—implications for Martian gypsum-rich dunes of Olympia Undae. Geomorphology 121:69. https://doi.org/10.1016/j.geomorph.2009.02.017

Thompson JB, Ferris FG (1990) Cyanobacterial precipitation of gypsum, calcite, and magnesite from natural alkaline lake water. Geology 18:995–998

Turner EC, Narbonne GM, James NP (2000) Framework composition of early Neoproterozoic calcimicrobial reefs and associated microbialites, Mackenzie Mountains, N.W.T., Canada. In: Grotzinger JP, James NP (eds) Carbonate sedimentation and diagenesis in the evolving Precambrian world. Society for Sedimentary Geology, Tulsa

Vasconcelos C, McKenzie JA (1997) Microbial mediation of modern dolomite precipitation and diagenesis under anoxic conditions (Lagoa Vermelha, Rio de Janeiro, Brazil). J Sediment Res 67:378–390

Visscher PT, Stolz JF (2005) Microbial mats as bioreactors: populations, processes, and products. Palaeogeogr Palaeoclimatol Palaeoecol 219:87–100. https://doi.org/10.1016/j.palaeo.2004.10.016

Vuille M, Bradley R, Keimig F (2000) Interannual climate variability in the Central Andes and its relation to tropical Pacific and Atlantic forcing. J Geophys Res 150:12447–12460. https://doi.org/10.1029/2000JD900134

Vuille M, Keimig F (2004) Interannual variability of summertime convective cloudiness and precipitation in the Central Andes derived from ISCCP-B3 data. J Clim 17:3334. https://doi.org/10.1175/1520-0442(2004)017<3334:IVOSCC>2.0.CO;2

Walter MR (1976) Stromatolites. Elsevier, Amsterdam

Warren-Rhodes KA, Rhodes KL, Pointing SB et al (2006) Hypolithic cyanobacteria, dry limit of photosynthesis, and microbial ecology in the hyperarid Atacama Desert. Microb Ecol 52:389–398

Watanabe Y, Martini JE, Ohmoto H (2000) Geochemical evidence for terrestrial ecosystems 2.6 billion years ago. Nature 408:574–578. https://doi.org/10.1038/35046052

Weber S (2005) Light-driven enzymatic catalysis of DNA repair: a review of recent biophysical studies on photolyase. Biochim Biophys Acta 1707(1):1–23

Wetzel RG, Diego S, Francisco S, et al. (2001) Limnology: lake and river ecosystems. 3rd edn. San Diego: Academic

Whittaker W, Bapna D, Maimone M, Rollins E (1997) Atacama Desert trek: a planetary analog field experiment. In: Proceedings of the 7th International Symposium for Artificial Intelligence, Robotics, and Automation for Space, Tokyo, July 1997

Wierzchos J, Ascaso C, McKay CP (2006) Endolithic cyanobacteria in halite rocks from the hyperarid core of the Atacama Desert. Astrobiology 6:415–422

Wierzchos J, Cámara B, De los Ríos A et al (2011) Microbial colonization of Ca-sulfate crusts in the hyperarid core of the Atacama Desert: implications for the search for life on Mars. Geobiology 9:44–60. https://doi.org/10.1111/j.1472-4669.2010.00254.x

Wolfe-Simon F, Blum JS, Kulp TR et al (2011) A bacterium that can grow by using arsenic instead of phosphorus. Science 332:1163–1166

Part II
Preandean and Andean Atacama Desert: Life at Limits

Chapter 3
The Desert Polyextreme Environment and Endolithic Habitats

Jacek Wierzchos, Carmen Ascaso, Octavio Artieda, and María Cristina Casero

3.1 Introduction

According to the Dutch botanist and microbiologist Lourens Baas-Becking, "Everything is everywhere, but, the environment selects" (Baas-Becking 1934). His idea was that while microbial life occurs worldwide, in a given environmental setting, most microbial species are only latently present (De Wit and Bouvier 2006). The combined extreme environmental conditions of hyperarid deserts give rise to perhaps the harshest setting faced by microbial life. As such, desert microbial ecosystems are excellent models to address the environmental selection of a terrestrial biome and the limits of life on our planet. The aridity of the desert environment implies a scarcity of water. Water is the single most important requirement for life on Earth, and theoretically, there is a threshold in the natural environment—the dry limit—where liquid water is too scarce for the full range of necessary functions required to sustain viable populations of organisms. In fact, there are some known environments with multiple and/or simultaneous forms of stress that determine the limits of life on Earth. While these environments can be considered polyextreme, surprisingly, they may be inhabited by microbiota able to survive such multiple extreme conditions. Hence, polyextreme environments are excellent microbial ecology models to study the biochemical resistance mechanisms of microorganisms.

This chapter describes the endolithic microbial communities discovered in the past decade within the hyperarid zone of the Atacama Desert (north Chile). Considering the scope of this book, we focused our attention on the Preandean and

J. Wierzchos (✉) · C. Ascaso · M. C. Casero
Grupo de Ecología y Geomicrobiología, Dpto. Biogeoquímica y Ecología Microbiana, Museo Nacional de Ciencias Naturales, Madrid, Spain
e-mail: j.wierzchos@mncn.csic.es

O. Artieda
Dpto. Biología Vegetal, Ecología y Ciencias de la Tierra, Centro Universitario de Plasencia, Universidad de Extremadura, Plasencia, Spain

M. E. Farías (ed.), *Microbial Ecosystems in Central Andes Extreme Environments*, https://doi.org/10.1007/978-3-030-36192-1_3

Andean zones of the Atacama Desert, excluding zones of the hyperarid core (Longitudinal Valley) harboring endolithic microbial communities colonizing the interior of halite (NaCl) rocks (Wierzchos et al. 2006; Robinson et al. 2015). After a description of the polyextreme environment of this part of the desert, we focus on its lithobiontic microbial habitats and their characteristics.

3.2 Characterizing the Polyextreme Environment of the Atacama Desert

While polyextremophilic/polyextremotolerant microorganisms as a group can survive individual extreme conditions, there are some effects of polyextreme stress combinations that have yet to be characterized. Where do we find one of the most polyextreme environments on Earth? Arid and hyperarid environments comprise over 30% of the Earth's land and as such represent the most extensive terrestrial biome (see Pointing and Belnap 2012). Among the deserts, the Atacama Desert is perhaps the most challenging polyextreme environment on Earth and is the most barren region imaginable. Its hyperarid climate is the result of the confluence of a subtropical high-pressure zone, the cold Humboldt Current on the coast, offshore winds, the Andean rain-shadow effect, and the latitudinal position of the region (Houston and Hartley 2003).The Longitudinal Valley of the Atacama Desert is the driest place on Earth (Hartley et al. 2005; Houston and Hartley 2003) with scarce precipitation of about 3 mm year^{-1} (McKay et al. 2003; Wierzchos et al. 2012a) and an extremely low mean annual relative humidity (RH) of 17.3% reported by Azúa-Bustos et al. (2015). Figure 3.1 shows the general physiography of the central part of the Atacama Desert.

The Preandean Depression of the Atacama Desert (up to 3300 m a.s.l.) represents a unique environment where there is synergy among various extreme climate and physical factors. The region's geological setting is also diverse, spanning from volcanic to evaporitic rocks that evolved during a long-lasting period of aridity (Hartley et al. 2005; Bull et al. 2018). The Preandean Atacama Desert experiences extremely dry conditions, with a mean annual RH of 16.5% in the Cordón de Lila area (Wierzchos et al. 2015), which is even lower than the humidity recorded in the Longitudinal Valley. Precipitation is extremely scarce, the mean annual value being 24–27 mm year^{-1}. Although this level of precipitation might seem higher than for the Longitudinal Valley of the Atacama Desert, Aridity Index (AI = annual rainfall precipitation/mean annual potential evapotranspiration) values have been calculated at around 0.0093 and 0.0075 for the Valle de la Luna and Cordón Lila areas, respectively (DiRuggiero et al. 2013; Wierzchos et al. 2013, 2015). These values are far below the threshold of 0.05 used to define the world's hyperarid zones (Houston and Hartley 2003). The extreme aridity of this location is also reflected by long timespans between scarce precipitations.

As expected, rainfall in the Western Cordillera, Altiplano, and Andean "Puna" areas is higher and has been reported below 200 mm year^{-1} for altitudes above

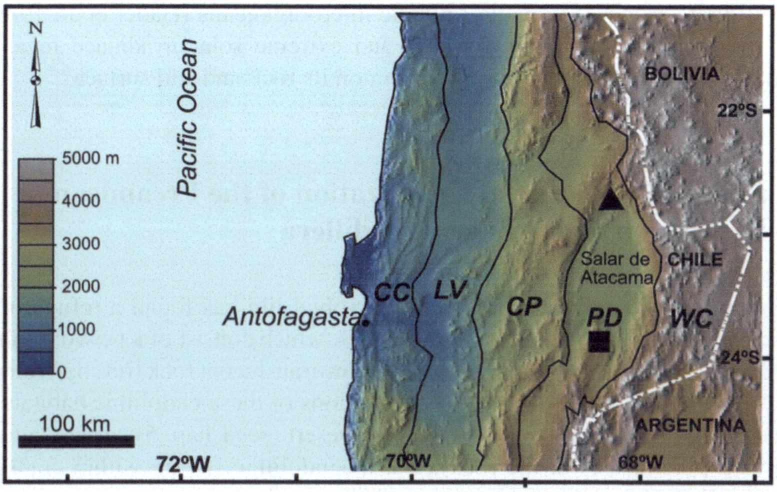

Fig. 3.1 General physiography of the central part of the Atacama Desert (Chile): *CC* Coastal Cordillera, *LV* Longitudinal Valley, *CP* Chilean Precordillera, *PD* Preandean Depression, *WC* Western Cordillera. Key field locations: black triangle = Valle de la Luna area, black square = Cordón de Lila area

4000 m a.s.l. and below 150 mm year^{-1} for altitudes between 3000 and 4000 m a.s.l., respectively, for latitudes between 27°S and 21°S (Tapia et al. 2018). Nevertheless, the vegetation in these areas is extremely sparse, with a maximum of 30% cover from 3800 to 4000 m (Núñez et al. 2002). Springs and a few perennial creeks are focal points for any life present. These parts of the Atacama Desert show mean annual temperatures of 12–17.3 °C (Meslier et al. 2018) and about 0 °C for the Preandean Depression and Western Cordillera, respectively.

The Atacama Desert, including Preandean, Western Cordillera, and Altiplano areas, holds other word records for an extreme environment. These zones are known as the places on Earth with the highest surface UV irradiance, where ground-based measurements have shown the UV index to exceed 20 (Cordero et al. 2014) and 43.3 (Cabrol et al. 2014), respectively. However, this last value was questioned by McKenzie et al. (2015). Annual UV-B (UV-A) doses have been estimated at about 5 kWh/m^2 (160 kWh/m^2) on the Andean plateau (Cordero et al. 2018). Moreover, the most extreme values for total solar and UV irradiances (1528 W m^{-2}) have been reported for "Puna de Atacama" by Piacentini et al. (2003). All these mentioned zones of the Atacama Desert are characterized by their locations (relatively close to the equator), high altitudes, relatively low total ozone column values, low relative humidity, prevalent cloudless conditions, and low aerosol loading (McKenzie et al. 2015). Further, UV enhancement was observed through the cloud-scattering effect (Piacentini et al. 2003; Cordero et al. 2014). While solar radiation is essential for phototrophic life, extremely high levels of UV radiation are harmful to cellular components (Jeffrey et al. 1996; Phoenix et al. 2006), and excessive photosynthetic active radiation (PAR) leads to photo-inhibition through damage to the reaction center of photosystem II in photosynthetic microorganisms (Häder et al. 1986). In effect, the combination of hyperaridity and extreme solar irradiance means the Preandean area virtually lacks any life form on its rock and soil surfaces.

3.3 Endolithic Microbial Colonization of the Preandean Depression and Western Cordillera

In this inhospitable polyextreme desert, microbial life has found a refuge in very specific endolithic (inside rocks) microhabitats, which consist of a network of pores and fissures connected to the surface within semi-translucent rock (rev. by Wierzchos et al. 2012b, 2018). Three different rock positions of these endolithic habitats have been described within rocks of the Atacama Desert (see Chap. 5): cryptoendolithic (occupying pore spaces in the rock), chasmoendolithic (living within cracks and fissures in the rock), and hypoendolithic (living inside the rock but close to the soil interface at the bottom). Endolithic colonization can be viewed as a stress avoidance strategy, whereby the overlying mineral substrate provides certain protection from incident lethal UV and PAR radiation and also offers enhanced moisture availability (Walker and Pace 2007; Wierzchos et al. 2012b). We would like to emphasize our discovery, characterization, and detailed description of novel endolithic microbial communities and their habitats in the hyperarid zone of the Preandean Depression of the Atacama Desert. These habitats are as diverse as the insides of gypsum (DiRuggiero et al. 2013; Ziolkowski et al. 2013), gypcrete (Meslier et al. 2018; Vítek et al. 2016; Wierzchos et al. 2015), calcite (Meslier et al. 2018; DiRuggiero et al. 2013), volcanic rocks (ignimbrite and rhyolite) (Meslier et al. 2018; Vítek et al. 2017; Crits-Christoph et al. 2016; Cámara et al. 2015; Wierzchos et al. 2013; DiRuggiero et al. 2013), and granite (Meslier et al. 2018). Communities are chlorophototrophic-based, with primary producers supporting a diversity of heterotrophic microorganisms (rev. in Wierzchos et al. 2018). Molecular and microscopy characterization of these endolithic microbial communities shows that the interior of the majority of rocks is colonized by communities dominated by *Cyanobacteria*, mostly of the Chroococcidiopsis genus, and Actinobacteria, Proteobacteria, Chloroflexi, *Bacterioidetes*, *Euryarchaeota* phyla and other heterotrophic microorganisms (Meslier et al. 2018). Eukaryotic algae have only been found in the gypcrete substrate (Wierzchos et al. 2015). Fungi as representative of another Eukaryotic biota were isolated from the Western Cordillera region close to the Sairecabur volcano (Pulschen et al. 2015). These fungi were described as *Cryptococcus friedmannii*, *Exophiala* sp., *Holtermanniella watticus*, *Rhodosporidium toruloides*. Despite higher rainfall in the Western Cordillera, unexpectedly, endolithic colonization is extremely scarce. We found very few examples of endolithic habitats within ignimbrite harboring fungi, phototrophic and heterotrophic bacteria in the Los Moaís de

Tara area and close to the Miscanti volcano (J. Wierzchos, pers. com.). The explanation for this paradox could be high weathering and exfoliation rates of rock surfaces induced by moisture and strong winds. We can hypothesis that endolithic colonization will thrive even in this very harsh polyextreme environment provided lithic substrates are stable over long time periods. Lower rates of weathering and exfoliation of rock surfaces can be expected in more arid environments, such as the Preandean Depression, when climate turned from relatively humid conditions during late glacial and early Holocene times to hyperarid conditions. In northern Chile, this change began around 14,000 year B.P. and culminated between 13,000 and 9500 year B.P. (Núñez et al. 2002).

3.4 Drivers of Endolithic Microbial Colonization: Rock Physical Properties and Their Architecture

3.4.1 Rock Porosity

The porosity of rocks connected to rock surface pores and fissures is an essential condition for endolithic colonization. However, absolute values of porosity measured by mercury intrusion porosimetry (MIP) or porosity measured by total immersion of rocks in water (TIW) gives only general information about the total volume of pores. For instance, reported porosity measured by MIP is 25.6–56% for gypcrete (Wierzchos et al. 2015) and 14.4% for ignimbrite (Wierzchos et al. 2013). The TIW method gives total porosity values of 12.15% for gypcrete, 17% for ignimbrite, 1.65% for calcite, and 0.3% for granite (Meslier et al. 2018). This span of absolute values of porosity clearly indicates that not only the total volume of pores should be taken into account for a better description of the bioreceptivity of endolithic habitats and rocks but also the pore shape, pore size distribution, pore distribution within the rock, and pore connectivity. X-ray computed tomography (CTScan) is a promising method that allows for porosity quantification within a whole piece of rock and also three-dimensional (3-D) distribution and visualization of the pore system within rocks (Needham et al. 2013). Figure 3.2 shows some examples of this approach for the above-mentioned endolithically colonized lithic substrates. All rocks show a porosity, and it plays an important role in water retention capacity, water absorption, and capillary condensation (Wierzchos et al. 2012a) acting as a moisture reservoir. Of note, in CTScan images, all these rocks displayed a significantly lower porosity close to the rock surface, which in the case of gypcrete was also confirmed by MIP (Wierzchos et al. 2015). This feature might indicate what we call the "eggshell" shield effect in which moisture within the rocks is protected against rapid evaporation rates, promoting endolithic colonization.

Fig. 3.2 Cross sections of endolithically colonized rocks (**a–d**) and corresponding CTScan 3-D reconstruction images (**a′–d′**). (**a**) calcite (close to VL) with black arrow, chasmoendolithic habitat and blue arrow, cryptoendolithic habitat; (**b**) gypcrete with arrow, cryptoendolithic habitat (CL); (**c**) ignimbrite with arrows, cryptoendolithic habitat (CL); (**d**) granite with arrows chasmoendolithic habitat (CL). (**a′–d′**) CTScan images for calcite, gypcrete, ignimbrite, and granite, respectively, where yellow signal shows the 3-D distribution of pores and fissures. CTScan images were obtained using Nikon Xtech 160 equipment (MNCN-CSIC, Madrid, Spain). Scale bar = 1 cm for all images

3.4.2 Water

Water is the most important resource for endolithic microbial communities and the most significant factor shaping community structure and composition (Meslier et al. 2018). Precipitation in the form of rain seems to be an essential source of liquid water for endolithic ecosystems. However, in the Preandean Depression area, rainfall is scarce (up to 27 mm year^{-1}) with long intervals between rain and evaporation rates as high as 2920 mm year^{-1} (Houston 2006). Our results indicate that water availability might depend on sources other than rainfall. One of the most important moisture sources for endolithic communities is water vapor condensation at night, namely dewfall, and absorption of this liquid water by the pore system. It is commonly accepted that dewfall can be a substantial source of moisture in many of the world's deserts (Büdel et al. 2008; Kidron et al. 2000; Kidron and Temina 2013). The frequency, duration, and water abundance of dewfall depends on atmospheric conditions such air T and air RH (air dew point conditions). However, dewfall on rock surfaces also depends on rock surface thermal conductivity (DiRuggiero et al. 2013). This property may induce a significantly lower rock surface temperature than air temperature and the formation of dewfall. This phenomenon for calcite colonized by chasmoendolithic microbial communities was reported by (DiRuggiero et al. 2013). These authors found microkarstic features on calcite rock surfaces attributed to frequent dewfall, as a common and important source of liquid water. Recently, through electrical conductivity measurements of the calcite surface (from VL) over a 2 year period, we were able to clearly show the presence of liquid water formed by dew during more than 1400 h year^{-1} (J. Wierzchos, pers. comm.).

Similarly, more than 300 h year^{-1} liquid water was deposited by dew on the surface of endolithically colonized gypcrete (CL) (J. Wierzchos, pers. comm.).

Because water is essential for endolithic microorganisms, they also suffer the most serious stress for all organisms—dehydration stress. These microorganisms can be considered as poikilohydric, as they faithfully combine features of poikilohydrous microorganisms such as: (1) the capacity to tolerate extreme and prolonged desiccation; (2) the capacity to enter into an anabiotic state when desiccated; (3) extreme heat resistance in the anabiotic state; and (4) the capacity to take up water as vapor, or liquid and instantaneously switch on metabolic activity (and vice versa).

3.4.3 Rock Light Properties

Extremely high UV and PAR irradiance along with desiccation could arrest the growth and proliferation of epilithic (inhabiting the rock surface) communities in extremely dry deserts (Cockell et al. 2008; Quesada et al. 1999) such as the Preandean Depression zone of the Atacama Desert. In such an environment, endoliths are better adapted to tolerate UV radiation by virtue of their strategic location and production of UV-screening compounds such as scytonemin and carotenoids (Cockell et al. 2003; Vítek et al. 2014a, b, 2016, 2017). Within endolithic habitats, the significantly attenuated solar irradiance allows for photosynthetic cyanobacteria to thrive as primary producers of these ecosystems. The position of the photosynthetic endolithic habitat is conditioned by light penetration, which depends on rock transparency, light reflection, absorption and dispersion by minerals, as well as by the presence, thickness, and optical properties of rock varnish. According to Walker and Pace (2007) and Matthes et al. (2001), no significant relationship exists between light penetration and depth of colonization. However, recent comprehensive studies of this relationship indicate clear correlation between rock transmittance (data based on light transmission spectra) and the depth of endoliths within the endolithically colonized rocks of the Preandean Depression (Meslier et al. 2018). These authors reported that spectral transmission varied greatly across substrates with values ranging from 7% to less than 0.001% of incident solar light at 2 mm below the rock surface. Calcite emerged as the most translucent substrate, with an average depth of the photosynthetic endolithic zone of 7 mm, followed by granite, gypsum and by ignimbrite as the least translucent, with a 2 mm deep photosynthetic endolithic zone. In addition, significantly higher transmission was reported in the red and infrared part of the spectrum (>700 nm) than for blue wavelengths (400–500 nm) (Meslier et al. 2018). It was also noted that all lithic substrates blocked UV radiation at a 2 mm depth. These data reinforce the hypothesis that light is the major driver for the localization of the endolithic colonization habitat but does not affect community richness or diversity. Similar avoidance of UV exposures relying on colonization of deeper positions by cyanobacteria in aragonite ($CaCO_3$) composed benthonic stromatolites from high-altitude volcanic lakes (Andes) was reported by Farías et al. (2013) and Toneatti et al. (2017).

3.4.4 Rock Thermal Properties

Several authors consider endolithic colonization as a strategy to avoid thermal stress by ameliorating or buffering the effects of temperature fluctuations within the endolithic microenvironment (Walker and Pace 2007; Gómez-Silva 2018; Wierzchos et al. 2012b; Nienow 2009). Perhaps this is true for polar deserts, but in the light of our recent measurements of air and endolithically colonized rock surface temperatures, the endolithic habitats of the Preandean Atacama Desert are extremely hot during mid-day solar irradiance, significantly warmer than the environmental temperature. Figure 3.3a–d shows two-dimensional (2-D) mid-day austral summer

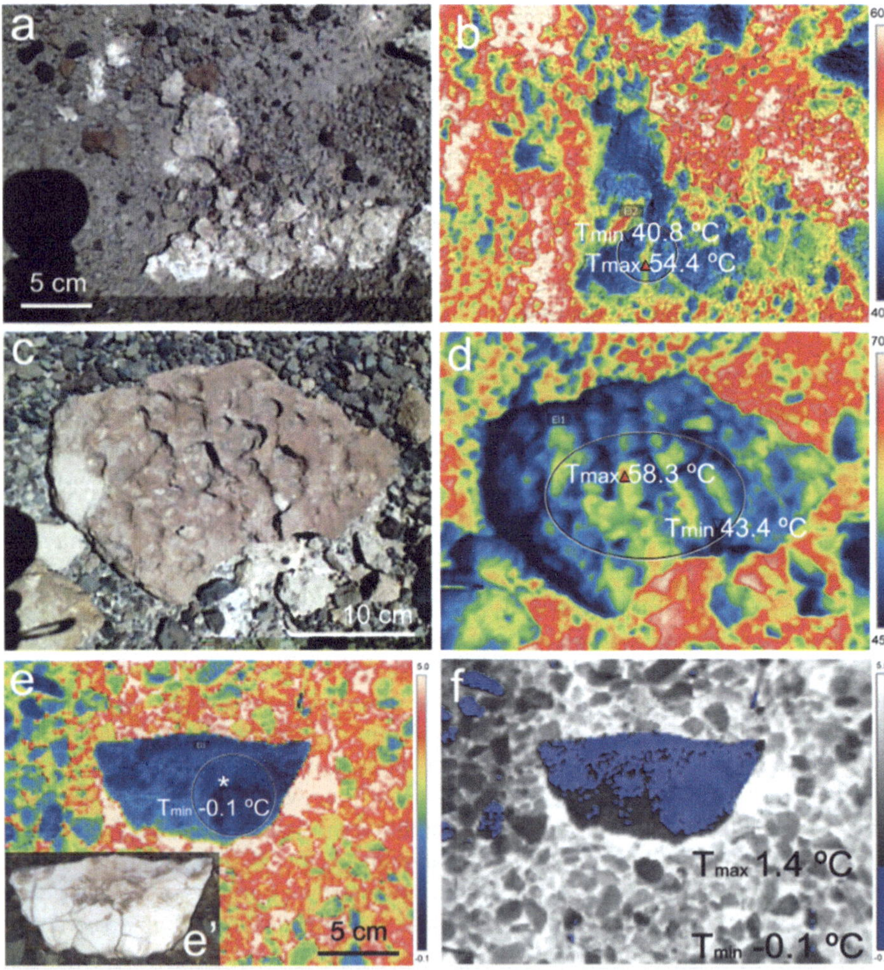

Fig. 3.3 Gypcrete (**a**, **b**), ignimbrite (**c**, **d**), and calcite (**e**, **f**) images and 2-D temperature distributions on rock surfaces measured by infrared radiation; FLIR, E40 camera system

temperature distributions on endolithically colonized gypcrete (CL) and ignimbrite (CL), and early morning temperature distributions on endolithically colonized calcite (VL). Note that maximum recorded temperatures for gypcrete and ignimbrite were 54.4 and 58.3 °C, respectively, when the ambient air temperature recorded was 34 °C.

These data clearly demonstrate that even in the temperate Atacama Desert, temperatures in the endolithic habitat are extremely high, much higher than air *T*, and rocks seem unable to mitigate such high temperatures. Although desert microorganisms can tolerate temperature extremes in excess of 60 °C when dormant (Schlesinger et al. 2003; Tracy et al. 2010), high temperatures could be an important limiting factor for endolithic growth, as temperatures in excess of 50 °C can quickly vaporize all the available water. Cooler early morning hours may be favorable for the metabolic activities of such endolithic communities, as host rocks could absorb water formed by the condensation of dew. This scenario is illustrated in Fig. 3.3e, f, where the calcite (VL) rock surface temperature was measured at 6 a.m. At this time, air temperature was 17 °C, air RH was 35%, and the dew point was 1.4 °C, meaning that any material with a temperature lower than 1.4 °C should condensate liquid dewfall water. The minimum T of this rock surface was −0.1 °C and *T* ranged between 1.4 °C and −0.1 °C, shown in blue in Fig. 3.3f. These 2-D blue spots reveal the presence of liquid water available for endolithic communities due to dewfall. Note that the mid-day *T* of this rock was 39 °C (data not shown) when air temperature was 32 °C, clearly indicating no buffering effect of the calcite rock. In summary, the extremely high temperatures of the endolithic habitat in the Atacama Desert could be another extreme factor for microbial life. However, the high thermal conductivity of colonized rocks might facilitate water vapor condensation due to dewfall condensation.

3.4.5 Rock Architecture

Rocks may be characterized by several physical properties such as: porosity, pore shape, pore size distribution, pore distribution within the rock, pore connectivity, thermal conductivity, permeability, density, water retention capacity, light transmittance, light absorbance, light reflectance, light scattering, dissolution, and mineralogical and chemical properties. Synergy among all these properties could promote the rock's bioreceptivity to endolithic colonization in the polyextreme environment of the Atacama Desert. We suggest the term "rock architecture" (sensu (Wierzchos et al. 2015) rather than "rock structure" to emphasize the functional role of the rock interior as a complex but realistic driver of the success of endolithic colonization. All porous rocks have structure, yet very few show a suitable architecture for endolithic microbial colonization, even in the polyextreme environment of the Atacama Desert.

3.5 Concluding Remarks

The hyperarid Preandean Depression of the Atacama Desert and its arid Western Cordillera is one of the most abiotic places on Earth with life-threatening polyextreme conditions. Despite water scarcity, extreme UV and PAR radiation, and the high temperatures of its rock substrates, microbial life, mostly represented by primary producers, has survival strategies that enable it to colonize endolithic habitats within porous rocks. Microbial life in these microhabitats thrives because of the maximum availability of scarce water and adaptive strategies to avoid excess solar irradiance and extreme evapotranspiration rates. This is made possible by the complex structural and mineralogical characteristics of the lithic substrate conceptually called the rock's "habitable architecture". Atacama's endolithic microbial communities are ubiquitous and provide tractable model systems for extreme environment microbial ecology studies.

References

Azúa-Bustos A, Caro-Lara L, Vicuña R (2015) Discovery and microbial content of the driest site of the hyperarid Atacama Desert, Chile. Environ Microbiol Rep 7(3):388–394. https://doi.org/10.1111/1758-2229.12261

Baas-Becking LGM (1934) Geobiologie of Inleiding tot de Milieukunde. WP Van Stockum & Zoon, The Hague, The Netherlands

Büdel B, Bendix J, Bicker FR, Allan Green TG (2008) Dewfall as a water source frequently activates the endolithic cyanobacterial communities in the granites of Taylor Valley, Antarctica1. J Phycol 44(6):1415–1424. https://doi.org/10.1111/j.1529-8817.2008.00608.x

Bull AT, Andrews BA, Dorador C, Goodfellow M (2018) Introducing the Atacama Desert. Antonie Van Leeuwenhoek 111(8):1269–1272. https://doi.org/10.1007/s10482-018-1100-2

Cabrol NA, Feister U, Häder DP, Piazena H, Grin EA, Klein A (2014) Record solar UV irradiance in the tropical Andes. Front Environ Sci 2. https://doi.org/10.3389/fenvs.2014.00019

Cámara B, Suzuki S, Nealson KH, Wierzchos J, Ascaso C, Artieda O, de los Ríos A (2015) Ignimbrite textural properties as determinants of endolithic colonization patterns from hyperarid Atacama Desert. Int Microbiol 17(4):235–247. https://doi.org/10.2436/20.1501.01.226

Cockell CS, McKay CP, Warren-Rhodes K, Horneck G (2008) Ultraviolet radiation-induced limitation to epilithic microbial growth in arid deserts-dosimetric experiments in the hyperarid core of the Atacama Desert. J Photochem Photobiol B 90:79–87

Cockell C, Rettberg EP, Horneck EG, Scherer K, Stokes EMD (2003) Measurements of microbial protection from ultraviolet radiation in polar terrestrial microhabitats. Photochem Photobiol 26:62–69

Cordero RR, Damiani A, Jorquera J, Sepúlveda E, Caballero M, Fernandez S, Feron S, Llanillo PJ, Carrasco J, Laroze D, Labbe F (2018) Ultraviolet radiation in the Atacama Desert. Antonie Van Leeuwenhoek 111(8):1301–1313. https://doi.org/10.1007/s10482-018-1075-z

Cordero RR, Seckmeyer G, Damiani A, Riechelmann S, Rayas J, Labbe F, Laroze D (2014) The world's highest levels of surface UV. Photochem Photobiol Sci 13(1):70–81. https://doi.org/10.1039/c3pp50221j

Crits-Christoph A, Robinson CK, Ma B, Ravel J, Wierzchos J, Ascaso C, Artieda O, Souza-Egipsy V, Casero MC, DiRuggiero J (2016) Phylogenetic and functional substrate specificity for endolithic microbial communities in hyper-arid environments. Front Microbiol 7:301. https://doi.org/10.3389/fmicb.2016.00301

De Wit R, Bouvier T (2006) 'Everything is everywhere, but, the environment selects'; what did Baas Becking and Beijerinck really say? Environ Microbiol 8(4):755–758. https://doi.org/10.1111/j.1462-2920.2006.01017.x

DiRuggiero J, Wierzchos J, Robinson CK, Souterre T, Ravel J, Artieda O, Souza-Egipsy V, Ascaso C (2013) Microbial colonisation of chasmoendolithic habitats in the hyper-arid zone of the Atacama Desert. Biogeosciences 10(4):2439–2450. https://doi.org/10.5194/bg-10-2439-2013

Farías ME, Rascovan N, Toneatti DM, Albarracín VH, Flores MR, Poiré DG, Collavino MM, Aguilar OM, Vazquez MP, Polerecky L (2013) The discovery of stromatolites developing at 3570 m above sea level in a high-altitude volcanic Lake Socompa, Argentinean Andes. PLoS One 8(1):e53497. https://doi.org/10.1371/journal.pone.0053497

Gómez-Silva B (2018) Lithobiontic life: "Atacama rocks are well and alive". Antonie Van Leeuwenhoek 111:1333. https://doi.org/10.1007/s10482-018-1033-9

Häder DP, Watanabe M, Furuya M (1986) Inhibition of motility in the cyanobacterium, Phormidium uncinatum, by solar and monochromatic UV irradiation. Plant Cell Physiol 27(5):887–894

Hartley AJ, Chong G, Houston J, Mather AE (2005) 150 million years of climatic stability: evidence from the Atacama Desert, northern Chile. J Geol Soc 162(3):421–424. https://doi.org/10.1144/0016-764904-071

Houston J (2006) Variability of precipitation in the Atacama desert: its causes and hydrological impact. Int J Climatol 26(15):2181–2198. https://doi.org/10.1002/joc.1359

Houston J, Hartley AJ (2003) The central Andean west-slope rainshadow and its potential contribution to the origin of hyper-aridity in the Atacama Desert. Int J Climatol 23(12):1453–1464. https://doi.org/10.1002/joc.938

Jeffrey WH, Aas P, Lyons MM, Coffin RB, Pledger RJ, Mitchell DL (1996) Ambient solar radiation-induced photodamage in marine bacterioplankton. Photochem Photobiol 64(3):419–427

Kidron GJ, Barzilay E, Sachs E (2000) Microclimate control upon sand microbiotic crusts, western Negev Desert, Israel. Geomorphology 36(1–2):1–18

Kidron GJ, Temina M (2013) The effect of dew and fog on lithic lichens along an altitudinal gradient in the Negev desert. Geomicrobiol J 30(4):281–290. https://doi.org/10.1080/01490451.2012.672542

Matthes U, Turner SJ, Larson DW (2001) Light attenuation by limestone rock and its constraint on the depth distribution of endolithic algae and cyanobacteria. Int J Plant Sci 162(2):263–270. https://doi.org/10.1086/319570

McKay CP, Friedmann EI, Gómez-Silva B, Cáceres-Villanueva L, Andersen DT (2003) Temperature and moisture conditions for life in the of observations including the El Niño of 1997–1998. Astrobiology 3:393–406

McKenzie RL, Bernhard G, Madronich S, Zaratti F (2015) Comment on "Record solar UV irradiance in the tropical Andes, by Cabrol et al.". Front Environ Sci 3:26. https://doi.org/10.3389/fenvs.2015.00026

Meslier V, Casero MC, Dailey M, Wierzchos J, Ascaso C, Artieda O, McCullough PR, DiRuggiero J (2018) Fundamental drivers for endolithic microbial community assemblies in the hyperarid Atacama Desert. Environ Microbiol 20(5):1765–1781. https://doi.org/10.1111/1462-2920.14106

Needham AW, Abel RL, Tomkinson T, Grady MM (2013) Martian subsurface fluid pathways and 3D mineralogy of the Nakhla meteorite. Geochim Cosmochim Acta 116:96–110. https://doi.org/10.1016/j.gca.2012.07.004

Nienow JA (2009) Extremophiles: dry environments (including Cryptoendoliths). In: Encyclopedia of microbiology. Elsevier, Oxford, pp 159–173

Núñez L, Grosjean M, Cartajena I (2002) Human occupations and climate change in the Puna de Atacama, Chile. Science 298(5594):821–824. https://doi.org/10.1126/science.1076449

Phoenix VR, Bennett PC, Engel AS, Tyler SW, Ferris FG (2006) Chilean high-altitude hot-spring sinters: a model system for UV screening mechanisms by early Precambrian cyanobacteria. Geobiology 4(1):15–28

Piacentini RD, Cede A, Bárcena H (2003) Extreme solar total and UV irradiances due to cloud effect measured near the summer solstice at the high-altitude desertic plateau Puna of Atacama (Argentina). J Atmos Sol Terr Phys 65(6):727–731

Pointing SB, Belnap J (2012) Microbial colonization and controls in dryland systems. Nat Rev Microbiol 10(8):551–562

Pulschen AA, Rodrigues F, Duarte RTD, Araujo GG, Santiago IF, Paulino-Lima IG, Rosa CA, Kato MJ, Pellizari VH, Galante D (2015) UV-resistant yeasts isolated from a high-altitude volcanic area on the Atacama Desert as eukaryotic models for astrobiology. Microbiologyopen 4(4):574–588. https://doi.org/10.1002/mbo3.262

Quesada A, Vincent WF, Lean DRS (1999) Community and pigment structure of Arctic cyanobacterial assemblages: the occurrence and distribution of UV-absorbing compounds. FEMS Microbiol Ecol 28(4):315–323. https://doi.org/10.1016/s0168-6496(98)00114-7

Robinson CK, Wierzchos J, Black C, Crits-Christoph A, Ma B, Ravel J, Ascaso C, Artieda O, Valea S, Roldan M, Gomez-Silva B, DiRuggiero J (2015) Microbial diversity and the presence of algae in halite endolithic communities are correlated to atmospheric moisture in the hyper-arid zone of the Atacama Desert. Environ Microbiol 17(2):299–315. https://doi.org/10.1111/1462-2920.12364

Schlesinger WH, Pippen JS, Wallenstein MD, Hofmockel KS, Klepeis DM, Mahall BE (2003) Community composition and photosynthesis by photoautotrophs under quartz pebbles, southern Mojave Desert. Ecology 84(12):3222–3231

Tapia J, González R, Townley B, Oliveros V, Álvarez F, Aguilar G, Menzies A, Calderón M (2018) Geology and geochemistry of the Atacama Desert. Antonie Van Leeuwenhoek 111(8):1273–1291. https://doi.org/10.1007/s10482-018-1024-x

Toneatti DM, Albarracín VH, Flores MR, Polerecky L, Farías ME (2017) Stratified bacterial diversity along physico-chemical gradients in high-altitude modern stromatolites. Front Microbiol 8:646. https://doi.org/10.3389/fmicb.2017.00646

Tracy CR, Streten-Joyce C, Dalton R, Nussear KE, Gibb KS, Christian KA (2010) Microclimate and limits to photosynthesis in a diverse community of hypolithic cyanobacteria in northern Australia. Environ Microbiol 12(3):592–607. https://doi.org/10.1111/j.1462-2920.2009.02098.x

Vítek P, Ascaso C, Artieda O, Casero MC, Wierzchos J (2017) Discovery of carotenoid red-shift in endolithic cyanobacteria from the Atacama Desert. Sci Rep 7(1):11116. https://doi.org/10.1038/s41598-017-11581-7

Vítek P, Ascaso C, Artieda O, Wierzchos J (2016) Raman imaging in geomicrobiology: endolithic phototrophic microorganisms in gypsum from the extreme sun irradiation area in the Atacama Desert. Anal Bioanal Chem 408(15):4083–4092. https://doi.org/10.1007/s00216-016-9497-9

Vítek P, Jehlička J, Ascaso C, Mašek V, Gómez-Silva B, Olivares H, Wierzchos J (2014a) Distribution of scytonemin in endolithic microbial communities from halite crusts in the hyperarid zone of the Atacama Desert, Chile. FEMS Microbiol Ecol 90(2):351–366. https://doi.org/10.1111/1574-6941.12387

Vítek P, Jehlicka J, Edwards HGM, Hutchinson I, Ascaso C, Wierzchos J (2014b) Miniaturized Raman instrumentation detects carotenoids in Mars-analogue rocks from the Mojave and Atacama deserts. Philos Trans R Soc A Math Phys Eng Sci 372(2030):20140196

Walker JJ, Pace NR (2007) Endolithic microbial ecosystems. Annu Rev Microbiol 61:331–347. https://doi.org/10.1146/annurev.micro.61.080706.093302

Wierzchos J, Ascaso C, McKay CP (2006) Endolithic cyanobacteria in halite rocks from the hyperarid core of the Atacama Desert. Astrobiology 6(3):415–422

Wierzchos J, Casero MC, Artieda O, Ascaso C (2018) Endolithic microbial habitats as refuges for life in polyextreme environment of the Atacama Desert. Curr Opin Microbiol 43:124–131. https://doi.org/10.1016/j.mib.2018.01.003

Wierzchos J, Davila AF, Artieda O, Cámara-Gallego B, de los Ríos A, Nealson KH, Valea S, Teresa García-González M, Ascaso C (2013) Ignimbrite as a substrate for endolithic life in the hyperarid Atacama Desert: implications for the search for life on Mars. Icarus 224(2):334–346. https://doi.org/10.1016/j.icarus.2012.06.009

Wierzchos J, Davila AF, Sánchez-Almazo IM, Hajnos M, Swieboda R, Ascaso C (2012b) Novel water source for endolithic life in the hyperarid core of the Atacama Desert. Biogeosciences 9(6):2275–2286

Wierzchos J, de los Ríos A, Ascaso C (2012a) Microorganisms in desert rocks: the edge of life on Earth. Int Microbiol 15(4):173–183

Wierzchos J, DiRuggiero J, Vitek P, Artieda O, Souza-Egipsy V, Skaloud P, Tisza M, Davila AF, Vilchez C, Garbayo I, Ascaso C (2015) Adaptation strategies of endolithic chlorophototrophs to survive the hyperarid and extreme solar radiation environment of the Atacama Desert. Front Microbiol 6:934. https://doi.org/10.3389/fmicb.2015.00934

Ziolkowski LA, Wierzchos J, Davila AF, Slater GF (2013) Radiocarbon evidence of active endolithic microbial communities in the hyperarid core of the Atacama desert. Astrobiology 13(7):607–616. https://doi.org/10.1089/ast.2012.0854

Chapter 4
Preandean Atacama Desert Endolithic Microbiology

María Cristina Casero, Victoria Meslier, Jacek Wierzchos, and Jocelyne DiRuggiero

4.1 Introduction

Desert microbial communities are adapted to extreme environmental conditions and are particularly sensitive to climate change (IPCC 2012). In both hot and cold deserts, these communities are subjected to high ultraviolet (UV) and solar radiation, scarcity of water, intense desiccation, strong temperature fluctuations, and oligotrophic conditions (Wierzchos et al. 2012; Billi et al. 2017). As such, they are highly sensitive to perturbation and therefore of great interest for predicting the impacts of today's changing climate. Microorganisms from desert communities are known as extremophiles. They have been used as model systems to investigate the limits of life (Dassarma 2006; Pikuta et al. 2007), and a number of studies have addressed their metabolic diversity and survival strategies (Dassarma 2006; Pointing and Belnap 2012; Wierzchos et al. 2018). Additionally, because the most arid deserts around the world (Atacama, Antarctica, Mojave, The Qaidam Basin) are analogues for Mars' environment, the study of desert extremophiles might help guide our search for life elsewhere (Fairén et al. 2010; Foing et al. 2011; Smith et al. 2014; Xiao et al. 2017; Bull et al. 2018). More recently, extremophiles have gained interest in applied research as potential sources for high-value bioactive compounds due to their ability to resist extreme environmental conditions (Finore et al. 2016; Stan-Lotter and Fendrihan 2017; Neifar et al. 2015).

M. C. Casero (✉) · J. Wierzchos
Grupo de Ecología y Geomicrobiología, Dpto. Biogeoquímica y Ecología Microbiana, Museo Nacional de Ciencias Naturales, CSIC, Madrid, Spain
e-mail: mcristina.casero@mncn.csic.es

V. Meslier· J. DiRuggiero (✉)
Department of Biology, Johns Hopkins University, Baltimore, MD, USA
e-mail: jdiruggiero@jhu.edu

M. E. Farías (ed.), *Microbial Ecosystems in Central Andes Extreme Environments*, https://doi.org/10.1007/978-3-030-36192-1_4

In the world's arid and hyperarid deserts, microorganisms find refuge inside rock substrates as a survival strategy, colonizing what is known as the endolithic habitat (Golubic et al. 1981; Wierzchos et al. 2012). There are many examples of these endolithic microbial communities (EMCs), which are photosynthetic-based with primary producers supporting a diversity of heterotrophic microorganisms (Friedmann 1980; Friedmann et al. 1988; De Los Ríos et al. 2005, 2014; Dong et al. 2007; Schmidt et al. 2011; Wei et al. 2015) and more specifically in the Atacama Desert (Wierzchos et al. 2006, 2011, 2013, 2015; DiRuggiero et al. 2013; Robinson et al. 2015; Crits-Christoph et al. 2016a, b; Meslier et al. 2018).

Several ecological properties of EMCs, initially proposed by (Friedmann and Ocampo-Friedmann 1984) and summarized by (Walker and Pace 2007) include:

(a) EMCs are among the simplest microbial ecosystems.
(b) EMCs are characterized by a core group of microorganisms that co-occur within a defined habitat.
(c) The extreme endolithic environment is seeded from a relatively small reservoir of microorganisms highly adapted to this unique environment.
(d) The composition of EMCs is influenced by biogeography and environmental factors such as the physical and chemical properties of substrates and climate.

Most studies of EMCs from the Preandean Depression of the Atacama Desert have focused on the determination of (1) the diversity, structure, and composition of the communities (who is there?) (Crits-Christoph et al. 2016b; Meslier et al. 2018; DiRuggiero et al. 2013), (2) the physicochemical structure of the substrate, also called the architecture, and the spatial organization of the community within that substrate (where are they?) (Wierzchos et al. 2018; Meslier et al. 2018; Cámara et al. 2015), and (3) adaptation strategies at the community and/or cellular levels (how are they able to survive?) (Wierzchos et al. 2015; Vítek et al. 2016, 2017).

In this chapter we will focus on the diversity, structure, and composition of EMCs from the Preandean zone of the hyperarid Atacama Desert, emphasizing the need to comprehensively address the diversity of these communities while considering the principles underlying their assembly, such as low-complexity, highly adapted microbiota, and influences of the biogeography, climate, and rock properties.

4.2 A Multidisciplinary Approach for Describing EMCs Structure, Composition, and Spatial Arrangement

The first reports on EMCs from hyperarid deserts involved the use of direct microscopy methods to visualize the microbe-rock interface and the endolithic settings (Wierzchos and Ascaso 2001 and references therein). At the time, EMCs were mostly characterized by their phototrophic members (prokaryotic- or eukaryotic-based communities), because of the difficulty in identifying heterotrophic members solely based on their morphology (Friedmann et al. 1988; De Los Ríos et al. 2014).

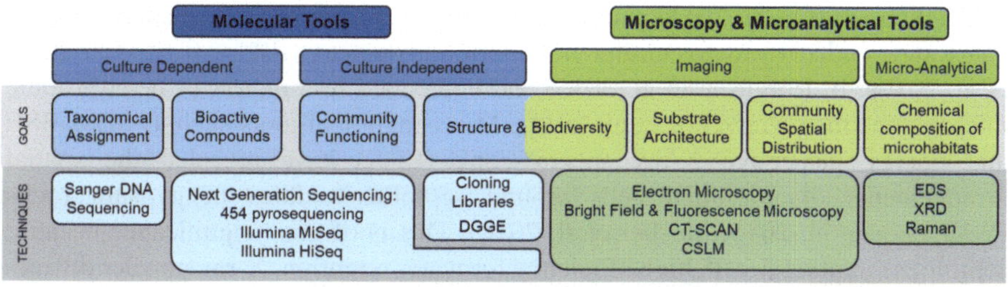

Fig. 4.1 Research tools used to study EMCs classified by techniques and specific goals

Early identifications of the heterotrophic component of EMCs were performed using culture-based methods (Hirsch et al. 1988; Siebert and Hirsch 1988), but because of the limitations of these methods, their diversity remained largely unexplored at the time. The emergence of high-throughput sequencing tools has deeply changed our view of microbial diversity across ecosystems, including that of the endolithic habitat (Walker and Pace 2007). Nowadays, characterization of EMCs, including those found in the Preandean area of the Atacama Desert, is routinely done with a multidisciplinary approach that involves sophisticated microscopy and molecular tools (Fig. 4.1).

The main contributions of microscopy and microanalytical tools to the study of EMCs have been the visualization and characterization of the microhabitat, including the spatial distribution of the microorganisms within. Scanning electron microscopy in backscattered electron mode (SEM-BSE) was essential in defining the different endolithic microhabitats and type of colonization, enabling one to distinguish (1) the cryptoendolithic habitat (Golubic et al. 1981), where microorganisms colonized the pore spaces of the lithic substrate, (2) the chasmoendolithic habitat (Golubic et al. 1981) characterized by colonized cracks and fissures, and (3) the hypoendolithic habitat (Wierzchos et al. 2011), where the colonization is located underneath the lithic substrate. This method has also provided invaluable information on the distribution of the microorganisms within each of these microhabitats (Wierzchos et al. 2011, 2013, 2018; Cámara et al. 2015; Crits-Christoph et al. 2016b; DiRuggiero et al. 2013; Meslier et al. 2018) and, together with computed tomography scanning (CT-Scan), allowed the description of the substrate's architecture (Wierzchos et al. 2018). The concept of the substrate's architecture has been defined as the space available for colonization and includes the pores, fissures, and cracks of the substrate and how they are connected to the surface (Wierzchos et al. 2015). The use of SEM at low temperatures (LTSEM) brought additional information, such as cytological identification of cells in situ and the characterization of their ultrastructural features (Wierzchos and Ascaso 2001). Other microscopy methods such as bright field microscopy provided the identification of the major phototrophic members (Wierzchos and Ascaso 2001; Wierzchos et al. 2013, 2015), while fluorescent microscopy and confocal laser scanning microscopy (CSLM) gave us essential information about the metabolic status of the microorganisms inhabiting EMCs. In particular, these methods revealed assemblages of live and intact dead cells (Wierzchos et al. 2011) and the spatial organization of cyanobacterial aggregates surrounded by extracellular polymeric substances (EPSs) (Crits-Christoph et al. 2016a, b; Robinson et al. 2015) and embedded in a matrix of heterotrophic microorganisms (Wierzchos et al. 2011). Microanalytical tools such as SEM-BSE and energy dispersive X-ray spectroscopy (EDS) have revealed the spatial arrangements of microbial cells around sepiolite nodules in gypsum rocks (Wierzchos et al. 2015; Meslier et al. 2018). This is of great significance because sepiolite is a mineral with high efficiency for water retention. X-ray powder diffraction (XRD) has also provided the mineral composition of many different lithic substrates, such as calcite, ignimbrite, and gypsum (Wierzchos et al. 2015; Meslier et al. 2018), and Raman spectroscopy has been used to characterize the distribution of pigments in the endolithic microhabitat of gypsum, ignimbrite, and halite (Vítek et al. 2010, 2013, 2014a, b, 2016, 2017).

Microscopy tools combined with culture-independent methods such as denaturing gradient gel electrophoresis (DGGE) (Wierzchos et al. 2013; Cámara et al. 2015) and, more recently, next generation sequencing (NGS) (DiRuggiero et al. 2013; Meslier et al. 2018; Crits-Christoph et al. 2016b; Wierzchos et al. 2015) have been used to characterize the phylogenetic diversity of EMCs. Pioneer studies revealed the discrepancies of biomass estimates between microscopy and molecular methods. For example, Dong et al. (2007) reported that cyanobacteria in gypsum endolithic communities represented 95% of the community when using microscopy, whereas the number fell to 40% when molecular methods were used. They suggested that those differences were potentially the result of contamination, the difficulty in isolating DNA from cyanobacteria, and the bias introduced by amplification of the DNA by PCR (Dong et al. 2007). Today, molecular tools such as amplicon sequencing of marker genes (i.e., 16S rRNA gene) and whole genome shotgun sequencing (WGS), combined with microscopy methods, have led to a more complete description of the endolithic microbiome. Habitats for these microbiomes include (1) the chasmoendolithic habitat of calcite (DiRuggiero et al. 2013; Meslier et al. 2018; Crits-Christoph et al. 2016b) and granite (Meslier et al. 2018) and (2) the cryptoendolithic habitat found in gypsum (Wierzchos et al. 2015; Dong et al. 2007; Meslier et al. 2018) and ignimbrite (Wierzchos et al. 2012; Cámara et al. 2015; Crits-Christoph et al. 2016b; Meslier et al. 2018) (Table 4.1).

Some of the most diverse EMCs found in the Preandean Atacama Desert are that of gypsum from Cordón de Lila and Tilocalar (Table 4.1). Phototrophs in these communities included *Cyanobacteria* (36–83%) and algae belonging to the *Chlorophyta* class. The algae were only detected at very low abundance, first by microscopy (Wierzchos et al. 2015), and their occurrence was later confirmed by cloning of the 18S rRNA gene and by metagenome sequencing (Meslier, pers. com.). Major heterotrophic bacteria of the gypsum EMCs included *Actinobacteria* (10–25%) and *Proteobacteria* (13–30%) and at lower relative abundance *Chloroflexi* (0–11%) and *Gemmatimonadetes* (0–6%) (Table 4.1).

Another high diversity EMC of the Preandean area of the Atacama Desert is the chasmoendolithic community of calcite from the nearby Valle de la Luna area. In this substrate, primary producers were exclusively *Cyanobacteria* with a relative

Table 4.1 Endolithic microbial communities found in lithic substrates from the Preandean area of the Atacama Desert: main members and relative abundances in the community

Substrate	Shannon Index	Main phyla relative abundance (%)					Other phyla detected	References
		Cyanobacteria	*Actinobacteria*	*Proteobacteria*	*Chloroflexi*	*Gemmatimonadetes*	*Chlorophyta*	
Gypsum	2.2–6.1	36–83	10–25	13–30	nd–11	<5–6	ns	[1] [4] [5] [7]
Calcite	3.7–6.1	50–60	10–20	3–5	nd–10.9	<5–15	nd	[1] [2] [5]
Ignimbrite	4.3–4.9	ns - 80	5–14	< 5	<5–11	<1	nd	[1] [2] [3] [6]
Granite	3.8	77	17	0.3	4	<1	nd	[1]

Abbreviations: *ns* not specified, *nd* not detected. [1] Meslier et al. (2018), [2] Crits-Christoph et al. (2016b), [3] Cámara et al. (2015) [4] Wierzchos et al. (2015), [5], DiRuggiero et al. (2013), [6] Wierzchos et al. (2012), [7] Dong et al. (2007)

abundance of 50 to 60%, while the heterotrophs included *Actinobacteria* (10–20%), *Proteobacteria* (3–5%), *Chloroflexi* (0–11%), and *Gemmatimonadetes* (0–15%) (Table 4.1). Using WGS Crits-Christoph et al. (2016b) identified additional heterotrophic bacteria with significant occurrence, such as *Deinococcus–Thermus* and *Bacteroidetes*.

EMCs from ignimbrite and granite were found to harbor significantly less diverse communities than other EMCs of the Preandean Atacama. In these substrates, *Cyanobacteria* were the only primary producers, reaching relative abundances of 80% and 77% in ignimbrite and granite, respectively, while *Proteobacteria* relative abundances dropped below 5% in the ignimbrite community and below 1% in the granite community (DiRuggiero et al. 2013; Crits-Christoph et al. 2016b; Meslier et al. 2018). The low relative abundance of heterotrophic bacteria in these EMCs might be the result of phototrophs' low metabolic activity in the harshest environments, leading to a reduced amount of fixed inorganic carbon and limiting, in turn, the abundance and diversity of the heterotrophic component of the community (Wierzchos et al. 2018; Meslier et al. 2018).

Abiotic factors promoting the diversity and composition of EMCs in the Preandean Atacama have been investigated by a number of multidisciplinary approaches (DiRuggiero et al. 2013; Wierzchos et al. 2015; Crits-Christoph et al. 2016b; Meslier et al. 2018). These studies have shown that the rock architecture, i.e., the space available for colonization, embodied by the size of the cracks, fissures, and pores and their connection to the surface, which is tightly linked to substrate water retention capacities, were the main drivers of community structure and diversity. In addition, specific properties of the substrates were also found to confer beneficial advantages to the EMCs; these include sepiolite nodules in gypsum, which considerably increase the water retention capability of the substrate, or the high thermal conductivity of calcite, promoting dewfall formation (DiRuggiero et al. 2013; Wierzchos et al. 2015; Crits-Christoph et al. 2016b; Meslier et al. 2018).

By colonizing rock substrates, EMCs find protection from the extremely intense solar irradiance found in hyperarid deserts (Rondanelli et al. 2015). Additional adaptation strategies, in particular against the deleterious effects of UV, include protective cell-layering, an array of screening pigments, and lipid production by phototrophs (see below; Vítek et al. 2013, 2016, 2017; Wierzchos et al. 2015, 2018). While UV can be significantly attenuated by the substrate, the decline in visible light transmission occurs at a much lower rate, providing sufficient light for photosynthesis (Hughes and Lawley 2003; Amaral et al. 2006). Meslier et al. (2018) measured the light transmittance in several substrates from the Preandean Atacama and found a direct relationship between the spectral properties of the substrate and the depth of the colonization zone; higher light transmitting substrates (calcite, gypsum, and granite) showed EMCs located deeper in the substrate, while EMCs from ignimbrite were located closer to the surface (Meslier et al. 2018).

The recent use of metagenomics has brought insights into the adaptation of EMCs to their unique environment. The functional analysis of calcite and ignimbrite EMCs revealed a broad diversity of stress response pathways, especially linked to survival under harsh conditions (Crits-Christoph et al. 2016b). These pathways were involved in (1) carbon starvation and low-nutrient stress, (2) cold shock genes, (3) oxidative stress genes related to osmotic stress/desiccation tolerance, and (4) secondary metabolites production (Crits-Christoph et al. 2016b). Using comparative metagenome analysis, the authors found that the ignimbrite community was significantly enriched in non-ribosomal peptide synthase (NRPS) and polyketide synthase (PKS) genes, suggesting a highly competitive environment for resources and space. The large number of gene clusters related to iron acquisition in the ignimbrite community also suggested iron starvation, while the presence of mycosporine-like gene clusters in the calcite community, but not in the ignimbrite community, indicated possible differences in UV radiation exposure of the EMCs (Crits-Christoph et al. 2016b). The differential abundances of secondary metabolites demonstrate the key role played by the substrate in the molecular adaptations of community members. Surprisingly, pathways for nitrogen fixation were not found in the metagenome of any of the Atacama EMCs investigated so far (Crits-Christoph et al. 2016a, b; Finstad et al. 2017). It is likely that the long-term accumulation of nitrate in the Atacama Desert, via atmospheric deposition, provides a major source of nitrogen to microbial communities in the form of nitrate and ammonium (Michalski et al. 2004; Crits-Christoph et al. 2016b; Finstad et al. 2017).

As demonstrated by the discussion above, only a comprehensive approach to the study of EMCs, using a combination of methods and tools, will allow for the elucidation of the mechanisms that generate and maintain their diversity.

4.3 Supporting the Microbial Community: Photoautotrophs as Primary Producers

Phototrophs are essential for the survival of EMCs because they are the only primary producers in a system where the import of exogenous organic carbon is negligible. As such, *Cyanobacteria* and microalgae carry essential functions in EMCs.

Phototrophic microorganisms perform oxygenic photosynthesis via two photosystems, PSI and PSII, connected by an electron transfer chain, similar to plants (Falkowski and Raven 2013). In oxygenic photosynthesis, photons collected by antenna complexes coupled to photosystems are transferred to chlorophyll molecules located in the photosystem core. This photon energy is used to break water molecules producing reduced nicotinamide adenine dinucleotide phosphate (NADPH) with oxygen as a by-product. In a subsequent step, NADPH is used to synthesize organic carbon from carbon dioxide via the Calvin cycle. The two main elements required for oxygenic photosynthesis, liquid water and light, are often limiting factors for the chlorophototrophs inhabiting endolithic substrates in hyper-arid deserts. Endolithic phototrophs can only perform photosynthesis during periods of time when liquid water is available and, because of high solar and UV radiations, they also need to use strategies to prevent photo-inhibition and photo-oxidative damage to their photosystems (Vítek et al. 2013, 2016, 2017; Wierzchos et al. 2015, 2018). While substrate colonization at greater depths inside the rock might give access to more retained water and increase protection against damaging UV irradiation, it might also decrease essential requirements for photosynthesis such as photosynthetic active radiation and CO_2 exchange (Boison et al. 2004; Rothschild et al. 1994). An example of such a strategy is the spatial arrangement of *Cyanobacteria* within Preandean EMCs habitats (Meslier et al. 2018) and will be discussed below.

4.3.1 Cyanobacteria

Cyanobacteria are found in most types of illuminated environment and were responsible for the "Great Oxidation Event" 2.4–2.1 billion years ago (Lyons et al. 2014). Their success as primary producers is due to several essential features (Whitton and Potts 2000).

(a) Their temperature optimum is higher by several degrees than that of most eukaryotic algae (Castenholz and Waterbury 1989), allowing them to colonize warmer environments.
(b) Desiccation and water stress tolerance made them some of the most successful organisms in hypersaline environments (Hu et al. 2012; Oren 2012).
(c) They display high tolerance to high levels of UV light radiation (Castenholz and Garcia-Pichel 2012).
(d) They can perform efficient photosynthetic CO_2 reduction with low concentrations of inorganic carbon (Pierce and Omata 1988; Raven 2012).

Most of the Cyanobacteria in EMCs from the Atacama Desert are members of *Chroococcidiopsis* (Wierzchos et al. 2011, 2015, 2018; Vítek et al. 2013, 2016, 2017; DiRuggiero et al. 2013; Cámara et al. 2015; Crits-Christoph et al. 2016b; Meslier et al. 2018). This cyanobacterial genus from the *Chroococcidiopsiales* order (Komarek et al. 2014) is the most abundant cyanobacteria in hyperarid environments where its adaptability to extreme conditions has been widely demonstrated (Smith et al. 2014). *Chroococcidiopsis* are often accompanied by other cyanobacterial taxa including members of other unicellular orders such as *Chroococales* and *Synechococcales*, and even members from filamentous orders such as *Oscillatoriales* and *Nostocales* (Table 4.2). Despite the detection of other cyanobacterial genus, *Chroococcidiopsis* is the only genus that has been consistently detected in all EMCs using microscopy approaches (Table 4.2). This discrepancy between molecular and microscopy methods with regard to cyanobacterial diversity is the result of a number of factors.

First, there is a technical factor associated with the evolution of research tools over the past decade, from DGGE to clone libraries and more recently to high-throughput sequencing platforms (454 pyrosequencing, Illumina-MiSeq, Illumina HiSeq). Additionally, all these methods have intrinsic limitations with DNA isolation, the selection of marker genes and their primers, library preparation, read

Table 4.2 Cyanobacterial taxa in endolithic microbial communities from the Preandean area of the Atacama Desert and the approaches used for their detection

Order	Genus	Tools used for cyanobacterial detection		Substrate	References
		Molecular	Microscopy		
Chroococcidiopsiales	*Chroococcidiopsis*	X	X	Gyp Ca Ign Gr	[1] [2] [3] [4] [5] [6] [7]
Chroococcales	*Gloeocapsa*	X		Ca	[2]
	Halothece	X		Gyp Ca	[1]
Synechococcales	*Acaryochloris*	X		Ca	[2]
	Synechococcus	X		Gyp Ca	[1]
Oscillatoriales	*Aerosakkonema*	X		Gyp Ca Ign Gr	[1]
	Phormidium	X		Ca	[5]
Nostocales	*Anabaena*	X		Gyp Ca	[5]

Abbreviations: *Gyp* gypsum, *Ca* calcite, *Ign* ignimbrite, *Gr* granite. [1] Meslier et al. (2018), [2] Crits-Christoph et al. (2016b), [3] Cámara et al. (2015), [4] Wierzchos et al. (2015), [5] DiRuggiero et al. (2013), [6] Wierzchos et al. (2012) [7] Dong et al. (2007)

length, and sequencing depth. Each of these steps can introduce significant biases that make it very difficult to compare studies (Rastogi and Sani 2011). The analysis of the sequencing data may also introduce significant biases with, in particular, the choice of parameters in defining unique taxa and the type of database (and its version) used for taxonomic annotation. While culture-independent methods have provided a large amount of sequencing information, especially during the past few years with NGS platforms, the increasing number of sequences in databases that belong to "uncultured cyanobacterium clone" seriously hinders accurate taxonomical assignments of this phylum. On the other hand, taxonomical assignment using microscopic methods is limited to morphologically different *Cyanobacteria*, their relative abundances in the sample, and requires a great deal of experience.

Another issue is the fact that the taxonomy and phylogeny of *Cyanobacteria* is an ongoing discussion, especially because of their antiquity, existing fossil representatives with very similar morphology to present-day species (William Schopf 1974; Knoll 2008), and a complex evolutionary history. Several features, in addition to genetic sequences, should be taken into account when defining cyanobacterial taxa, including morphological characteristics, ultrastructural details such as thylakoid structure, and type of cell division (Komarek et al. 2014). This is essential for accurate *Cyanobacteria* taxonomy assignment but is not always practical, in particular, in studies with large numbers of samples.

Several adaptation strategies to water stress have been described for *Chroococcidiopsis*. For one, *Chroococcidiopsis* belongs to the group of anhydrobiotic cyanobacteria. These cyanobacteria cope with the lack of water by entering an ametabolic state involving structural, physiological, and molecular changes (Feofilova 2003). Another adaptation to the scarcity of water is the production of

EPSs providing a depository for water and stabilizing desiccation-related enzymes and molecules (Wright et al. 2006).

To protect themselves from the damaging effects of UV radiation, *Chroococcidiopsis* from EMCs of the Preandean Depression synthesize UV-absorbing and/or screening compounds. One such compound, the sunscreen pigment scytonemin, has been detected in the EPS sheath of EMCs from gypsum from the Preandean Depression and other hyperarid regions of the Atacama Desert (Wierzchos et al. 2015; Vítek et al. 2014a, 2016). It has also been hypothesized that the simultaneous exposure to both desiccation and UV radiation may further induce scytonemin biosynthesis (Fleming and Castenholz 2007). Recently, a novel light-adaptation strategy was discovered in *Cyanobacteria* inhabiting ignimbrite rocks (Vítek et al. 2017). The authors attributed the shift in carotenoid composition (red-shift of υ1 (C=C) band) in the cyanobacteria to a light-dependent change in carotenoid conjugation that would mediate a non-photochemical quenching mechanism (Kirilovsky and Kerfeld 2016). Since *Cyanobacteria* are major components of EMCs, and are most often located in the upper part of the endolithic microhabitat, the strategies developed by this phylum to deal with extreme environmental conditions constitute a benefit for the entire community.

4.3.2 Eukaryotic Phototrophs

Eukaryotic microalgae constitute a basic component of the world's ecosystems as they contribute to about 40% of global productivity (Andersen 1992). Apart from their important role in marine ecosystems, they are found as important primary producers in desert soils (Lewis and Lewis 2005) and in lithic microhabitats as free-living organisms (Meyer et al. 1988; Robinson et al. 2015; Wierzchos et al. 2015; Vítek et al. 2016) or as photobionts of lichens (Palmer Jr and Friedmann 1990; Wierzchos et al. 2011).

While most endolithic communities in the Preandean Depression do not harbor microalgae, their occurrence has been reported in gypsum from the Cordón de Lila depression (Wierzchos et al. 2015). The authors suggested that while these algae have morphological similarity with three genera of *Chlorophyta* (*Spongiochloris*, *Deasonia*, and *Neochloris*), they also have novel features, in situ and in culture, and therefore should be classified as a new genus of the *Chlorophyceae* class. Similar to *Cyanobacteria*, EMCs' algae show special strategies to deal with environmental stresses such as the accumulation of photoprotective carotenoids and storage lipids (Wierzchos et al. 2015; Vítek et al. 2016). In gypsum, both carotenoid and storage lipids increased in algal cells with their closeness to the surface of the substrate where solar radiation is the highest (Wierzchos et al. 2015; Vítek et al. 2016). Carotenogenesis has also been attributed to photoprotection against high solar radiation (Oren et al. 1995) and also as a strategy to mitigate high salinity and nutrition stresses in other aerial microalgae (Aburai et al. 2015; Oren et al. 1995; Hanagata and Dubinsky 1999). In gypsum of Cordón de Lila, the algae were located in the upper part of the substrate (with first red and then green cells) followed by *Cyanobacteria*. This very special arrangement suggests that they may act as a protective barrier for the community against the harmful effects of high levels of solar radiation because of their high carotenoids content (Vítek et al. 2017).

4.4 Diversity of Heterotrophs in Communities of the Preandean Depression of the Atacama Desert

Before the use of molecular tools to characterize EMCs, heterotrophic members of these communities received little attention compared to their phototrophic counterparts. This was most likely because of the difficulty in characterizing heterotrophic bacteria morphologically and also because of the idea that the main contributors to the activity and resilience of EMCs were primary producers. Indeed, heterotrophic microorganisms benefit from photosynthetic conversion of CO_2 into organic carbon, provided either by *Cyanobacteria* or microalgae. Nonetheless, investigations of the diversity and community structure of EMCs have since revealed far more diverse heterotrophic assemblages than previously thought (Walker and Pace 2007; Dong et al. 2007; Lacap et al. 2011; DiRuggiero et al. 2013; Crits-Christoph et al. 2016b; Meslier et al. 2018), emphasizing the idea that primary producers can nurture and support the development of a high diversity of heterotrophic microorganisms.

In EMCs of the Preandean Atacama Desert, the most abundant heterotrophic phyla have been assigned to *Actinobacteria*, *Chloroflexi*, and *Proteobacteria*, although additional phyla such as *Bacteroidetes*, *Gemmatimonadetes*, *Deinococcus*–Thermus, *Firmicutes*, *Planctomycetes*, and *Verrucomicrobia* have also been consistently reported in several lithic substrates (Table 4.3) (Dong et al. 2007; DiRuggiero et al. 2013; Rasuk et al. 2014; Wierzchos et al. 2015; Crits-Christoph et al. 2016b; Meslier et al. 2018). In contrast, arid and hyperarid soils are typically dominated by sparse heterotrophic microorganisms with mostly members of the *Actinobacteria* phylum (Costello et al. 2009; Bull 2011; Lynch et al. 2012, 2014; Neilson et al. 2012; Crits-Christoph et al. 2013; Schulze-Makuch et al. 2018). While the recent sequencing of the genome of *Pseudonocardia* sp. from a volcanic soil showed that chemoautotrophic microbes may provide organic carbon to a community, it was a rather specific case where trace gases from volcanic activity provided the energy sources for carbon fixation (Lynch et al. 2014).

In the *Actinobacteria* phylum, the main identified taxa included *Blastococcus*, *Geodermatophilus*, *Modestobacter*, *Jatrophihabitans*, *Marmoricola*, *Microlunatus*, *Pseudonocardia*, *Euzebya*, *Solirubrobacter*, *Conexibacter*, and *Rubrobacter* genera, for which the relative abundances could vary greatly between substrates (Table 4.3). For the most part, these genera were also consistently recovered in lithic substrates and soils of hyperarid deserts around the world (Connon et al. 2007; Lacap et al. 2011; DiRuggiero et al. 2013; Crits-Christoph et al. 2013, 2016b; Wierzchos et al. 2015; Meslier et al. 2018; Bull et al. 2018) and recognized for their

Table 4.3 Major heterotrophic taxa detected in calcite (Ca), gypsum (Gy), ignimbrite (Ig), and granite (Gr) substrates from Preandean Depression of the Atacama Desert

Phylum	Class	Order	Family	Ca	Gy	Ig	Gr	References
Actinobacteria	Acidimicrobia	Acidimicrobiales	Other/uncultured	**d(+)**	**d(+)**	d(−)	d(−)	[1]
	Actinobacteria	Frankiales	Frankiaceae	**d(+)**	d(−)	d(−)	**d(+)**	[1] [2]
			Geodermatophilaceae	**d(+)**	**d(+)**	**d(+)**	**d(+)**	[1]
			Other/uncultured	**d(+)**	**d(+)**	**d(+)**	**d(+)**	[1]
		Micrococcales	Intrasporangiaceae	nd	d(−)	d(−)	**d(+)**	[1]
		Propionibacteriales	Nocardioidaceae	**d(+)**	**d(+)**	**d(+)**	**d(+)**	[1]
			Propionibacteriaceae	**d(+)**	**d(+)**	d(−)	**d(+)**	[1] [2]
		Pseudonocardiales	Pseudonocardiaceae	**d(+)**	**d(+)**	d(−)	d(−)	[1]
	Nitriliruptoria	Euzebyales	Euzebyaceae	**d(+)**	**d(+)**	**d(+)**	d(−)	[1]
	Rubrobacteria	Rubrobacterales	Rubrobacteriaceae	**d(+)**	**d(+)**	**d(+)**	**d(+)**	[1] [2] [4]
	TakashiAC-B11	uncultured bacterium		**d(+)**	**d(+)**	d(−)	d(−)	[1]
	Thermoleophilia	Gaiellales	uncultured bacterium	d(−)	**d(+)**	d(−)	nd	[1]
		Solirubrobacterales	Patulibacteraceae	d(−)	**d(+)**	d(−)	nd	[1]
			Solirubrobacteraceae	**d(+)**	**d(+)**	**d(+)**	**d(+)**	[1]
			Conexibacteraceae	d	nd	d	nd	[1]
			Other(0319-6M6, S1-80, 480-2)	**d(+)**	**d(+)**	**d(+)**	**d(+)**	[1]
Chloroflexi	Chloroflexi Incertae Sedis	Unknown Order	Unknown family	**d(+)**	**d(+)**	**d(+)**	**d(+)**	[1]
	Chloroflexia	Kallotenuales	AKIW781/Other	**d(+)**	**d(+)**	**d(+)**	**d(+)**	[1]
	Thermomicrobia	AKYG1722, JG30-KF-CM45	Other/uncultured bacterium	**d(+)**	**d(+)**	**d(+)**	**d(+)**	[1] [2] [4]
	Other[a]			**d(+)**	**d(+)**	d(−)	d(−)	[1]

Proteobacteria	α-Proteobacteria	Caulobacterales	Caulobacteraceae	**d(+)**	**d(+)**	d(−)	nd	[1] [4]
		Rhizobiales	Methylobacteriaceae	**d(+)**	**d(+)**	d(−)	d(−)	[1] [4]
			Rhizobiaceae	d(−)	**d(+)**	d(−)	nd	[1] [3]
			Other (D05-2, JG34-KF-361, P-102)	**d(+)**	**d(+)**	d(−)	d(−)	[1] [4] [5]
		Sphingomonadales	Sphingomonadaceae	**d(+)**	**d(+)**	d(−)	d(−)	[1], [3] [4] [5]
			Other (Ellin6055, JG34-KF-161)	**d(+)**	**d(+)**	**d(+)**	**d(+)**	[1]
	β-Proteobacteria	Burkholderiales	Burkholderiaceae	d	**d(+)**	d	nd	[1] [4]
	γ-Proteobacteria	Methylococcales/Other		d	d	d	nd	[4] [5]

Abbreviations: *d* detected taxa, *nd* not detected; (+) relative abundance average > 1.10E−3; (−), relative abundance average < 1.10E−3. [a]Other Chloroflexi comprising taxa belonging to TK10, Gitt-GS-136, JG30-KF-CM66 and KD4-96. Data covering the Preandean Depression of the Atacama Desert are from Meslier et al. (2018) [1], completed by Crits-Christoph et al. (2016b) [2], Wierzchos et al. (2015) [3], DiRuggiero et al. (2013) [4] and Dong et al. (2007) [5]

ability to resist extreme desiccation (*Geodermatophilus*, *Blastococcus*, *Rubrobacter*), high UV and ionizing radiation (*Geodermatophilus*, *Rubrobacter*, *Modestobacter*), oligotrophic conditions (*Modestobacter*, *Rubrobacter*), temperature fluctuations (*Modestobacter*, *Rubrobacter*), high salinity (*Marmoricola*, *Pseudonocardia*), and high concentrations of metals (*Rubrobacter*) (Bull 2011; Mohammadipanah and Wink 2016; Idris et al. 2017; Schulze-Makuch et al. 2018; Bull et al. 2018). The identification and characterization of *Actinobacteria* from the extreme biosphere have been of particular interest over the past decade because of their incredible potential as sources of novel bioactive compounds for agricultural, pharmaceutical, and industrial applications (Bull 2011; Mohammadipanah and Wink 2016; Idris et al. 2017; Rateb et al. 2018). Indeed, some researchers argue that most of the novel chemical diversity will likely be discovered in the rare actinobacterial biosphere (Mohammadipanah and Wink 2016). Although culture-dependent methods are challenging, and only reveal a fraction of the microbial diversity in any given environment, they are necessary for the discovery and production of bioactive compounds.

Members of the *Chloroflexi* phyla, such as *Thermobaculum* and *Thermomicrobia*, and various genera of *Proteobacteria*, mainly from the *alpha*-, *beta*-, and *gamma-proteobacteria* (*Methylobacterium*, *Ensifer*, *Sphingomonas*, and *Burkholderia*) are known to be anoxygenic photosynthesizers (Table 4.3). However, no anoxygenic photosynthesizers have been found so far in EMCs from the Preandean area of the Atacama Desert (Crits-Christoph et al. 2016b; Frigaard 2016; Meslier et al. 2018). *Chloroflexi* relative abundances in various desert soils and lithic habitats are highly variable (Lacap et al. 2011; Neilson et al. 2012) and, although unique adaptive traits have been reported, such as atypical peptidoglycan contents, their functional role remains to be elucidated (Neilson et al. 2012).

Variations in the detection and relative abundances of members of the *Proteobacteria* have also been reported in desert soils and lithic communities, yet, some genera display several putative functional advantages for survival in hyperarid deserts (Drees et al. 2006; Dong et al. 2007; Connon et al. 2007; Neilson et al. 2012; Crits-Christoph et al. 2013; Rasuk et al. 2014; Makhalanyane et al. 2015; Van Goethem et al. 2017). Some examples include (1) members of the *Methylobacterium* genus known to use methanol as both carbon and energy source (Makhalanyane et al. 2015), (2) members of the *Burkholderia* genus, a common soil inhabitant that promotes plant growth via nitrogen fixation (Coenye and Vandamme 2003; Suárez-Moreno et al. 2012; Stopnisek et al. 2016), and (3) members of the *Ensifer* genus that harbor mineral weathering activity, such as potassium feldspar-solubilization, and potential nitrogen fixation (Rogel et al. 2001; Wang et al. 2016; Peng et al. 2017). Recently, the combined use of metagenomics and metatranscriptomics in hypoliths from the Namib Desert demonstrated the close relationship between active members of *Cyanobacteria* and *alpha-Proteobacteria* and their key role in maintaining and facilitating nutrient cycling in the community (Van Goethem et al. 2017). While the functional contribution of *Proteobacteria* in EMCs of hyperarid deserts needs to be further investigated, it is likely that their characterization and isolation will be of great interest, notably for their potential production of secondary metabolites (Suárez-Moreno et al. 2012).

4.5 Future Directions of Research

Endolithic microbial communities are excellent models for microbial ecology studies because of their low complexity and containment. However, they are challenging to study because of the difficulties in collecting samples and because of their low biomass. Nevertheless, the combination of state-of-the-art microscopy and molecular methods has revealed novel and diverse communities with unique adaptations to their extreme environments. Despite this progress, many knowledge gaps still remain.

In particular, very little is known about the interactions between the biotic and abiotic components of these unique ecosystems. Greater efforts, using sophisticated microscopy, are needed to characterize substrate architecture, define the different microhabitats within each substrate, and how these are colonized by different members of the community. Interactions between functional groups in the community and the role of viruses in shaping these communities should also be investigated to obtain a holistic picture of EMCs.

Cultivation efforts and single cell genomics will provide the opportunity to study the genetic diversity of EMCs members, refine taxonomic annotations, and investigate the potential role of biogeography in shaping these communities. Additionally, having isolates at hand will allow for the production of secondary metabolites for agricultural, pharmaceutical, and industrial applications.

Finally, the use of omics strategies and longitudinal studies will give us insights into the functioning of these communities at the ecosystem level, their response to environmental stresses, and help us predict how resistant and resilient these microbial communities might be to climate change. Because arid and hyperarid deserts, at the dry limit for life, are fragile ecosystems, their inhabitants will be on the front line of the major changes in climate ahead of us.

Acknowledgments Part of this work was funded by NASA grant NNX15AP18G and NSF grant DEB1556574 to JDR, and by MCIU/AEI (Spain) and FEDER (UE) grant PGC2018-094076-B-I00 to JW.

References

Aburai N, Sumida D, Abe K (2015) Effect of light level and salinity on the composition and accumulation of free and ester-type carotenoids in the aerial microalga Scenedesmus sp. (Chlorophyceae). Algal Res 8:30–36. https://doi.org/10.1016/j.algal.2015.01.005

Amaral G, Martinez-Frias J, Vazquez L (2006) Astrobiological significance of minerals on Mars surface environment: UV-shielding properties of Fe (jarosite) vs. Ca (gypsum) sulphates. World Appl Sci J 2:112–116

Andersen RA (1992) Diversity of eukaryotic algae. Biodivers Conserv 1(4):267–292. https://doi.org/10.1007/BF00693765

Billi D, Baqué M, Verseux C, Rothschild L, de Vera JP (2017) Desert cyanobacteria: potential for space and earth applications. In: Adaption of microbial life to environmental extremes:

novel research results and application, 2nd edn. Springer, New York, pp 133–146. https://doi.org/10.1007/978-3-319-48327-6_6
Boison G, Mergel A, Jolkver H, Bothe H (2004) Bacterial life and dinitrogen fixation at a gypsum rock. Appl Environ Microbiol 70(12):7070–7077. https://doi.org/10.1128/aem.70.12.7070-7077.2004
Bull AT (2011) Actinobacteria of the extremobiosphere. In: Horikoshi K (ed) Extremophiles handbook. Springer, Japan, pp 1204–1231
Bull AT, Andrews BA, Dorador C, Goodfellow M (2018) Introducing the Atacama Desert. Antonie van Leeuwenhoek 111(8):1269–1272. https://doi.org/10.1007/s10482-018-1100-2
Cámara B, Suzuki S, Nealson KH, Wierzchos J, Ascaso C, Artieda O, de los Ríos A (2015) Ignimbrite textural properties as determinants of endolithic colonization patterns from hyper-arid Atacama Desert. Int Microbiol 17(4):235–247. https://doi.org/10.2436/20.1501.01.226
Castenholz RW, Garcia-Pichel F (2012) Cyanobacterial responses to UV radiation. In: Whitton BA (ed) Ecology of cyanobacteria II: their diversity in space and time. Springer Netherlands, Dordrecht, pp 481–499. https://doi.org/10.1007/978-94-007-3855-3_19
Castenholz RW, Waterbury JB (1989) Cyanobacteria. In: Staley JT, Bryant MP, Pfennig N, Holt JG (eds) Bergey's manual of systematic bacteriology, vol 3. Williams & Wilkins, Baltimore, pp 1710–1727
Coenye T, Vandamme P (2003) Diversity and significance of Burkholderia species occupying diverse ecological niches. Environ Microbiol 5:719–729. https://doi.org/10.1046/j.1462-2920.2003.00471.x
Connon SA, Lester ED, Shafaat HS, Obenhuber DC, Ponce A (2007) Bacterial diversity in hyperarid Atacama desert soils. J Geophys Res Biogeo 112(4). https://doi.org/10.1029/2006JG000311
Costello EK, Halloy SRP, Reed SC, Sowell P, Schmidt SK (2009) Fumarole-supported islands of biodiversity within a hyperarid, high-elevation landscape on Socompa Volcano, Puna de Atacama, Andes. Appl Environ Microbiol 75(3):735–747. https://doi.org/10.1128/AEM.01469-08
Crits-Christoph A, Gelsinger DR, Ma B, Wierzchos J, Ravel J, Davila A, Casero MC, DiRuggiero J (2016a) Functional interactions of archaea, bacteria and viruses in a hypersaline endolithic community. Environ Microbiol 18(6):2064–2077. https://doi.org/10.1111/1462-2920.13259
Crits-Christoph A, Robinson CK, Barnum T, Fricke WF, Davila AF, Jedynak B, McKay CP, DiRuggiero J (2013) Colonization patterns of soil microbial communities in the Atacama Desert. Microbiome 1(1):28. https://doi.org/10.1186/2049-2618-1-28
Crits-Christoph A, Robinson CK, Ma B, Ravel J, Wierzchos J, Ascaso C, Artieda O, Souza-Egipsy V, Casero MC, DiRuggiero J (2016b) Phylogenetic and functional substrate specificity for endolithic microbial communities in hyper-arid environments. Front Microbiol 7:301. https://doi.org/10.3389/fmicb.2016.00301
Dassarma S (2006) Extreme halophiles are models for astrobiology. Microbe 1. https://doi.org/10.1128/microbe.1.120.1
De Los Ríos A, Wierzchos J, Ascaso C (2014) The lithic microbial ecosystems of Antarctica's McMurdo Dry Valleys. Antarct Sci 25(4):459. https://doi.org/10.1017/S0954102014000194
De Los Ríos A, Wierzchos J, Sancho LG, Green TGA, Ascaso C (2005) Ecology of endolithic lichens colonizing granite in continental Antarctica. Lichenologist 37(5):383–395. https://doi.org/10.1017/S0024282905014969
DiRuggiero J, Wierzchos J, Robinson CK, Souterre T, Ravel J, Artieda O, Souza-Egipsy V, Ascaso C (2013) Microbial colonisation of chasmoendolithic habitats in the hyper-arid zone of the Atacama Desert. Biogeosciences 10(4):2439–2450. https://doi.org/10.5194/bg-10-2439-2013
Dong H, Rech JA, Jiang H, Sun H, Buck BJ (2007) Endolithnic cyanobacteria in soil gypsum: occurrences in Atacama (Chile), Mojave (United States), and Al-Jafr Basin (Jordan) Deserts. J Geophys Res Biogeosci 112(2). https://doi.org/10.1029/2006JG000385
Drees KP, Neilson JW, Betancourt JL, Quade J, Henderson DA, Pryor BM, Maier RM (2006) Bacterial community structure in the hyperarid core of the Atacama Desert, Chile. Appl Environ Microbiol 72(12):7902–7908. https://doi.org/10.1128/AEM.01305-06

Fairén AG, Davila AF, Lim D, Bramall N, Bonaccorsi R, Zavaleta J, Uceda ER, Stoker C, Wierzchos J, Dohm JM, Amils R, Andersen D, McKay CP (2010) Astrobiology through the ages of Mars: the study of terrestrial analogues to understand the habitability of Mars. Astrobiology 10(8):821–843. https://doi.org/10.1089/ast.2009.0440

Falkowski PG, Raven JA (2013) Aquatic photosynthesis, 2nd edn. Princeton University Press, Princeton

Feofilova EP (2003) Deceleration of vital activity as a universal biochemical mechanism ensuring adaptation of microorganisms to stress factors: a review. Appl Biochem Microbiol 39:1–18. https://doi.org/10.1023/A:1021774523465

Finore I, Lama L, Poli A, Di Donato P, Nicolaus B (2016) Biotechnology implications of extremophiles as life pioneers and wellspring of valuable biomolecules. In: Microbial factories: biodiversity, biopolymers, bioactive molecules, vol 2. Springer India, New Delhi, pp 193–216. https://doi.org/10.1007/978-81-322-2595-9_13

Finstad KM, Probst AJ, Thomas BC, Andersen GL, Demergasso C, Echeverría A, Amundson RG, Banfield JF (2017) Microbial community structure and the persistence of cyanobacterial populations in salt crusts of the hyperarid Atacama desert from genome-resolved metagenomics. Front Microbiol 8:1435. https://doi.org/10.3389/fmicb.2017.01435

Fleming ED, Castenholz RW (2007) Effects of periodic desiccation on the synthesis of the UV-screening compound, scytonemin, in cyanobacteria. Environ Microbiol 9(6):1448–1455. https://doi.org/10.1111/j.1462-2920.2007.01261.x

Foing BH, Stoker C, Ehrenfreund P (2011) Astrobiology field research in Moon/Mars analogue environments. Int J Astrobiol 10(3):137–139. https://doi.org/10.1017/S1473550411000139

Friedmann EI (1980) Endolithic microbial life in hot and cold deserts. Orig Life 10(3):223–235. https://doi.org/10.1007/bf00928400

Friedmann EI, Hua M, Ocampo-Friedmann R (1988) Cryptoendolithic lichen and cyanobacterial communities of the Ross Desert, Antarctica. Polarforschung 58(2–3):251–259

Friedmann EI, Ocampo-Friedmann R (1984) Endolithic microorganisms in extreme dry environments: analysis of a lithobiontic microbial habitat. In: Current perspectives in microbiology. Proc 3rd international symposium. ASM Press, Washington, DC, pp 177–185

Frigaard NU (2016) Biotechnology of anoxygenic phototrophic bacteria. Adv Biochem Eng Biotechnol 156:139–154. https://doi.org/10.1007/10_2015_5006

Golubic S, Friedmann I, Schneider J (1981) The lithobiontic ecological niche, with special reference to microorganisms. J Sediment Petrol 51(2):475–478

Hanagata N, Dubinsky Z (1999) Secondary carotenoid accumulation in Scenedesmus komarekii (Chlorophyceae, Chlorophyta). J Phycol 35(5):960–966. https://doi.org/10.1046/j.1529-8817.1999.3550960.x

Hirsch P, Hoffmann B, Gallikowski CC, Mevs U, Siebert J, Sittig M (1988) Diversity and identification of heterotrophic bacteria from Antarctic Rocks of the McMurdo Dry Valleys (Ross Desert). Supplement to: Hirsch, P et al. (1988): 3.7 Diversity and identification of heterotrophs from Antarctic Rocks of the McMurdo Dry Valleys (Ross Desert). Polarforschung 58(2/3):261–269, hdl:10013/epic.29622.d001. PANGAEA. https://doi.org/10.1594/PANGAEA.763324

Hughes KA, Lawley B (2003) A novel Antarctic microbial endolithic community within gypsum crusts. Environ Microbiol 5(7):555–565

Hu C, Gao K, Whitton BA (2012) Semi-arid regions and deserts. In: Ecology of cyanobacteria II: their diversity in space and time, vol 9789400738553. Springer Netherlands, Dordrecht, pp 345–369. https://doi.org/10.1007/978-94-007-3855-3_12

Idris H, Goodfellow M, Sanderson R, Asenjo JA, Bull AT (2017) Actinobacterial rare biospheres and dark matter revealed in habitats of the Chilean Atacama Desert. Sci Rep 7:8373. https://doi.org/10.1038/s41598-017-08937-4

IPCC (2012) Summary for policy makers. In: Field CB, Barros V, Stocker TF, Qin D, Dokken DJ, Ebi KL et al (eds) Managing the risks of extreme events and disasters to advance climate change adaptation: a special report of working groups I and II of the intergovernmental panel on climate change. Cambridge University Press, Cambridge, pp 1–19

Kirilovsky D, Kerfeld CA (2016) Cyanobacterial photoprotection by the orange carotenoid protein. Nat Plants 2:16180. https://doi.org/10.1038/nplants.2016.180
Knoll A (2008) Cyanobacteria and Earth history. In: The cyanobacteria: molecular biology, genomics and evolution. Caister Academic Press, UK
Komarek J, Kastovsky J, Mares J, Johansen JR (2014) Taxonomic classification of cyanoprokaryotes (cyanobacterial genera) 2014, using a polyphasic approach. Preslia 86:295–335
Lacap DC, Warren-Rhodes KA, McKay CP, Pointing SB (2011) Cyanobacteria and chloroflexi-dominated hypolithic colonization of quartz at the hyper-arid core of the Atacama Desert, Chile. Extremophiles 15(1):31–38. https://doi.org/10.1007/s00792-010-0334-3
Lewis LA, Lewis PO (2005) Unearthing the molecular phylodiversity of desert soil green algae (Chlorophyta). Syst Biol 54(6):936–947. https://doi.org/10.1080/10635150500354852
Lynch RC, Darcy JL, Kane NC, Nemergut DR, Schmidt SK (2014) Metagenomic evidence for metabolism of trace atmospheric gases by high-elevation desert Actinobacteria. Front Microbiol 5:698. https://doi.org/10.3389/fmicb.2014.00698
Lynch RC, King AJ, Farías ME, Sowell P, Vitry C, Schmidt SK (2012) The potential for microbial life in the highest-elevation (>6000 m a.s.l.) mineral soils of the Atacama region. J Geophys Res 117:G02028. https://doi.org/10.1029/2012JG001961.
Lyons TW, Reinhard CT, Planavsky NJ (2014) The rise of oxygen in Earth's early ocean and atmosphere. Nature 506(7488):307–315. https://doi.org/10.1038/nature13068
Makhalanyane TP, Valverde A, Gunnigle E, Frossard A, Ramond JB, Cowan DA (2015) Microbial ecology of hot desert edaphic systems. FEMS Microbiol Rev 39(2):203–221. https://doi.org/10.1093/femsre/fuu011
Meslier V, Casero MC, Dailey M, Wierzchos J, Ascaso C, Artieda O, McCullough PR, DiRuggiero J (2018) Fundamental drivers for endolithic microbial community assemblies in the hyperarid Atacama Desert. Environ Microbiol 20(5):1765–1781. https://doi.org/10.1111/1462-2920.14106
Meyer MA, Huang GH, Morris GJ, Friedmann EI (1988) The effect of low temperature on Antarctic endolithic green algae. Polarforschung 58(2–3):113–119
Michalski G, Böhlke JK, Thiemens M (2004) Long term atmospheric deposition as the source of nitrate and other salts in the Atacama Desert, Chile: new evidence from mass-independent oxygen isotopic compositions. Geochim Cosmochim Acta 68(20):4023–4038. https://doi.org/10.1016/j.gca.2004.04.009
Mohammadipanah F, Wink J (2016) Actinobacteria from arid and desert habitats: diversity and biological activity. Front Microbiol 6:1541. https://doi.org/10.3389/fmicb.2015.01541
Neifar M, Maktouf S, Ghorbel RE, Jaouani A, Cherif A (2015) Extremophiles as source of novel bioactive compounds with industrial potential. In: Biotechnology of bioactive compounds: sources and applications, vol 9781118733493. Wiley, Hoboken, NJ, pp 245–267. https://doi.org/10.1002/9781118733103.ch10
Neilson JW, Quade J, Ortiz M, Nelson WM, Legatzki A, Tian F, LaComb M, Betancourt JL, Wing RA, Soderlund CA, Maier RM (2012) Life at the hyperarid margin: novel bacterial diversity in arid soils of the Atacama Desert, Chile. Extremophiles 16(3):553–566. https://doi.org/10.1007/s00792-012-0454-z
Oren A (2012) Salts and brines. In: Whitton BA (ed) Ecology of cyanobacteria II: their diversity in space and time. Springer Netherlands, Dordrecht, pp 401–426. https://doi.org/10.1007/978-94-007-3855-3_15
Oren A, Kühl M, Karsten U (1995) An endoevaporitic microbial mat within a gypsum crust: Zonation of phototrophs, photopigments, and light penetration. Mar Ecol Prog Ser 128:151–159. https://doi.org/10.3354/meps128151
Palmer RJ Jr, Friedmann EI (1990) Water relations and photosynthesis in the cryptoendolithic microbial habitat of hot and cold deserts. Microb Ecol 19:111–118
Peng Q, Yi L, Peng Q, Peng Y (2017) Draft genome sequence of the potassium feldspar-solubilizing bacterium Ensifer adhaerens L18. Genome Announc 5(17):e00199–e00117. https://doi.org/10.1128/genomeA.00199-17

Pierce J, Omata T (1988) Uptake and utilization of inorganic carbon by cyanobacteria. Photosynth Res 16:141–154

Pikuta EV, Hoover RB, Tang J (2007) Microbial extremophiles at the limits of life. Crit Rev Microbiol 33(3):183–209. https://doi.org/10.1080/10408410701451948

Pointing SB, Belnap J (2012) Microbial colonization and controls in dryland systems. Nat Rev Microbiol 10(8):551–562. https://doi.org/10.1038/nrmicro2831

Rastogi G, Sani RK (2011) Molecular techniques to assess microbial community structure, function, and dynamics in the environment. In: Ahmad I, Ahmad F, Pichtel J (eds) Microbes and microbial technology: agricultural and environmental applications. Springer New York, New York, pp 29–57. https://doi.org/10.1007/978-1-4419-7931-5_2

Rasuk MC, Kurth D, Flores MR, Contreras M, Novoa F, Poire D, Farias ME (2014) Microbial characterization of microbial ecosystems associated to evaporites domes of gypsum in Salar de Llamara in Atacama Desert. Microb Ecol 68(3):483–494. https://doi.org/10.1007/s00248-014-0431-4

Rateb ME, Ebel R, Jaspars M (2018) Natural product diversity of actinobacteria in the Atacama Desert. Antonie Van Leeuwenhoek 111(8):1467–1477. https://doi.org/10.1007/s10482-018-1030-z

Raven JA (2012) Carbon. In: Whitton BA (ed) Ecology of cyanobacteria II: their diversity in space and time. Springer Netherlands, Dordrecht, pp 443–460. https://doi.org/10.1007/978-94-007-3855-3_17

Robinson CK, Wierzchos J, Black C, Crits-Christoph A, Ma B, Ravel J, Ascaso C, Artieda O, Valea S, Roldán M, Gómez-Silva B, Diruggiero J (2015) Microbial diversity and the presence of algae in halite endolithic communities are correlated to atmospheric moisture in the hyper-arid zone of the Atacama Desert. Environ Microbiol 17(2):299–315. https://doi.org/10.1111/1462-2920.12364

Rogel MA, Hernández-Lucas I, Kuykendall LD, Balkwill DL, Martinez-Romero E (2001) Nitrogen-fixing nodules with Ensifer adhaerens harboring rhizobium tropici symbiotic plasmids. Appl Environ Microbiol 67(7):3264–3268. https://doi.org/10.1128/AEM.67.7.3264-3268.2001

Rondanelli R, Molina A, Falvey M (2015) The Atacama surface solar maximum. Bull Am Meteorol Soc 96. https://doi.org/10.1175/BAMS-D-13-00175.1

Rothschild LJ, Giver LJ, White MR, Mancinelli RL (1994) Metabolic activity of microorganisms in evaporites. J Phycol 30(3):431–438

Schmidt SK, Lynch RC, King AJ, Karki D, Robeson MS, Nagy L, Williams MW, Mitter MS, Freeman KR (2011) Phylogeography of microbial phototrophs in the dry valleys of the high Himalayas and Antarctica. Proc R Soc B Biol Sci 278(1706):702–708. https://doi.org/10.1098/rspb.2010.1254

Schulze-Makuch D, Wagner D, Kounaves SP, Mangelsdorf K, Devine KG, de Vera J-P, Schmitt-Kopplin P, Grossart H-P, Parro V, Kaupenjohann M, Galy A, Schneider B, Airo A, Frösler J, Davila AF, Arens FL, Cáceres L, Cornejo FS, Carrizo D, Dartnell L, DiRuggiero J, Flury M, Ganzert L, Gessner MO, Grathwohl P, Guan L, Heinz J, Hess M, Keppler F, Maus D, McKay CP, Meckenstock RU, Montgomery W, Oberlin EA, Probst AJ, Sáenz JS, Sattler T, Schirmack J, Sephton MA, Schloter M, Uhl J, Valenzuela B, Vestergaard G, Wörmer L, Zamorano P (2018) Transitory microbial habitat in the hyperarid Atacama Desert. Proc Natl Acad Sci U S A 115(11):2670–2675. https://doi.org/10.1073/pnas.1714341115

Siebert J, Hirsch P (1988) Characterization of 15 selected coccal bacteria isolated from antarctic rock and soil samples from the McMurdo-Dry Valleys (South-Victoria Land). Polar Biol 9(1):37–44. https://doi.org/10.1007/BF00441762

Smith HD, Baqué M, Duncan AG, Lloyd CR, McKay CP, Billi D (2014) Comparative analysis of cyanobacteria inhabiting rocks with different light transmittance in the Mojave Desert: a Mars terrestrial analogue. Int J Astrobiol 17:271. https://doi.org/10.1017/S1473550414000056

Stan-Lotter H, Fendrihan S (2017) Adaption of microbial life to environmental extremes: novel research results and application, 2nd edn. Springer, New York. https://doi.org/10.1007/978-3-319-48327-6

Stopnisek N, Zuhlke D, Carlier A, Barberan A, Fierer N, Becher D, Riedel K, Eberl L, Weisskopf L (2016) Molecular mechanisms underlying the close association between soil Burkholderia and fungi. ISME J 10(1):253–264. https://doi.org/10.1038/ismej.2015.73

Suárez-Moreno ZR, Caballero-Mellado J, Coutinho BG, Mendonça-Previato L, James EK, Venturi V (2012) Common features of environmental and potentially beneficial plant-associated Burkholderia. Microb Ecol 63(2):249–266. https://doi.org/10.1007/s00248-011-9929-1

Van Goethem MW, Makhalanyane TP, Cowan DA, Valverde A (2017) Cyanobacteria and Alphaproteobacteria may facilitate cooperative interactions in niche communities. Front Microbiol 8:2099. https://doi.org/10.3389/fmicb.2017.02099

Vítek P, Ascaso C, Artieda O, Casero MC, Wierzchos J (2017) Discovery of carotenoid red-shift in endolithic cyanobacteria from the Atacama Desert. Sci Rep 7(1):11116. https://doi.org/10.1038/s41598-017-11581-7

Vítek P, Ascaso C, Artieda O, Wierzchos J (2016) Raman imaging in geomicrobiology: endolithic phototrophic microorganisms in gypsum from the extreme sun irradiation area in the Atacama Desert. Anal Bioanal Chem 408(15):4083–4092. https://doi.org/10.1007/s00216-016-9497-9

Vítek P, Cámara-Gallego B, Edwards HGM, Jehlička J, Ascaso C, Wierzchos J (2013) Phototrophic community in gypsum crust from the Atacama desert studied by Raman spectroscopy and microscopic imaging. Geomicrobiol J 30(5):399–410. https://doi.org/10.1080/01490451.2012.697976

Vítek P, Edwards HGM, Jehlička J, Ascaso C, De Los Ríos A, Valea S, Jorge-Villar SE, Davila AF, Wierzchos J (2010) Microbial colonization of halite from the hyper-arid Atacama desert studied by Raman spectroscopy. Philos Trans R Soc A Math Phys Eng Sci 368(1922):3205–3221. https://doi.org/10.1098/rsta.2010.0059

Vítek P, Jehlička J, Ascaso C, Mašek V, Gómez-Silva B, Olivares H, Wierzchos J (2014a) Distribution of scytonemin in endolithic microbial communities from halite crusts in the hyperarid zone of the Atacama Desert, Chile. FEMS Microbiol Ecol 90(2):351–366. https://doi.org/10.1111/1574-6941.12387

Vítek P, Jehlicka J, Edwards HGM, Hutchinson I, Ascaso C, Wierzchos J (2014b) Miniaturized Raman instrumentation detects carotenoids in Mars-analogue rocks from the Mojave and Atacama deserts. Philos Trans R Soc A Math Phys Eng Sci 372(2030):20140196. https://doi.org/10.1098/rsta.2014.0196

Walker JJ, Pace NR (2007) Endolithic microbial ecosystems. Annu Rev Microbiol 61:331–347. https://doi.org/10.1146/annurev.micro.61.080706.093302

Wang Y, Chen W, He L, Wang Q, Sheng X-F (2016) Draft genome sequence of Ensifer adhaerens M78, a mineral-weathering bacterium isolated from soil. Genome Announc 4(5):e00969–e00916. https://doi.org/10.1128/genomeA.00969-16.

Wei STS, Fernandez-Martinez MA, Chan Y, Van Nostrand JD, de los Rios-Murillo A, Chiu JMY, Ganeshram AM, Cary SC, Zhou J, Pointing SB (2015) Diverse metabolic and stress-tolerance pathways in chasmoendolithic and soil communities of Miers Valley, McMurdo Dry Valleys, Antarctica. Polar Biol 38(4):433–443. https://doi.org/10.1007/s00300-014-1598-3

Whitton B, Potts M (2000) Introduction to the cyanobacteria. In: Whitton B, Potts M (eds) Ecology of cyanobacteria: their diversity in time and space. Springer, New York

Wierzchos J, Ascaso C (2001) Life, decay and fossilisation of endolithic microorganisms from the Ross Desert, Antarctica. Polar Biol 24:863–868. https://doi.org/10.1007/s003000100296

Wierzchos J, Ascaso C, McKay CP (2006) Endolithic cyanobacteria in halite rocks from the hyperarid core of the Atacama Desert. Astrobiology 6(3):415–422. https://doi.org/10.1089/ast.2006.6.415

Wierzchos J, Cámara B, De los ríos A, Davila AF, sánchez almazo IM, Artieda O, Wierzchos K, Gómez-silva B, McKay C, Ascaso C (2011) Microbial colonization of Ca-sulfate crusts in the hyperarid core of the Atacama Desert: implications for the search for life on Mars. Geobiology 9(1):44–60. https://doi.org/10.1111/j.1472-4669.2010.00254.x

Wierzchos J, Casero MC, Artieda O, Ascaso C (2018) Endolithic microbial habitats as refuges for life in polyextreme environment of the Atacama Desert. Curr Opin Microbiol 43:124–131. https://doi.org/10.1016/j.mib.2018.01.003

Wierzchos J, Davila AF, Artieda O, Cámara-Gallego B, de los Ríos A, Nealson KH, Valea S, Teresa García-González M, Ascaso C (2013) Ignimbrite as a substrate for endolithic life in the hyper-arid Atacama Desert: implications for the search for life on Mars. Icarus 224(2):334–346. https://doi.org/10.1016/j.icarus.2012.06.009

Wierzchos J, de los Ríos A, Ascaso C (2012) Microorganisms in desert rocks: the edge of life on earth. Int Microbiol 15(4):173–183. https://doi.org/10.2436/20.1501.01.170

Wierzchos J, DiRuggiero J, Vítek P, Artieda O, Souza-Egipsy V, Škaloud P, Tisza MJ, Davila AF, Vílchez C, Garbayo I, Ascaso C (2015) Adaptation strategies of endolithic chlorophototrophs to survive the hyperarid and extreme solar radiation environment of the Atacama Desert. Front Microbiol 6:934. https://doi.org/10.3389/fmicb.2015.00934

William Schopf J (1974) The development and diversification of Precambrian life. Orig Life 5(1):119–135. https://doi.org/10.1007/BF00927018

Wright DJ, C Smith S, Joardar V, Scherer S, Jervis J, Warren A, Helm R, Potts M (2006) UV irradiation and desiccation modulate the three-dimensional extracellular matrix of Nostoc commune (cyanobacteria). J Biol Chem 280:40271–40281. https://doi.org/10.1074/jbc.M505961200

Xiao L, Wang J, Dang Y, Cheng Z, Huang T, Zhao J, Xu Y, Huang J, Xiao Z, Komatsu G (2017) A new terrestrial analogue site for Mars research: the Qaidam Basin, Tibetan Plateau (NW China). Earth Sci Rev 164:84–101. https://doi.org/10.1016/j.earscirev.2016.11.003

Chapter 5
Desert Biosignatures

Petr Vítek and Jacek Wierzchos

5.1 Introduction

Some biomolecular structures are known to exclusively originate through biological processes, such as the group of carotenoids. Conversely, the origin of some biomolecules (e.g., amino acids) can be either biogenic or abiogenic. The major group of biosignatures that are the subject of this chapter are characterized as biomolecules of living organisms or the molecular remnants of extinct biota stemming from the endolithic microbial communities harboring inside rocks from the Preandean Depression (Chile) (for a review, see Wierzchos et al. 2018 and references therein). For a detailed description of the polyextreme environment of this part of the Atacama Desert and the endolithic habitats found there, see Chap. 4 of this book; for characterization of the endolithic microbiota and their microbial ecology, see Chap. 5. This chapter includes identifications and descriptions of "biomarkers" that, from an organic geochemistry point of view, can be characterized as molecular compounds derived from organisms from all three domains of life, having the ability to be preserved in the sedimentary record (Vandenabeele et al. 2012).

Biomolecules in the geological environment can be detected using various analytical techniques, one of them being the conventional destructive "wet" techniques, such as analyses based on high-performance liquid chromatography (HPLC). On the other hand, Raman spectroscopic detection of biosignatures, as described in the following text, represents a nondestructive vibrational spectroscopic method based on the interaction of an incident laser beam with the sample. Moreover, it allows both organic and inorganic compounds to be detected at the same time. The pigments

P. Vítek (✉)
Global Change Research Institute of the Czech Academy of Sciences, Brno, Czech Republic
e-mail: vitek.p@czechglobe.cz

J. Wierzchos
Museo Nacional de Ciencias Naturales, Consejo Superior de Investigaciones Científicas (CSIC), Madrid, Spain

M. E. Farías (ed.), *Microbial Ecosystems in Central Andes Extreme Environments*, https://doi.org/10.1007/978-3-030-36192-1_5

associated with phototrophic endolithic microorganisms that represent these kinds of molecular biosignatures are valuable targets for Raman spectroscopic analysis (Jehlička et al. 2014). Apart from microbial pigments, the method has been applied for the detection of osmotic solutes in extremophilic bacteria (Jehlička et al. 2012). Raman spectroscopy allows nondestructive spectral discrimination between cyanobacteria and algae (Vítek et al. 2013) and a better understanding of the survival strategies of organisms in response to the harsh desert environment (Vítek et al. 2010, 2014; Wierzchos et al. 2015). Recently, relatively fast Raman imaging (mapping) has allowed the spatial distribution of microbially associated pigments in the rock matrix to be imaged (Vítek et al. 2014, 2016, 2017; Morillas et al. 2015). It is currently planned to integrate miniaturized Raman counterparts into the scientific payloads of robotic rovers that are part of both the 2020 ExoMars mission (European Space Agency (ESA)/Roscosmos) and the Mars 2020 mission (National Aeronautics and Space Administration (NASA)).

Specifically, the Atacama Desert environment is used as a model (analog) for testing of the instrumentation being suggested as a part of future payloads in astrobiologically targeted missions to Mars, including Raman spectroscopy (Vítek et al. 2012; Wei et al. 2015; Wilhelm et al. 2017). Single-point Raman detection, as well as the spatial imaging of these biosignatures, is presented here.

5.2 Microbial Pigments as Biosignatures

Organic pigments are compounds that cannot be abiotically synthesized in nature. Thus, their presence and identification in terrestrial sediments (and perhaps eventually in extraterrestrial sediments) unambiguously point to biological activity. All known photosynthetic organisms use either some form of chlorophyll for their oxygenic photosynthesis or bacteriochrorophyll for anoxygenic photosynthesis. The chlorophyll molecule is formed by a porphyrin (tetrapyrrole) ring with an Mg^{2+} ion in the middle and an alcohol (phytol) side chain.

Phycobiliproteins are accessory pigments that enable coverage of a broader spectrum of visible sunlight (and thus more efficient photosynthesis). Their molecular structure is formed by an open tetrapyrrole chain bonded to proteins. These blue or reddish pigments can be found in cyanobacteria (cyan = blue; the name of the family is derived from the blue pigmentation caused by phycobilins) and in some algae (though not in green algae and higher plants). Three types of phycobiliproteins can be found in the photosynthetic apparatus of these organisms: allophycocyanin, phycocyanin (blue), and phycoerythrin (reddish).

Carotenoids are also accessory pigments involved in photosynthesis, photoprotection, and membrane stabilization. The characteristic feature of carotenoid structures lies in the long conjugated double-bond system composed of isoprenoic units. As part of the photosynthetic apparatus, carotenoids play an important role in the harvesting of energy and the transfer of this energy to the chlorophyll.

There are a variety of ultraviolet (UV)-protective screening pigments. Two major groups of biomolecules (i.e., mycosporine-like amino acids and scytonemins) are described as UV-screening compounds in cyanobacteria (Cockell and Knowland 1999). Scytonemin is well known as a UV radiation–screening pigment and is characteristic of cyanobacteria (Proteau et al. 1993; Garcia-Pichel 1998; Dillon and Castenholz 1999; Edwards et al. 2000).

Other UV-protective substances can be found within lichens and have been described, for example, from Antarctic habitats (Edwards et al. 2004) or from the Atacama Desert (Villar et al. 2005), where parietin, usnic acid, and a few other protective compounds in lichens have been described.

5.3 Protective Role of Pigments

5.3.1 Stratification of Pigments in Preandean Depression Rocks

In studying cryptoendolithic phototrophic colonization, our team, to date, has recognized a variety of substrates in the Preandean Depression of the Atacama Desert, which have revealed different types of colonization patterns. These are interpreted as different adaptation strategies, reflected by specific stratification of the pigments, depending on the composition of the community (being dominated by algae, cyanobacteria, or a combination of both).

On the basis of Raman spectroscopy, three different adaptation strategy models were revealed in gypsum and ignimbrite (Fig. 5.1). The gypcretes from which the data originate come from the Cordon de Lila area (23°53′S, 68°08′W; 2720 m above sea level). The gypcretes are predominantly colonized by cyanobacteria, heterotrophic bacteria, and algae (Wierzchos et al. 2015). They were subjected to Raman imaging by Vítek et al. (2016). Imaging of the cryptoendolithic community, which is dominated by algae, revealed a gradient of an enhanced carotenoid Raman signal with closer proximity to the surface (Fig. 5.1a). Ubiquitous carotenoids, with nine or more conjugated double bonds, absorb in the range corresponding to blue-green irradiation (440–520 nm) (Cockell and Knowland 1999). These longer-chained carotenoids are able to cope with toxic reactive oxygen species, which can be formed during photosynthesis or as a result of UV irradiation, and would otherwise destroy the biomolecules forming the photosynthetic apparatus. The key photoprotective mechanism in algae (and higher plants) is dissipation of excess excitation energy in the xanthophyll cycle (Lunch et al. 2013). Hence, the increased carotenoid content at the upper positions is interpreted as a shield-like photoprotective function of the carotenoids (Wierzchos et al. 2015; Vítek et al. 2016).

Within some transects of the gypcrete (from the same area in the Cordon de Lila), cyanobacteria occur in greater numbers in cryptoendolithic positions. In those cases, the scenario depicted in Fig. 5.1b occurs. Stratification of the colo-

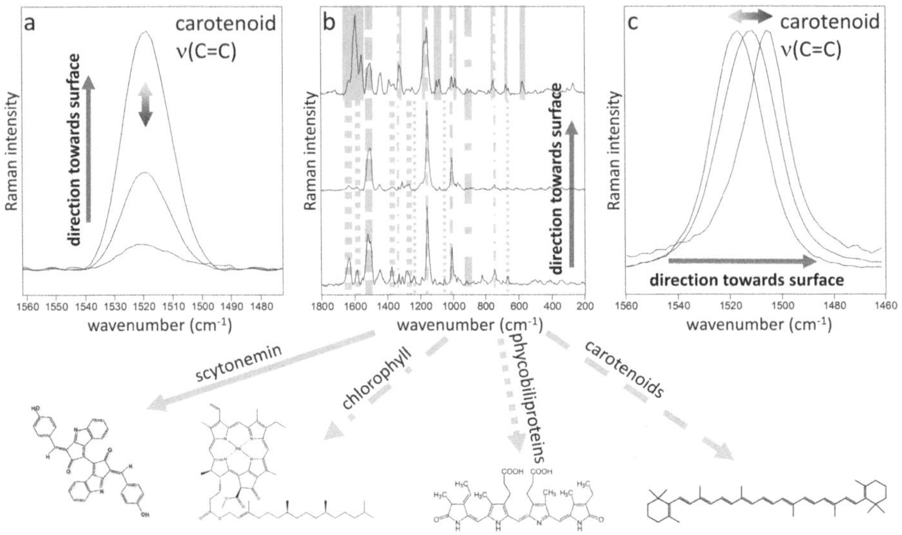

Fig. 5.1 Three different model situations, representing different strategies of adaptation to higher doses of light (including ultraviolet (UV) light), recognized by Raman spectroscopy. **a** Depth-dependent change in the carotenoid content in algae. **b** Depth-dependent vertical stratification of the composition of cyanobacterial pigments, including scytonemin. **c** Depth-dependent change in the carotenoid composition in cyanobacteria

nized zone may be both visually and spectrally described as follows: the coloration from the top to the bottom changes as brown–red–green. Carotenoids were detected within all of the layers studied by strong corroborative Raman bands around 1515 cm^{-1} due to ν(C=C) vibration, a ν(C–C) band at 1155 cm^{-1}, and a weak band at 1003 cm^{-1}, corresponding to the in-plane rocking modes of the CH_3 groups attached to the polyene chain (Fig. 5.1b). Features of chlorophyll at 1326, 910, and 745 cm^{-1} were also detected within all three layers. Strong scytonemin features were detected in the brown (uppermost) layer in the hypoendolithic colonization zone and represent the typical spectral feature of the uppermost layer. The underlying red layer is dominated only by carotenoids, with weaker features of chlorophyll. Below it is the green layer, which is rich in phycobiliproteins (Fig. 5.1b). The detection of scytonemin and phycobiliproteins points to a colony rich in cyanobacteria. Phycobiliprotein Raman features are enhanced toward the lower position, characterized by bands at 1635, 1585, 1373, 1285, 1239, 1112, 1055, 823, and 667 cm^{-1}. This is attributed to the effect of decreasing amounts of light getting into the deeper positions. The light-dependent occurrence of spectroscopic features of phycobiliproteins were previously observed by Villar et al. (2006), who examined Raman spectroscopic features of phycocyanin within cyanobacteria inhabiting volcanic rocks in the arctic environment of Svalbard (Norway).

Another pattern of stratification in the pigment composition (Fig. 5.1c) was observed within the cyanobacterial community from ignimbrites from a distinct location within the Cordon de Lila area, as studied by Vítek et al. (2017). A 1- to 2-mm-thick colonization zone, dominated by *Chroococcidiopsis* sp., was described, with a Raman spectroscopic record that points to changes in carotenoid composition within the transect through the colonization zone. A gradient of the carotenoid ν(C=C) band position was observed, with the lower wavenumber position detected closer to the surface. This is interpreted as the result of a change in carotenoid conjugation, with larger relative amounts of long-chained carotenoids increasing with proximity to the surface. This may possibly be related to nonphotochemical quenching (NPQ) of cyanobacteria via the orange carotenoid protein (OCP) transition to the red form of OCP. Carotenoids are also key molecular structures in this photoprotective mechanism of cyanobacteria, which works in a completely different manner from the xanthophyll cycle in algae and plants (see Kirilovsky and Kerfeld (2016)). The gradient of carotenoid composition was accompanied by an enhanced signal of phycobiliproteins toward deeper positions (Vítek et al. 2017). The Raman bands due to molecular biosignatures detected in this study are summarized in Table 5.1.

5.3.2 *Scytonemin*

Scytonemin is a yellow-brown lipid-soluble pigment that is produced as an extracellular sheath exclusively by certain strains of cyanobacteria. It acts as an effective UV radiation screening compound. Cyanobacterial pigmentation was first described by Nägeli in 1849 (Nägeli 1849) and was later named "scytonemin" (Nägeli and Schwenderer 1877). However, its exact structure was determined much later in 1993 (Proteau et al. 1993). It is widespread within cyanobacteria and may have originated in a common ancestor or in an early member of this lineage (Garcia-Pichel and Castenholz 1991). A genomic study revealed that the biosynthetic genes of scytonemin are highly conserved among evolutionarily diverse strains (Soule et al. 2009). It was demonstrated that the key parameter controlling the biosynthesis of scytonemin by cyanobacteria is the amount of UV-A irradiance (Garcia-Pichel and Castenholz 1991). Together with the UV radiation stress, other stress factors can affect the synthesis of this screening pigment (Dillon et al. 2002).

Table 5.1 Major Raman bands detected in this study

Biomolecules	Wavenumbers (cm^{-1})
Carotenoids	**1515–1527**, **1155–1157**, 1003
Chlorophyll	1555, **1326**, **910**, **745**
Phycobiliproteins	**1635**, **1585**, **1373**, 1285, 1239, 1112, 1053, 823, **667**
Scytonemin	1605, **1592**, **1555**, **1325**, **1176**, 1098, 1081, 985, 754, 577
Lipids	**1657**, **1445**

Key corroborative bands are in bold

The Raman signal of scytonemin is characterized by bands at 1595 and 1554 due to ν(CCH) vibrational modes; at 1326 (the ν(C=N) indole ring); and at 1173 cm^{-1} due to vibration of the ν(C=C–C=C) system (Edwards et al. 2000). Scytonemin has been detected in a variety of rock substrates in the Atacama Desert. Varnali and Edwards (2013a, b, 2014) presented the Raman spectroscopic features of iron(III) complexes, methoxy derivatives of scytonemin, and its reduced and oxidized forms. To date, scytonemin has been found to be associated with cyanobacterial communities in distinct areas of the Atacama Desert, and it was first detected within halite-inhabiting cyanobacteria (*Chroococcidiopsis* sp.) in the Yungay region (Vítek et al. 2010). A comprehensive Raman and microscopic study of scytonemin in Yungay halites has previously been presented by Vítek et al. (2014). More recently, the pigment was detected within cryptoendolithic cyanobacteria (close to the rock surface) and hypoendolithic cyanobacteria (close to the rock base) occupying gypcrete from the Cordon de Lila area (Vítek et al. 2016). There, the Raman imaging method revealed the spatial distribution of the pigment within the gypsum microhabitat.

Here we detail another two examples of cyanobacterial biosynthesis of scytonemin, detected by Raman spectroscopy (both in the classical point analysis mode and in the imaging mode) from the Preandean region of the Atacama Desert—specifically, from the Valle de la Luna and Cordon de Lila areas. Chasmoendolithic colonization is described from two different substrates: calcite and gypcrete. Calcium carbonate colonized by chasmoendolithic phototrophs was displayed in the Valle de la Luna area. The microscopic observations revealed large cyanobacterial aggregates, which is in accord with the work of DiRuggiero et al. (2013) and Meslier et al. (2018). The Raman imaging of these aggregates confirmed the presence of scytonemin through the entirety of the cell aggregates studied (see Fig. 5.2). The imaging acquisition was performed using the point-to-point mode. The Raman map is based on direct classical least squares (DCLS) analysis, showing the distribution of the two main spectral components: the cyanobacterial spectral record (purple color) including strong bands of scytonemin (corroborative bands at 1605, 1592, 1555, 1325, and 1176 cm^{-1}) and carbonate. The carbonate matrix around the cell aggregates is represented by a gray color and is characterized by the strong ν_1 carbonate band at 1087 cm^{-1}. The position of the weak carbonate ν_4 band around 712 cm^{-1} points to calcite, a trigonal form of $CaCO_5$. The Raman signal due to carotenoids is dominated by strong corroborative bands at 1510 cm^{-1} (the broader band probably splits from more carotenoid bands due to ν(C=C) vibration), at 1156 cm^{-1}, and at 1004 cm^{-1} (see Fig. 5.2f). The signal due to carotenoid ν(C=C) vibration is depicted by a red color separately in Fig. 5.2e. The calcite matrix around the cell aggregates is represented by a gray color. As chasmoendolithic colonization does not allow preparation of an absolutely flat surface without destruction of the colony, the relative intensity of the signal within the map is also slightly affected by the morphology of the sample. The detection of scytonemin within various endolithic positions in gypcrete is presented in Fig. 5.3.

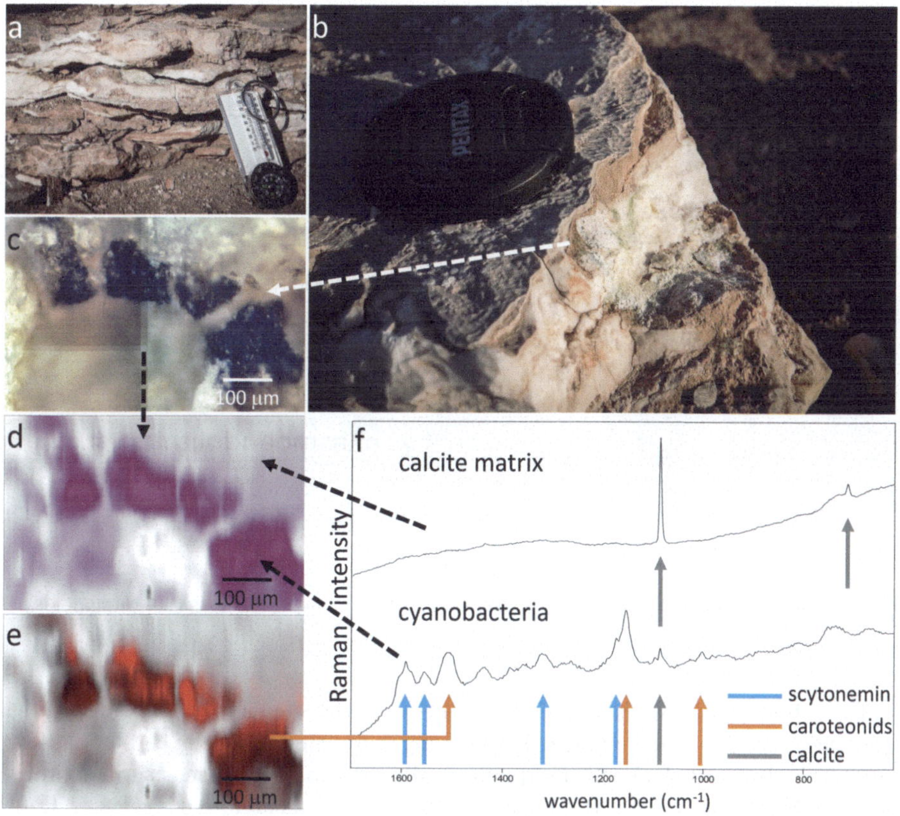

Fig. 5.2 **a**, **b** Calcite from the Valle de la Luna, with macroscopically visible chasmoendolithic colonization. **c** Optical microscopy image of cyanobacterial aggregate. **d** Raman image using point Raman imaging based on direct classical least squares (DCLS) analysis, showing two components: scytonemin (*magenta*) and calcite (*gray*). **e** Signal of carotenoid ν(C=C) band (*red*) and calcite (*gray*). **f** Raman spectra of calcite matrix and the cyanobacterial spectral record, which comprises strong scytonemin and carotenoids bands

5.4 Algal Lipids

Algae are known to contain lipid storage bodies, which can comprise a significant portion of the algal mass (e.g., see Hanagata and Dubinsky (1999)). In Raman spectra, lipids are characterized by a couple of corroborative Raman bands in the 1000–1800 cm^{-1} range, with the strongest bands located around 1657 cm^{-1} (assigned to the C=C stretching vibrational mode), as well as at 1445 cm^{-1} (because of scissoring of CH_2). The two bands are used as indicators of lipid unsaturation (C=C) or saturation (vibration of CH_2 groups), respectively (Samek et al. 2010; Wu et al. 2011). After correction of the carotenoid contribution to the band around 1445 cm^{-1}, the ratio of the unsaturated/saturated band indicator can be correlated with the number of double bonds in the lipid structure (see Samek et al. 2010; Wu et al. 2011).

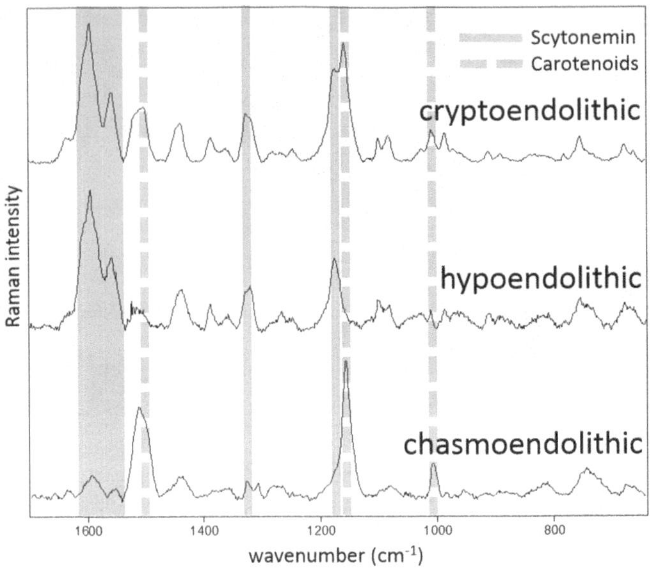

Fig. 5.3 Scytonemin detected at three different endolithic positions, together with carotenoids

Here, both of the bands assigned to lipids were detected as weak features in the spectral record, obtained on algae in gypcrete, dominated by strong carotenoid features. The observed algal cells morphologically resemble the algae described previously by Wierzchos et al. (2015) from the Cordon de Lila area.

5.5 Raman Spectroscopic Approaches to Desert Biosignature Detection

5.5.1 Some Aspects of Point Raman Analysis of Biomolecules in Preandean Depression Rocks

The strong advantage of conventional Raman spectroscopy lies in the minimal (if any) necessity for sample preparation, which favors the technique for applications in robotic instrumentation for planetary explorations as one of the first analytical tools to touch the selected samples.

It is generally known that the sensitivity of the Raman spectroscopic method is highly dependent on the particular molecular structure of the analyte being detected. Pigments are prominent molecular compounds to be detected by Raman spectroscopic analysis of phototrophic biota. Among these, carotenoids are unique and excellent Raman scatterers because of their polyenic structure. Very low concentrations of this biomarker can be detected by Raman microspectrometry (e.g., 0.1–1 mg kg^{-1} in resonance using a 514.5 nm laser, and 1–10 mg kg^{-1} using a 785 nm laser for excitation) with a randomly targeted laser beam (Vítek et al. 2009a, b). Employing the instrumentation with a larger laser spot diameter may even slightly improve the performance of the instrumentation during analysis of low concentrations of a solid analyte in a solid matrix (Vandenabeele et al. 2012). This aspect was found to be an advantage of the miniaturized Raman systems working with a large laser spot diameter in comparison with their benchtop counterparts (see the next paragraph).

On the other hand, objects focused at the micrometer level allow microorganisms to be analyzed in the context of the surrounding rock matrix. Moreover, selection of the appropriate area for analysis (e.g., cell, cell aggregate, accumulation of decayed biomass) leads to acquisition of the spectral record from the area of greater concentrations of the molecular compounds of interest. These may not otherwise be detectable when a highly dispersed biomolecule occurs in the surrounding matrix (for a more detailed discussion about the limits of detection, see Vandenabeele et al. (2012)).

5.5.2 *Miniaturized Raman Spectrometers and Desert Mars Analogs*

Desert rock materials are used for the testing of instrumentation aimed at detecting molecular traces of life as part of future payloads in astrobiological missions to Mars (e.g., the ExoMars or NASA mission in 2020). A miniaturized Raman spectrometer is planned as part of the complement of instrumentation on these missions, and representative instrumentation is tested for the detection of various biosignatures, including microbial pigments (Culka et al. 2014; Malherbe et al. 2017).

It has been demonstrated that miniaturized instruments using 532 and 785 nm excitation sources provide a more stable signal than their benchtop counterparts at the lowest concentrations of β-carotene in an evaporitic matrix. It was 0.1 mg kg^{-1} in the case of 532 nm (Vítek et al. 2014) and 1 mg kg^{-1} in the case of 785 nm (Vítek et al. 2011). In this case, comparable or slightly better performance than that of the benchtop counterparts is interpreted as the result of a larger sample volume being measured during use of a larger laser spot size. This may be a key factor during remote analysis of a solid analyte dispersed in a solid matrix, which can be inhomogeneous at the microscale (Vandenabeele et al. 2012). Recently, miniaturized Raman instrumentation using 532 nm for excitation was successfully applied for detection of carotenoids of halophilic prokaryotes in synthetic inclusions of NaCl, KCl, and various sulfates (Jehlička et al. 2018).

Scytonemin and carotenoids were detected during the analytical testing of miniaturized Raman spectrometers on a natural Atacama halite sample in the context of the future planned robotic missions mentioned above (Vítek et al. 2012). In another study, the carotenoids associated with different halophilic microbes were unique molecular compounds detected within different types of desert evaporites (i.e., gypsum, halite, thenardite, burkeite, and trona) with use of a miniaturized Raman system with a 532 nm excitation wavelength (Vítek et al. 2014).

5.5.3 Raman Imaging

Raman imaging brings yet another dimension to Raman spectroscopic analysis, allowing the two- or even three-dimensional distribution of an analyte within the scanned area to be obtained (Nasdala et al. 2012; Foucher et al. 2017). Contrary to conventional single point Raman analysis, there is the necessity for some more-or-less complex sample preparation for most of the Raman imaging applications. There is the need for the scanned surface to be as flat as possible to minimize bias of the signal intensity, caused by varying laser focus. The duration of the imaging acquisition is predominantly of the order of hours. Raman imaging of geomicrobiological samples of gypcrete and ignimbrite from the Preandean Cordon de Lila area has been applied by Vítek et al. (2016, 2017) and helped to reveal the significant biomolecular adaptations of algae and cyanobacteria to stress conditions in the Preandean region of the Atacama Desert. Apart from a conventional Raman imaging technique, coherent anti-Stokes Raman scattering (CARS) microscopy was used to obtain high-resolution images of scytonemin in the cyanobacterium *Nostoc commune* (Venckus et al. 2018).

There are various approaches on how to perform the imaging acquisition (Lee 2012; Wang et al. 2015). One of the rapid imaging methods uses line focus, where the laser beam is focused as a prolonged elliptical-shaped spot in combination with a microscopic stage that is synchronized with the continuous charge-coupled device (CCD) readout. Here (Fig. 5.2), point-to-point mapping was applied. It is more time consuming; nevertheless, a better signal-to-noise ratio can be obtained.

The spatial resolution of the method equipped with a confocal Raman microscope is diffraction limited; hence, only those objects that are not smaller than about half of the value of the excitation wavelength can be displayed (the Abbe limit). The modern benchtop Raman spectroscopic setups, combined with atomic force microscopy (AFM), allow for spatial resolution that is not diffraction limited, ranging from $\lambda/10$ to $\lambda/50$ (for more information, see Prats-Mateu and Gierlinger (2017)).

5.6 Conclusions

Pigments are molecular structures that are good Raman scatterers and that dominate the Raman signal of molecular biosignatures from phototrophic desert microorganisms. In addition, spectral signs of lipids have been recorded in algal cells in gypcretes from the Cordon de Lila area. The technique allowed observation of the spatial distribution of pigments within the microbial colonies. Accordingly, three different adaptation strategies were observed and defined within different rock habitats. Spatially resolved and depth-dependent changes in Raman features were observed and interpreted as:

- Changes in the relative carotenoid content
- Changes in the relative intensities of scytonemin, carotenoids, chlorophyll, and phycobiliproteins

- Changes in the composition (structure) of carotenoids

The imaging capability of Raman spectroscopy has played an important role in revealing some microbial adaptation strategies, and its development in the future represents a promising tool for in vivo detection of molecular biosignatures within rocks.

Acknowledgments This work was supported by the Ministry of Education, Youth and Sports of the Czech Republic within National Sustainability Program I (NPU I) grant number LO1415, INTER-COST grant number LTC18036 to P. Vítek, and grant number PGC2018-094076-B-I00 to all authors from the Ministry of Science, Innovation and Universities, Spain.

References

Cockell CS, Knowland J (1999) Ultraviolet radiation screening compounds. Biol Rev 74:311–345

Culka A, Osterrothová K, Hutchinson I, Ingley R, McHugh M, Oren A, Edwards HGM, Jehlička J (2014) Detection of pigments of halophilic endoliths from gypsum: Raman portable instrument and European Space Agency's prototype analysis. Philos Trans R Soc A 372:20140205

Dillon JG, Castenholz RW (1999) Scytonemin, a cyanobacterial sheath pigment, protects against UVC radiation implications for early photosynthetic life. J Phycol 35:673–681

Dillon JG, Tatsumi CM, Tandingan PG, Castenholz RW (2002) Effect of environmental factors on the synthesis of scytonemin, a UV-screening pigment, in a cyanobacterium (*Chroococcidiopsis* sp.). Arch Microbiol 177:322–331

DiRuggiero J, Wierzchos J, Robinson CK, Souterre T, Ravel J, Artieda O, Souza-Egipsy V, Ascaso C (2013) Microbial colonisation of chasmoendolithic habitats in the hyper-arid zone of the Atacama Desert. Biogeosciences 10:2439–2450

Edwards HGM, Garcia-Pichel F, Newton EM, Wynn-Williams DD (2000) Vibrational Raman spectroscopic study of scytonemin, the UV-protective cyanobacterial pigment. Spectrochim Acta A 56:193–200

Edwards HGM, Cockell ChS, Newton EM, Wynn-Williams DD (2004) Protective pigmentation in UVB-screened Antarctic lichens studied by Fourier transform Raman spectroscopy: an extremophile bioresponse to radiation stress. J Raman Spectrosc 35:463–469

Foucher F, Guimbretiére G, Bost N, Westall F (2017) Petrographical and mineralogical applications of Raman mapping. In: Maaz K (ed) Raman spectroscopy and applications. IntechOpen. https://www.intechopen.com/books/raman-spectroscopy-and-applications/petrographical-and-mineralogical-applications-of-raman-mapping. doi:10.5772/65112

Garcia-Pichel F (1998) Solar ultraviolet and the evolutionary history of cyanobacteria. Orig Life Evol Biosph 28:321–347

Garcia-Pichel F, Castenholz RW (1991) Characterization and biological implications of scytonemin, a cyanobacterial sheath pigment. J Phycol 27:395–409

Hanagata N, Dubinsky Z (1999) Secondary carotenoid accumulation in *Scendesmus* Komarekii (Chlorophyceae, Chlorophyta). J Phycol 35:960–966

Jehlička J, Oren A, Vítek P (2012) Use of Raman spectroscopy for identification of compatible solutes in halophilic bacteria. Extremophiles 16:507–514

Jehlička J, Edwards HGM, Oren A (2014) Raman spectroscopy of microbial pigments. Appl Environ Microbiol 80:3286–3295

Jehlička J, Culka A, Mana L, Oren A (2018) Using a portable Raman spectrometer to detect carotenoids of halophilic prokaryotes in synthetic inclusions in NaCl, KCl, and sulfates. Anal Bioanal Chem 410:4437–4445

Kirilovsky D, Kerfeld CA (2016) Cyanobacterial photoprotection by the orange carotenoid protein. Nat Plants 2:16180

Lee E (2012) Imaging modes. In: Zoubir A (ed) Raman imaging, techniques and applications, Springer series in optical sciences, vol 168. Berlin: Springer, pp 1–37

Lunch CK, LaFountain AM, Thomas S, Frank HA, Lewis LA, Cardon ZG (2013) The xanthophyll cycle and NPQ in diverse desert and aquatic green algae. Photosynth Res 115:139–151

Malherbe C, Hutchinson IB, McHugh M, Ingley R, Jehlička J, Edwards HGM (2017) Accurate differentiation of carotenoid pigments using flight representative Raman spectrometers. Astrobiology 17:351–362

Meslier V, Cristina Casero M, Dailey M, Wierzchos J, Ascaso C, Artieda O, McCullough PR, DiRuggiero J (2018) Fundamental drivers for endolithic microbial community assemblies in the hyperarid Atacama Desert. Environ Microbiol 20:1765–1781

Morillas H, Maguregui M, Marcaida I, Trebolazabala J, Salcedo I, Madariaga JM (2015) Characterization of the main colonizer and biogenic pigments present in the red biofilm from La Galea Fortress sandstone by means of microscopic observations and Raman imaging. Microchem J 121:48–55

Nägeli C (1849) Gattungen einzelliger Algen, physiologisch und systematisch bearbeitet. Neue Denkschrift Allg Schweiz Natur Ges 10:1–138

Nägeli C, Schwenderer S (1877) Das Mikroskop, 2. Leipzig: Wilhelm Engelmann Verlag

Nasdala L, Beyssac O, Schopf JW, Bleisteiner B (2012) Application of Raman-based images in the Earth sciences. In: Zoubir A (ed) Raman imaging, techniques and applications. Springer series in optical sciences 168. Berlin: Springer, pp 145–187

Prats-Mateu B, Gierlinger N (2017) Tip in-light on: advantages, challenges, and applications of combining AFM and Raman microscopy on biological samples. Microsc Res Tech 80:30–40

Proteau PJ, Gerwick WH, Garcia-Pichel F, Castenholz R (1993) The structure of scytonemin, an ultraviolet sunscreen pigment from the sheaths of cyanobacteria. Experientia 49:825–829

Samek O, Jonáš A, Pilát Z, Zemánek P, Nedbal L, Tříska J, Kotas P, Trtílek M (2010) Raman microspectroscopy of individual algal cells: sensing unsaturation of storage lipids in vivo. Sensors 10:8635–8651

Soule T, Palmer K, Gao Q, Potrafka RM, Stout V, Garcia-Pichel G (2009) A comparative genomics approach to understanding the biosynthesis of the sunscreen scytonemin in cyanobacteria. BMC Genomics 10:336

Vandenabeele P, Jehlička J, Vítek P, Edwards HGM (2012) On the definition of Raman spectroscopic detection limits for the analysis of biomarkers in solid matrices. Planet Space Sci 62:48–54

Varnali T, Edwards HGM (2013a) A potential new biosignature of life in iron-rich extreme environments: an iron(III) complex of scytonemin and proposal for its identification using Raman spectroscopy. Planet Space Sci 82–83:128–135

Varnali T, Edwards HGM (2013b) Theoretical study of novel complexed structures for methoxy derivatives of scytonemin: potential biomarkers in iron-rich stressed environments. Astrobiology 13:861–869

Varnali T, Edwards HGM (2014) Reduced and oxidized scytonemin: theoretical protocol for Raman spectroscopic identification of potential key biomolecules for astrobiology. Spectrochim Acta A 117:72–77

Venckus P, Paliulis S, Kostkevičiene J, Dementjev A (2018) CARS microscopy of scytonemin in cyanobacteria *Nostoc commune*. J Raman Spectrosc 49:1333–1338

Villar SEJ, Edwards HGM, Seaward MRD (2005) Raman spectroscopy of hot desert, high altitude epilithic lichens. Analyst 130:730–737

Villar SEJ, Edwards HGM, Benning LG (2006) Raman spectroscopic and scanning electron microscopic analysis of a novel biological colonisation of volcanic rocks. Icarus 184:158–169

Vítek P, Osterrothová K, Jehlička J (2009a) Beta-carotene—a possible biomarker in Martian evaporitic environment: Raman micro-spectroscopic study. Planet Space Sci 57:454–459

Vítek P, Jehlička J, Edwards HGM, Osterrothová K (2009b) Identification of β-carotene in an evaporitic matrix—evaluation of Raman spectroscopic analysis for astrobiological research on Mars. Anal Bioanal Chem 393:1967–1975

Vítek P, Edwards HGM, Jehlicka J, Ascaso C, De los Rıos A, Valea S, Villar SEJ, Davila AF, Wierzchos J (2010) Microbial colonization of halite from the hyper-arid Atacama Desert studied by Raman spectroscopy. Philos Trans R Soc A 368:3205–3221

Vítek P, Edwards HGM, Jehlička J, Cox R (2011) Evaluation of portable Raman instrumentation for identification of β-carotene and mellitic acid in two-component mixtures with halite. Spectrochim Acta A 80:32–35

Vítek P, Jehlicka J, Edwards HGM, Hutchinson I, Ascaso C, Wierzchos J (2012) Miniaturized Raman system and the detection of traces of life in halite from the Atacama Desert: some considerations for the search for life signatures on Mars. Astrobiology 12:1095–1099

Vítek P, Camara-Gallego B, Edwards HGM, Jehlička J, Ascaso C, Wierzchos J (2013) Phototrophic community in gypsum crust from the Atacama Desert studied by Raman spectroscopy and microscopic imaging. Geomicrobiol J 30:399–410

Vítek P, Jehlička J, Ascaso C, Mašek V, Gómez-Silva B, Oliváres H, Wierzchos J (2014) Distribution of scytonemin in endolithic microbial communities from halite crusts in the hyperarid zone of the Atacama Desert, Chile. FEMS Microbiol Ecol 90:351–366

Vítek P, Ascaso C, Artieda O, Wierzchos J (2016) Raman imaging in geomicrobiology: endolithic phototrophic microorganisms in gypsum from the extreme sun irradiation area in the Atacama Desert. Anal Bioanal Chem 408:483–492

Vítek P, Ascaso C, Artieda O, Casero MC, Wierzchos J (2017) Discovery of carotenoid red-shift in endolithic cyanobacteria from the Atacama Desert. Sci Rep 7:11116

Wang A, Korotev RL, Jolliff BL, Ling Z (2015) Raman imaging of extraterrestrial materials. Planet Space Sci 112:23–34

Wei J, Wang AA, Lambert JL, Wettergreen D, Cabrol N, Warren-Rhodes K, Zacny K (2015) Autonomous soil analysis by the Mars micro-beam Raman spectrometer (MMRS) on-board a rover in the Atacama Desert: a terrestrial test for planetary exploration. J Raman Spectrosc 46:810–821

Wierzchos J, DiRuggiero J, Vítek P, Artieda O, Souza-Egipsy V, Škaloud P, Tisza M, Davila AF, Vílchez C, Garbayo I, Ascaso C (2015) Adaptation strategies of endolithic chlorophototrophs to survive the hyperarid and extreme solar radiation environment of the Atacama Desert. Front Microbiol 6:934

Wierzchos J, Cristina Casero M, Artieda O, Ascaso C (2018) Endolithic microbial habitats as refuges for life in polyextreme environment of the Atacama Desert. Curr Opin Microbiol 43:124–131

Wilhelm MB, Sanz-Arranz JA, Sobron P, Rull F, Davila AF (2017) Critical assessment of biosignature detection with Raman spectroscopy on biologically lean soils. Paper presented at the Astrobiology Science Conference 2017, Mesa, 24–28 Apr 2017. https://www.hou.usra.edu/meetings/abscicon2017/pdf/3424.pdf. Accessed 24 Nov 2019

Wu H, Volponi JV, Oliver AE, Parikh AN, Simmons BA, Singh S (2011) In vivo lipidomics using single-cell Raman spectroscopy. Proc Natl Acad Sci U S A 108:3809–3814

Chapter 6
Microbial Characterization of Andean Peatland's Soil

Carolina Belfiore, Ana Paula Santos, Alejandra Leonor Valdez, Manuel Contreras Leiva, and María Eugenia Farías

6.1 Introduction

Wetlands are defined as one of the most important terrestrial ecosystems, and they are distributed in all regions throughout the world from the tundra to the tropics (Jungblut et al. 2012). They serve as a focus of human activity in arid lands and support much of the region's biodiversity (Nicholas 1998; Ashley 2001; Ruthsatz 1993; Squeo et al., 2006b), maintain vertebrate populations (Halloy 1991; Seimon et al. 2007), and regulate hydrological resources (Reboratti 2006).

These aspects of wetland value led to the Ramsar Convention (Turner 2014). The Ramsar Convention is an international treatise for the conservation and sustainable use of wetlands; Article 1.1 defines wetlands, as "areas of marsh, fen, peatland or water, whether natural or artificial, permanent or temporary, with water that is static or flowing, fresh, brackish or salt, including areas of marine water the depth of which at low tide does not exceed six metres". This definition includes all lakes and rivers, swamps and marshes, wet grasslands, peatlands, oases, estuaries, deltas and low tides, mangroves and other coastal areas, coral reefs, and artificial sites such as fish ponds, rice fields, reservoirs and salt pans (Izquierdo et al. 2018). A particular case is the vegas or bofedales, which, although they originate from outcrops of the aquifer, are kept near or on the surface because the characteristic vegetation promotes the accumulation of organic matter and

C. Belfiore · A. P. Santos · A. L. Valdez
Laboratorio de Investigaciones Microbiológicas de Lagunas Andinas (LIMLA), Planta Piloto de Procesos Industriales Microbiológicos (PROIMI-CONICET), Tucumán, Argentina

M. C. Leiva
Centro de Ecología Aplicada (CEA), Santiago, Chile

M. E. Farías (✉)
LIMLA-PROIMI-CONICET, Tucuman, Argentina

M. E. Farías (ed.), *Microbial Ecosystems in Central Andes Extreme Environments*, https://doi.org/10.1007/978-3-030-36192-1_6

modifies the hydrology of the place with its ability to retain and regulate the flow of water. If the vegetation did not keep the water under its cover, it could circulate and form another type of wetland or aquatic system (Izquierdo et al. 2018).

6.2 Andean Wetlands: The Importance

Wetlands are found throughout the world, except in Antarctica. The world has 7–9 million km^2 of wetlands, which is 4–6% of the land surface. Approximately 56% of the 4–6% of the land surface is found in the tropical and subtropical regions; nevertheless, the wadis of southern Africa, the oases of the Middle East, and the billabongs of Australia are examples of arid land wetlands. Much less known are the peatlands in the high Andean arid zone of the central Andes, which have been referred to as bofedales, vegas, cushion bogs, and wet grasslands (Fig. 6.1). These wetlands of arid and semi-arid zones are strategic; annual rainfall is exclusively during the summer and on average less than 400 mm per year (Cabrera 1976; Morales et al. 2018). Despite hyper-aridity, intense solar radiation, high-velocity winds, hypoxia, daily frost, and a short growing season, bofedales are near the hydrological and altitudinal limits for plant life in the cold and arid grasslands of Perú, Bolivia, Chile, and Argentina (Ruthsatz 1993, 2000; Squeo et al. 1994, 2006a; Villagran and Castro 1997). These peatlands are like no other in the world; they have been referred to as "highland bogs" (Wilcox et al. 1986; Ruthsatz 1993).

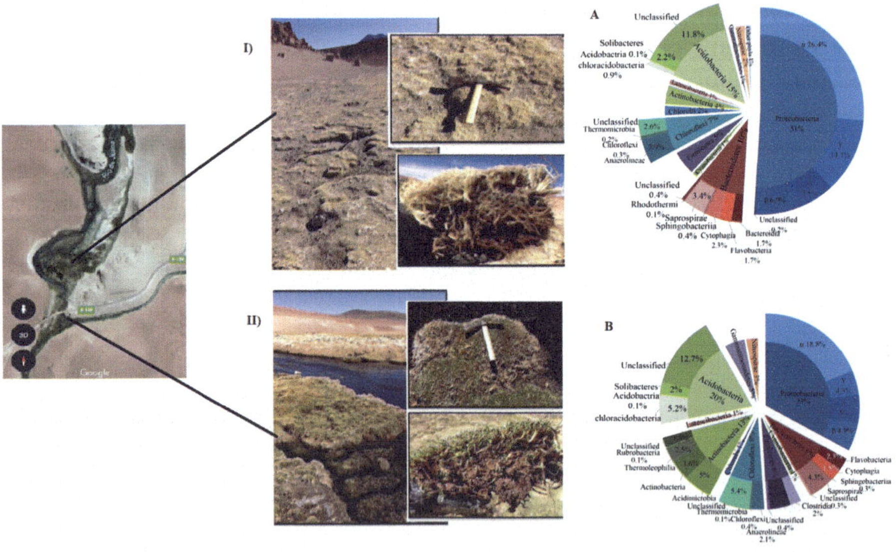

Fig. 6.1 Tocorpuri peatland in different states of conservation (I) dead and (II) active peatland, and comparison of the microbial soil diversity. (Adapted from Belfiore et al. (2018))

Wetlands are vital for human survival; they are among the world's most productive environments, are cradles of biological diversity, and provide water for countless species of plants and animals.

Wetlands reduce the impacts of floods, absorb pollutants, and improve water quality, and, in addition, an immense variety of species of plants, insects, amphibians, reptiles, birds, fish, mammals, and microbes can be a part of them. The complex, dynamic relationships among the organisms inhabiting the wetland environment are the reason they differ from one another.

With regard to plants, several thousand species grow in wetlands, ranging from mosses and grasses to shrubs and trees. Submerged plants, plants that float on the water's surface, and emergent plants can grow in the wetlands. In Andean peatlands, members of the Juncaceae are the most common species, Oxychloe andina and Patosia clandestina being the community dominants and primary peat-formers (Ruthsatz 1993, 2000; Squeo et al. 2006b).

The vegetation of each peatland is in close relation to the available water resource and contrasts sharply with surrounding terrestrial communities by having plant cover usually greater than 70% and high plant productivity (Squeo et al. 1993, 1994, 2006b). It is controlled by four main interacting ecological factors: (a) water quantity and seasonal availability, especially during dry periods, (b) favorable ambient temperatures and occurrence of frost events that control the duration of the growing season, (c) water pH, availability of nutrients (mainly, N, P, K, Ca, and Mg), and exposure to toxic elements such as As, B, Fe, and Al in the water, and (d) biotic factors, such as seed dispersion by animals, grazing, and human impacts (Villagran et al. 1983, 2003; Ruthsatz 1993, 2000; Squeo et al. 2006a, b).

With regard to wildlife, mammals that might live in wetlands include beavers, otters, bobcats, deer, and minks. Alligators, snakes, turtles, newts, and salamanders are among the reptiles and amphibians. Invertebrates, such as crayfish, shrimp, mosquitoes, snails, and dragonflies, also live in wetlands, along with birds, including plover, grouse, storks, herons, and other waterfowl (Gambrel 2017). The wetlands play a critical role in preservation of the diversity of the Cordillera de los Andes. They are areas of valuable threatened species conservation, including for vicuña, guanaco, llama, alpaca, among others, which depend upon the peatlands for grazing and water (Calvo et al. 2000). Migratory birds, such as *Phoenicopterus andinus*, *Ph. jamesi*, *Netta erythropthalma*, *Theristicus melanopis*, *Gallinula melanops*, and fish and amphibians, such as *Atelopus muisca* (http://ramsar.rgis.ch/pdf/cop9/cop9_doc26_s.pdf), use wetlands. They are refuge and reproduction areas.

6.3 Impact on Water Resource

Wetland areas are not only important to our environment but also directly benefit our society (Knutsen 2014). The peatlands have cultural, environmental, and economic social importance, since they are the livelihood for communities. Wetlands are formed and maintained with the conjunction and interaction of three elements:

water, soil, and plant or animal organisms. Since the 1970s, wetlands have been considered as the kidneys of the planet; however, during the past century, these valuable ecosystems were destroyed. The most relevant culprits responsible for wetland degradation and loss are: the extraction of water for agricultural and mining uses, overgrazing, burning, pollution, intense urbanization processes, the construction of large infrastructure works, sectionized and incomplete developmental planning, in addition to the rapid growth of the human population (Abramovitz 1996; Rangel 2000; Canevari et al. 2001; Hofstede 2003; Calvo et al. 2000; Holden et al. 2004). According to the Intergovernmental Panel on Climate Change (IPCC) of the United Nations, wetlands are among the ecosystems most vulnerable to climate change. It is mentioned, for example, that continental wetlands will be affected by changes in precipitation and experience increased frequency and intensity of droughts, storms, and floods. It is important to mention that in the case of the Andean wetlands, the surveys indicate that the state of conservation is vulnerable (WWF 2014).

6.4 Microbiology of the Soils

The loss of wetlands raises the need for their recovery; however, a problem arises when irrigation is not enough, and thus the present microbiota could be relevant. Soil is known as a complex habitat with a large number of microbial populations (Reyes et al. 2006). The microbial diversity in soil ecosystems exceeds, by far, that of the eukaryotic organisms. Less than 1% of the billion microorganisms of the soil can be cultivated and characterized; soil ecosystems are, generally, uncharted (Torsvik and Øvreas 2002). There are several studies about the flora and fauna of these wetlands; however, very few are about the microbiota and its role in these ecosystems. The activities of microorganisms in wetlands play an important role in biogeochemical processes, and they are key to the functions of wetlands (Farias et al. 2014; Hartman et al. 2008; van den Heuvel et al. 2010, 2012; Hu et al. 2014). The soil is considered ecologically to be an extremely stable habitat (Kloepper and Beuchamp 1992). The number and types of microorganisms present in a soil depend on various environmental factors, such as nutrients, moisture, aeration, temperature, pH, and agricultural practices (Chao et al. 1986). The populations and the structure of the community tend to remain constant; however, this homeostasis and dominance of certain populations can be disrupted when environmental conditions are altered.

The microbial diversity in peatlands is characterized by organisms that have developed physiological and metabolic adaptations to cope with the constraining conditions found in these ecosystems, such as low oxygen availability, cold temperature, acidity, and oligotrophy. *Proteobacteria*, *Bacteroidetes*, *Acidobacteria*, *Actinobacteria*, *Firmicutes*, and *Chloroflexi* seem to be the most relevant genera in wetlands (Belfiore et al. 2018; Hartman et al. 2008; Lv et al. 2014; Niederberger et al. 2015).

While there is not much knowledge on the subject, the response of the microbial communities to the disturbances could be the key to the recovery of these ecosystems, since the microbial activities influence the growth and performance of plants

(Lamers et al. 2012). The understanding of the microorganisms living in wetland ecosystems is far from complete and necessitates a more intensive and comprehensive assessment. A first contribution is given by Belfiore et al. (2018), who described that the loss of water in Andean peatland causes great changes in the physicochemical characteristics of the soil, which leads to a decrease of the microbiota *Proteobacteria* indicating it is more sensitive to extreme conditions. This phylum decreases in abundance with desiccation (Zhou et al. 2016). *Acidobacteria*, *Actinobacteria*, and *Gemmatinoidetes* increased in these samples showing a better adaptation to the change of conditions (low soil moisture) (Fig. 6.1). They have the ability to survive in adverse condition (Goodfellow and Williams 1983; Potts 1994; Griffin 1977), for example, *Acidobacteria* produce exopolysaccharides and Firmicutes produce endospores under stressful environmental conditions (Lv et al. 2014). Although cyanobacteria dominate the bacterial populations of many extreme environments (Witton and Potts 1999), and many studies reveal the ability of this phylum to withstand large periods of drying (Lange et al. 1989; Potts 1994, 2001; Kranner et al. 2003, 2008; Potts et al. 2005), its presence in Salar de Atacama is scarce (Farias et al. 2013, 2014; Rasuk et al. 2014, 2016; Fernandez et al. 2016).

Soil is considered an ecologically stable habitat; nevertheless, this homeostasis and dominance of certain populations can be disrupted when environmental conditions are altered. Available data about microbial communities in different types of wetlands is scarce, and therefore it is impossible to make any general conclusions about the dynamics of microbial community structure in each one of them.

Deeper knowledge of microbial communities for each particular wetland is required because the recovery of the wetland could depend not only on hydration of soils but also on the bioinoculation of the native organisms.

References

Abramovitz JN (1996) Imperiled waters, impoverished future: the decline of freshwater ecosystems. Worldwatch Paper 128. Worldwatch Institute, Washington, DC

Ashley GM (2001) Archeological sediments in springs and wetlands. In: Stein JK, Ferrand WR (eds) Sediments in archaeological contexts. University of Utah Press, Salt Lake City, UT, 218p

Belfiore C, Fernandez A, Santos AP, Contreras M, Farías ME (2018) Characterization and comparison of microbial soil diversity in two Andean peatlands in different states of conservation—Vega Tocorpuri. J Geosci Environ Protect 6:194–210

Cabrera AL (1976) Regiones Fitogeogra'ficas Argentinas. Editorial Acme, Buenos Aires

Calvo MAA, Pozo Torres VL, Rojas García MF, Zenteno A (2000) Protección de humedales (Vegas y Bofedales) en el Norte de Chile

Canevari P, Castro G, Sallaberry M, Naranjo LG (2001) Guía de los Chorlos y Playeros de la Región Neotropical. In: American Bird Conservancy, WWF-US, Humedales para las Américas y Manomet Conservation Science, Asociación Calidris. Colombia, Santiago de Cali

Chao WL, Nelson EB, Harman GE, Hoch HC (1986) Colonization of the rhizosphere by biological control agents applied to seeds. Phytopathology 76:60–65

Farias ME, Rascovan N, Toneatti DM, Albarracin VH, Flores MR, Poiré DG (2013) The discovery of stromatolites developing at 3570 m above sea level in a high-altitude volcanic lake Socompa, Argentinean Andes. PLoS One 8:e53497

Farias ME, Contreras M, Rasuk MC et al (2014) Characterization of bacterial diversity associated with microbial mats, gypsum evaporites and carbonate microbialites in thalassic wetlands: tebenquiche and La Brava, Salar de Atacama, Chile. Extremophiles 18(2):311–329
Fernandez AB, Rasuk MC, Visscher P, Contreras M, Novoa F, Poire D, Patterson MM, Ventosa A, Farias ME (2016) Microbial diversity in sediment ecosystems (evaporites domes, microbial mats, and crusts) of hypersaline laguna tebenquiche. Front Microbiol 7:1284
Gambrel E (2017) Animals that live in wetland. Wetland Plants & Wildlife. United States, Florida.
Goodfellow M, Williams S (1983) Ecology of actinomycetes. Annu Rev Microbiol 37:189–216
Griffin DM (1977) Water potential and wood-decay fungi. Annu Rev Phytopathol 15:319–329
Halloy S (1991) Islands of life at 6000 m altitude: the environment of the highest autotrophic communities on Earth (Socompa Volcano, Andes). Arctic Alpine Res 23:247–262. https://doi.org/10.1080/00040851.1991.12002843
Hartman WH, Richardson CJ, Vilgalys R, Bruland GL (2008) Environmental and anthropogenic controls over bacterial communities in wetland soils. Proc Natl Acad Sci U S A 105:17842–17847
Hofstede R (2003) Los Páramos en el Mundo: su diversidad y sus habitantes. In: Hofstede R, Segarra P, Mena P (eds) Los Páramos del Mundo Proyecto Atlas Mundial de Los Páramos. Global Peatland Initiative/NC-IUCN/EcoCiencia, Quito, pp 15–38
Holden J, Chapman PJ, Labadz JC (2004) Artificual drainage of Peatland: hydrological and hydrochemical process and wetland restoration. Prog Phys Geog 28:95–123
Hu B, Shen L, Lian X et al (2014) Evidence for nitrite-dependent anaerobic methane oxidation as a previously overlooked microbial methane sink in wetlands. Proc Natl Acad Sci U S A 111:4495–4500
Izquierdo AE, Aragon R, Navarro CJ and Casagranda E (2018) Humedales de la Puna: principals proveedores de servicios ecosistemicos de la region. Serie conservacion de la naturaleza 24: La Puna argentina: naturaleza y cultura
Jungblut AD, Wood SA, Hawes I, Webster-Brow J, Harris C (2012) The Pyramid Trough Wetland: environmental and biological diversity in a newly created Antarctic protected area. FEMS Microbiology Ecology, 82(2), 356–366
Kloepper J, Beuchamp C (1992) A review of issues related to measuring colonization of plant roots by bacteria. Can J Microbiol 38:1219–1232
Knutsen C (2014) Wetlands and mining: a case study of "NorthMet". DJUB 1(Spring 2014):1–9
Kranner I, Zorn M, Turk B, Wornik S, Beckett RP, Batic F (2003) Biochemical traits of lichens differing in relative desiccation tolerance. New Phytol 160:167–176
Kranner I, Beckett R, Hochman A, Nash TH III (2008) Desiccation-tolerance in lichens: a review. Bryologist 111:576–593
Lamers LPM, Van Diggelen JMH, Op Den Camp HJM, Visser EJW, Lucassen ECHET, Vile MA et al (2012) Microbial transformations of nitrogen, sulfur, and iron dictate vegetation composition in wetlands: a review. Front Microbiol 3:156. https://doi.org/10.3389/fmicb.2012.00156
Lange OL, Bilger W, Rimke S, Schreiber U (1989) Chlorophyll fluorescence of lichens containing green and blue-green algae during hydration by water vapor uptake and by addition of liquid water. Bot Acta 102:306–313
Lv X, Yu J, Fu Y, Ma B, Qu F, Ning K, Wu H (2014) A meta-analysis of the bacterial and archeal diversity observed in wetland soils. ScientificWorldJournal 2014:437684
Morales MS, Chritie DA, Neukom R, Rojas F, Villalba R (2018) Variabilidad hidroclimatica en el sur del Altiplano pasado presente y futuro. In: Grau HR, Babot MJ, Izquierdo A, Grau A (eds) La Puna argentina: Naturaleza y cultura, Serie conservación de la Naturaleza No. 24. Miguel Lillo Foundation, Argentina, pp 75–91
Nicholas GP (1998) Wetlands and hunter-gatherers: a global perspective. Curr Anthropol 39:720–731. https://doi.org/10.1086/204795
Niederberger TD, Sohm JA, Gunderson TE, Parker AE, Tirindelli J, Capone DG, Carpenter EJ, Cary SC (2015) Microbial community composition of transiently wetted Antartic dry valley soils. Front Microbiol 6:1–12
Potts M (1994) Desiccation tolerance in prokaryotes. Microbiol Rev 58:755–805
Potts M (2001) Desiccation tolerance: a simple process? Trends Microbiol 9:553–559
Potts M, Slaughter SM, Hunneke FU, Garst JF, Helm RF (2005) Desiccation tolerance of prokaryotes: application of principles to human cells. Integr Comp Biol 45(5):800–809

Rangel CH (2000) La Diversidad Beta: Tipos de vegetación. In: Rangel-Ch JO (ed) Colombia Diversidad Biótica III. La región de vida paramuna. Instituto de Ciencias Naturales-Instituto Alexander von Humboldt, Bogotá, pp 658–719
Rasuk MC, Kurth D, Flores MR, Contreras M, Novoa F, Poire D, Farias ME (2014) Microbial characterization of microbial ecosystems associated to evaporites domes of gypsum in Salar de Llamara in Atacama desert. Microb Ecol 68:483–494
Rasuk MC, Fernandez AB, Kurth D, Contreras M, Novoa F, Poiré D, Farias ME (2016) Bacterial diversity in microbial mats and sediments from the Atacama desert. Microb Ecol 71:44–56
Reboratti C (2006) La Argentina rural entre la modernización y la exclusión. En publicación: América Latina: cidade, campo e turismo. Amalia Inés Geraiges de Lemos, Mónica Arroyo, María Laura Silveira. CLACSO, Consejo Latinoamericano de Ciencias Sociales, San Pablo. ISBN: 978-987-1183-64-7
Reyes I, Valery A, Valduz Z (2006) Phosphate-solubilizing microorganisms isolated from rhizospheric and bulk soils of colonizer plants at an abandoned rock phosphate mine. Pant Soil 287:69–75
Ruthsatz B (1993) Flora and ecological conditions of high Andean peatlands of Chile between 18°00′ (Arica) and 40°30′ (Osorno) south latitude. Phytocoenologia 25:185–234
Ruthsatz B (2000) Die Hartpolstermoore der Hochanden und ihre Artenviefalt. Ber D Reinh-Tüxen-Ges 12:351–371
Seimon TA, Seimon A, Daszak P, Halloy SRP et al (2007) Upward range extension of Andean anurans and chytridiomycosis to extreme elevations in response to tropical deglaciation. Glob Chang Biol 13:288–299
Squeo FA, Veit H, Arancio G, Gutierrez JR, Arroyo MTK, Olivares N (1993) Spatial heterogeneity of high mountain vegetation in the andean desert zone of Chile (30°S). Mt Res Dev 13:203–209
Squeo FA, Osorio R, Arancio G (1994) Flora de los Andes de Coquimbo: cordillera de Doña Ana. Ediciones de la Universidad de La Serena, La Serena, Chile, 176pp
Squeo FA, Ibacache E, Warner BG, Espinoza D, Aravena R, Gutierrez JR (2006a) Productividad y diversidad florística de la Vega Los Tambos, Cordillera de Doña Ana: variabilidad interanual, herbivoría y nivel freático. In: Cepeda J (ed) Geoecología de la Alta Montaña del Valle del Elqui. Ediciones Universidad de La Serena, La Serena, Chile, pp 333–362
Squeo FA, Warner BG, Aravena R, Espinoza D (2006b) Bofedales: high altitude peatlands of the central Andes. Rev Chil Hist Nat 79:1579–1600
Torsvik and Øvreas (2002) Microbial diversity and function in soil: from genes to ecosystems. Curr Opin Microbiol 5:240–245
Turner K (2014) Environmental economics. Ambio 20(2):59–63
Van den Heuvel RN, van der Biezen E, Jetten MS, Hefting MM, Kartal B (2010) Denitrification at pH 4 by a soil-derived rhodanobacter-dominated community. Environ Microbiol 12:3264–3271
Van den Heuvel MP, van Soelen IL, Stam CJ, Kahn RS, Boomsma DI, Hulshoff Pol HE (2012) Genetic control of functional brain network efficiency in children. Eur Neuropsychopharmacol 23:19–23
Villagran C, Castro V (1997) Etnobotánica y manejo ganadero de las vegas, bofedales y quebradas en el Loa superior, Andes de Antofagasta, Segunda Región, Chile. Chungará (Chile) 29:275–304
Villagran C, Arroyo MTK, Marticorena C (1983) Efectos de la desertización en la distribución de la flora andina de Chile. Rev Chil Hist Nat 56:137–157
Villagran C, Romo M, Castro V (2003) Etnobotánica del sur de los Andes de la Primera Región de Chile: un enlace entre las culturas altiplánicas y las de quebradas altas del Loa superior. Chungará (Chile) 35:73–124
Wilcox BP, Bryant FC, Wester D, Allen BL (1986) Grassland communities and soils on a high elevation grassland of central Peru. Phytologia 61:231–250
Witton BA, Potts M (eds) (1999) The Ecology of Cyanobacteria; Their Success in Time and Space. Kluwer Academic, Dordrecht.
WWF (2014) Living Planet Report 2014: species and spaces, people and places. WWF, Gland, Switzerland
Zhou X, Fornara D, Ikenaga M, Akagi I, Zhang R, Jia Z (2016) The resilience of microbial community under drying and rewetting cycles of three forest soils. Front Microbiol 19:1101

Part III
Life in Andean Volcanoes and Fumaroles

Chapter 7
Of Microbes and Mummies: Tales of Microbial Activity and Inactivity at 6000 m a.s.l.

Steven K. Schmidt, Preston Sowell, Zachary R. Schubert, Lara Vimercati, Adam J. Solon, Dorota L. Porazinska, Pacifica Sommers, John L. Darcy, and Eli M. S. Gendron

7.1 Introduction

In this chapter, we review recent microbiological studies of the highest elevation sites in the Atacama region. Whereas studies of lower elevations of the hyper-arid region of the Atacama Desert have illuminated our understanding of how microbial life has adapted to extremely dry conditions (see Chaps. 3–6), much less work has been done at high elevations. The work reviewed in this chapter focuses on areas above 6000 m a.s.l. on volcanoes in the southern portion of the Andean Central Volcanic Zone (14–28°S) near where Bolivia, Argentina, and Chile meet (around 24°S, 68°W) in the high-elevation region known as the Altiplano or the Puna de Atacama (Costello et al. 2009). The Puna is dry (0–50 mm annual precipitation), cold (0 °C mean annual air temperature), and has a mean elevation of about 4400 m a.s.l. (Allmendinger et al. 1997). Rising from the Puna are the highest volcanoes on Earth, some of which are just under 7000 m a.s.l. in elevation (Allmendinger et al. 1997; Lynch et al. 2012). Figure 7.1 shows one of the highest volcanoes in this region, Volcán Llullaillaco (elevation 6739 m a.s.l.).

In many respects, conditions above 6000 m a.s.l. are even more extreme than lower elevation regions of the Atacama (Solon et al. 2018). As one ascends in elevation from the arid core of the Atacama, the climate initially becomes slightly wetter allowing for the existence of sparse plant communities at elevations between 3500

S. K. Schmidt · Z. R. Schubert · L. Vimercati · A. J. Solon · D. L. Porazinska · P. Sommers
J. L. Darcy · E. M. S. Gendron
Department of Ecology and Evolutionary Biology, University of Colorado, Boulder, CO, USA

P. Sowell (✉)
Geotic Solutions, Boulder, CO, USA
e-mail: psowell@geoticsolutions.com

M. E. Farías (ed.), *Microbial Ecosystems in Central Andes Extreme Environments*, https://doi.org/10.1007/978-3-030-36192-1_7

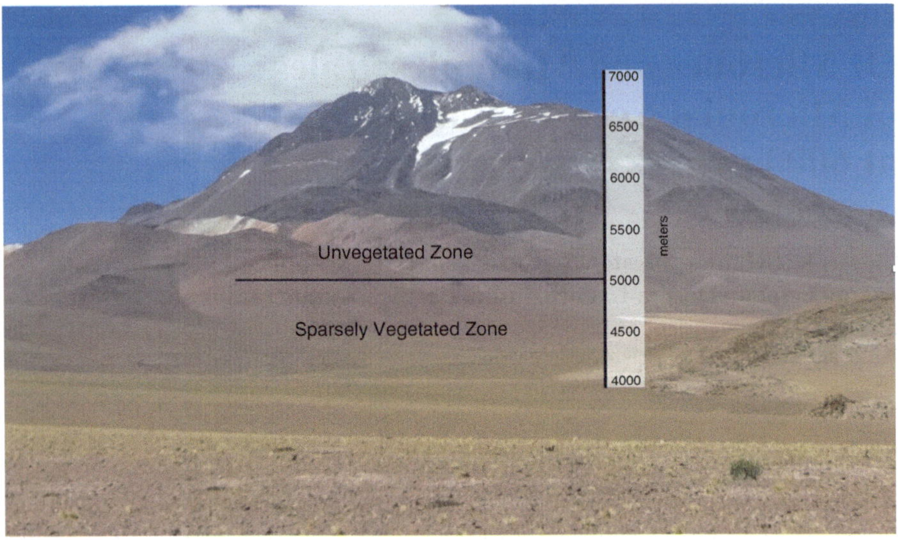

Fig. 7.1 Photo taken in March of 2016 from the vegetated zone northwest of Volcán Llullaillaco showing that most of the prominence of the mountain lies above the vegetated zone. (Photo credit D. Porazinska)

and 5000 m a.s.l. Above about 5000 m a.s.l. there are no plants (except some mosses near fumaroles) and even hardy macro-organisms such as lichens are very rare. Therefore, most of the surface area of the massive stratovolcanoes in this region is completely devoid of plant life (Fig. 7.1). Besides extreme dryness, organisms at high elevations must contend with a thin atmosphere and extreme cold—and more importantly—the highest UV radiation (Cabrol et al. 2014; Pulschen et al. 2015) and the most extreme daily freeze–thaw cycles (Lynch et al. 2012; Vimercati et al. 2016) yet measured on Earth. In addition, all soil-like material (tephra) so far studied above 6000 m a.s.l. has very low pH values ranging from 4.2 to 4.8 (Schmidt et al. 2017a, b).

We start this chapter by reviewing some archaeological studies which demonstrated that mummies buried for 500 years at high elevations on Volcán Llullaillaco did not decay even in the absence of preserving chemicals. This finding was our original inspiration for visiting this volcano, because from a microbiological perspective it indicated an environment where microbes were completely inhibited from growing despite the presence of an abundant food source. Such a situation is rare within the biosphere and indicated an unusually extreme environment. Our examination of the environmental conditions that completely inhibited microbial growth on the mummies of Llullaillaco is followed by a discussion of evidence (pro and con) for the presence of active microorganisms in surface soils elsewhere on Llullaillaco and other volcanoes in the region. We end with a discussion of oases of life, such as fumaroles and ice fields (Nieves Penitentes), at extreme elevations on Volcán Llullaillaco and Volcán Socompa and the implications of our findings for the fields of microbiology and astrobiology.

7.2 Preservation of Human Remains at Extreme Elevations

Mountains have played a prominent role in the Andean cultures for millennia, a tradition that extends to the present day (Bastien 1978). The view of mountains as deities (*Apus*) likely originated from their perceived control of meteorological phenomena. Life in the Andes has long been ruled by climactic extremes resulting in drought or flooding (D'Altroy 2002), a result of the El Niño Southern Oscillation and La Niña cycles (Garreaund 2009). Those events wreaked havoc on early Andean civilizations such as the Inca. Fearing that angry *Apus* were the cause of their suffering, the Inca often took desperate measures to appease them (Cieza de Leon 1864).

The Incan Empire proliferated between 1438 and 1532 A.D. (D'Altroy 2002) throughout the Andes, and during that period their veneration of mountains led the Inca to build shrines known as "high-altitude sanctuaries" (Reinhard and Ceruti 2010) on the very summits of over 100 mountains exceeding 5200 m (Reinhard 2014). As snow and ice limited the elevations that could feasibly be reached in more northern latitudes of their empire (Reinhard and Ceruti 2010), the highest, accessible mountain summits to the Inca were the unglaciated volcanoes of the Central and Southern Andean Volcanic Zones (Ramos and Aleman 2000). One purpose for the high-altitude sanctuaries was to facilitate *capacocha* offering ceremonies, which included the sacrifice of children (de Betanzos 1996). The victims were buried on the summit along with other offerings of pottery, textiles, and figurines. To date, the remains of at least 27 sacrificial victims have been discovered on the summits of high peaks in the former Incan Empire (Besom 2009), many of them in various states of natural mummification (Medina Rojos 1958). In 1999, archaeologists recovered the 500-year-old mummified remains of two girls and a boy, from near the summit of Llullaillaco (Reinhard and Ceruti 2010). At 6712 m, the Llullaillaco shrine is the highest archaeological site in the world and its mummies are the best naturally preserved mummies yet discovered (Wilson et al. 2013). The artifacts buried with them included nearly pristine textiles and feathered headdresses (Fig. 7.2, Reinhard and Ceruti 2010). The remarkable preservation of the bodies and other organic material at the site without the use of artificial preservatives implied a lack of microbial activity at the site, which is what originally inspired our microbiological expeditions to this area (Costello et al. 2009; Lynch et al. 2012).

Mummification is the process of preserving non-bony tissue in human remains, whether by artificial or natural processes. The most well-known artificially preserved mummies are from the ancient Egyptian culture, but only 400 km northwest of Llullaillaco, the Chinchorro culture began artificially preserving their dead along the coastal desert of southern Peru and northern Chile by 5050 BC (Arriaza et al. 1998), up to 2000 years before the Egyptians. Marquet et al. (2012) have suggested that the practice may have been inspired by seeing the bodies of their ancestors accumulating in the landscape through the natural process of mummification, which was induced by the hyperarid environment. In either artificial or natural mummification, the usual decay of flesh, including by microbial activity, is deterred or prevented. Shortly after death, the body's own cells release enzymes that begin a

Fig. 7.2 Two of the mummies recovered from near the top of Volcán Llullaillaco in 1999. These children were buried alive without any preserving chemicals. Note the pristine condition of the fabrics and artifacts buried with these 500-year-old mummies. (Photo courtesy of J. Reinhard)

process of auto-digestion (Lynnerup 2007). In temperate conditions, the body will then begin to decay due to the activity of enteric microorganisms (Lynnerup 2007) and from colonization by environmental microorganisms (Micozzi 1991). The presence of liquid water is essential for both the enzymatic process of auto-digestion and for microbial growth, which is why desiccation is such an efficient preservative (Lynnerup 2007). Cold temperatures are also an impediment to microbial growth (and therefore decomposition), but microbes can be active at temperatures well below 0 °C as long as some liquid water is available (Schmidt et al. 2009; Vimercati et al. 2016). Therefore, the most effective preservative of human bodies is freezing at temperatures below about −13 °C under dry conditions (Previgliano et al. 2005)—conditions that are present on the summits of high-elevation volcanoes (see Fig. 7.3).

Reinhard and Ceruti (2010) noted that the Llullaillaco mummies were likely frozen immediately after their death and their remarkable preservation indicates that they remained frozen for the 500 years before they were discovered. Furthermore, the summit of Llullaillaco where the bodies were excavated is frequently covered with snow, yet the soil surrounding the bodies was dry and excavating the site was surprisingly easy (Reinhard and Ceruti 2010). The lack of moisture at the depth that the mummies were buried (1.2–2.2 m, Reinhard and Ceruti 2010) is due to the fact that at those elevations, snowfall sublimates back to the atmosphere and therefore liquid water never penetrated to that depth. Furthermore, due to the insulating properties of the soil (tephra), the temperature at depth can remain very constant despite

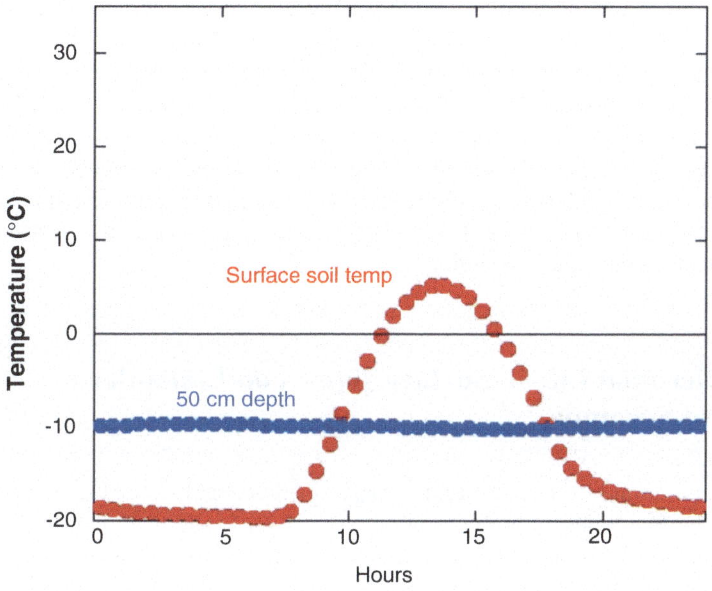

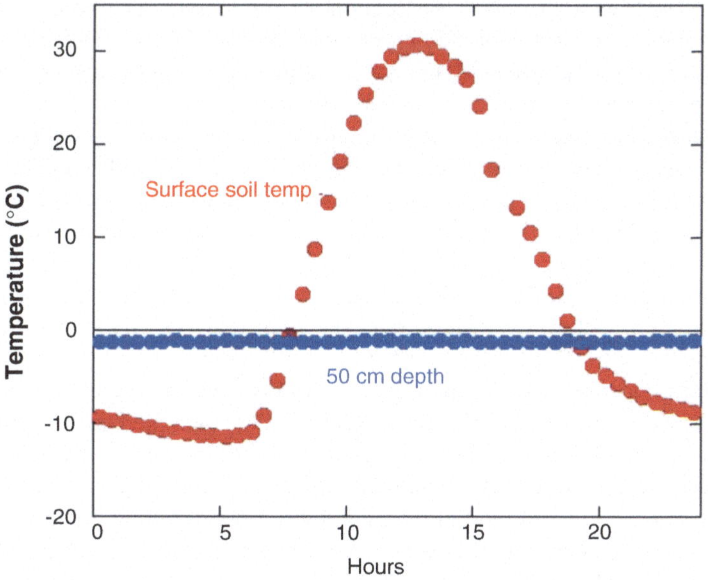

Fig. 7.3 *Left panel*: Diurnal temperature fluctuations on an average (mean for the month of July) winter day at 5820 m a.s.l. on Volcán Saírecabur. Note the constant −10 °C at 50 cm depth (blue dots), whereas soil surface temperatures fluctuated across the freezing point (red dots). *Right panel*: An average summer (December) day at 5820 m a.s.l. on Volcán Saírecabur. Note the constant temperature just under 0 °C at 50 cm depth, whereas soil surface temperatures fluctuated wildly across the freezing point. (Data are redrawn from Schubert (2014) and Schmidt (1999))

extreme temperature fluctuations at the soil surface. This phenomenon is illustrated in Fig. 7.3, which shows constant, cold, soil temperatures at 50 cm depth (at 5820 m a.s.l.), while temperatures were fluctuating dramatically at the surface. Soil temperatures at almost a thousand meters higher and 50 cm deeper on Llullaillaco are likely much colder than those shown in Fig. 7.3. Thus, the remarkable lack of decay in the mummies of Llullaillaco was due to the unique combination of cold and dry conditions. These conditions effectively stopped all microbial activity and growth, leading us to wonder if microbes are active anywhere at elevations above 6000 m a.s.l. in the Atacama region.

7.3 Microbial Life in Surface Tephra on Llullaillaco and Socompa

The apparent total lack of microbial activity (as indicated by lack of decomposition) at deeper tephra depths at high elevations on Volcán Llullaillaco led us to the question of whether microorganisms can function at shallower soil depths at high elevations. Over the past 10 years we have sampled shallower soil depths on both Llullillaco and Socompa, and in this section we summarize our findings indicating that transiently active microbial communities may exist in the top 5 cm of tephra material at elevations above 6000 m a.s.l., even in areas away from sources of water such as ice fields and fumaroles (which are discussed in the next section).

If constant cold and dryness inhibit microbial life at deeper soil depths at high elevations, this is not the case near the soil surface where soil temperatures can fluctuate wildly on a daily basis (Fig. 7.3). In addition to these freeze–thaw cycles, life on the soil surface is subjected to intense UV radiation and aridity (Cabrol et al. 2014; Lynch et al. 2012; Pulschen et al. 2015). Given these stressors, we initially expected that surface soils would be almost sterile at the highest elevations in the Atacama region. Below we discuss relevant work that has challenged this initial assumption.

The first molecular study of microbial life in high-elevation tephra material was the pioneering work of Costello et al. (2009). They sampled tephra material at 5285 m a.s.l. on Volcán Socompa and showed that these "soils" had total organic carbon (TOC) levels (0.03%) on a par with the most oligotrophic soils of Antarctica and lower Atacama Desert, and nitrogen and water levels below the limit of detection. Despite these extreme conditions, they detected (using cloning and Sanger sequencing) simple microbial communities dominated by Actinobacteria, a Basidiomyceteous yeast in the genus *Naganishia* (previously known as *Cryptococcus*), and dark-pigmented members of the Dothideomycetes (e.g. *Cladosporium* and *Ulocladium*).

Since the work of Costello et al. (2009), there have been a number of other studies using modern approaches (e.g., metagenomics, functional gene libraries, high-throughput amplicon sequencing, and culturing approaches) to characterize the microbial communities on these mountains (Lynch et al. 2012, 2014; Solon et al. 2018; Vimercati et al. 2016). Here, we summarize just the work examining microbial communities at elevations above 6000 m a.s.l., where Lynch et al. (2012) found soil conditions even more oligotrophic than those reported by Costello et al. (2009) with TOC levels of 0–0.005%. To our knowledge these are among the lowest TOC values ever reported for a terrestrial ecosystem (cf. Crits-Christoph et al. 2013; Goordial et al. 2016). This high-elevation site (6330 m a.s.l.) was dominated by members of the enigmatic Chlorofexi group that were not closely related to any cultured bacteria, but one sub-group likely is in the *Ktedonobacter* clade (Lynch et al. 2012). However, most of these high-elevation Chloroflexi are most closely related to the B12-WMSP1 clade that includes environmental sequences from a dark, volcanic ice cave on Mt. Erebus (KJ625634, Tebo et al. 2015), snow-covered alpine tundra (DQ450730, Costello and Schmidt 2006), and desert rock varnish (FJ595536).

At slightly lower elevations (6030 m a.s.l.) on Llullaillaco, dry tephra soils were dominated by sequences related to the actinobacterium *Pseudonocardia*. Metagenomic and functional gene analyses of DNA, from 6030 m a.s.l., showed that this organism is capable of using trace levels of gases such as H_2, CO, and CH_4 as energy sources (Lynch et al. 2012, 2014). Use of gases (especially CO) by bacteria in oligotrophic volcanic sediments has also been noted in other studies (King and King 2014; Tebo et al. 2015; Weber and King 2010) perhaps indicating that organisms in the *Pseudonocardia* group are active at 6030 m a.s.l. on Llullaillaco. It is noteworthy that members of the *Ktedonobacter* group of the Chloroflexi also have the ability to use CO as an energy source (King and King 2014). Thus, one hypothesis for how microbes might function in extremely oligotrophic volcanic tephra is through the utilization of volcanic and atmospheric gases, but a direct test of this hypothesis has yet to be done in the field.

Further evidence that some microbes can function in the extremely dry and oligotrophic tephra above 6000 m a.s.l. comes from recent studies of eukaryotic phylotypes from high volcanoes (Pulschen et al. 2015; Schmidt et al. 2017a, b; Vimercati et al. 2016). The dominant eukaryote in high-elevation tephra (Lynch et al. 2012; Solon et al. 2018) is closely related to *Naganishia friedmannii* (previously *Cryptococcus friedmannii*) originally described from endolithic communities of the Dry Valleys of Antarctica (Friedmann 1982; Vishniac 1985). Unlike *Ktedonobacter* and *Pseudonocardia*, *N. friedmannii* is easy to grow in culture and can use a broad array of organic compounds for growth (Vimercati et al. 2016), which has led researchers to hypothesize that this organism is an opportunitroph that utilizes Aeolian deposited organic matter to rapidly grow during rare snowmelt events (Schmidt et al. 2017a, b; Vimercati et al. 2016).

In order to grow in surface soils at elevations above 6000 m a.s.l., *N. friedmannii* and numerous fungi in the Dothideomycetes would also need to withstand high levels of UV radiation and the extreme freeze–thaw cycles (Fig. 7.3). Recent research indicates that *N. friedmannii* is able to withstand both of these stressors. Pulschen et al. (2015) demonstrated that an *N. friedmannii* isolate from Volcán Saírecabur could withstand levels of UV radiation as high as *Deinococcus radiodurans*, the most radiation-resistant bacterium known. Likewise, Vimercati et al. (2016) demonstrated that the *N. friedmannii* isolate from soils above 6000 m a.s.l. on Llullaillaco could grow exponentially during diurnal freeze–thaw cycles (−10 °C to +30 °C) that would be typical of shallow soils at high elevations (Fig. 7.3). This was the first definitive determination that any organism can grow during extreme freeze–thaw cycles and provides strong evidence that *N. friedmannii* may be able to grow during rare periods of snowmelt and carbon availability as hypothesized above.

Despite recent evidence that some organisms such as *Pseudonocardia,* B12-WMSP1, and *Naganishia* have phenotypic traits that would allow them to function in surface soils above 6000 m a.s.l., it should be noted that the communities of these extreme soils are among the simplest yet described for a terrestrial ecosystem (Costello et al. 2009; Lynch et al. 2012; Schmidt et al. 2018; Solon et al. 2018). The diversity and structure of these communities are discussed in the following two sections where they are compared to the more robust microbial communities found near fumaroles and ice fields (penitentes).

7.4 Microbial Communities of Fumarolic Soils Above 6000 m a.s.l.

In parallel with recent work on the potential for microbial life in dry tephra on Socompa and Llullaillaco, we have also been exploring potential "oases" of abundant life in this otherwise austere landscape. This work was inspired by the pioneering studies of Stephan Halloy who originally described remarkable moss-dominated communities, supported by water vapor and heat from fumaroles, at elevations up to 6050 m a.s.l. on Volcán Socompa (Halloy 1991). During two recent expeditions, we resampled some of the same fumaroles studied by Halloy (1991) and compared these fumarolic microbial communities to communities in dry tephra soils nearby (Costello et al. 2009; Solon et al. 2018). For example, Fig. 7.4 shows alpha diversity plots of the bacterial (16S) and eukaryotic (18S) communities of fumarolic and non-fumarolic tephra at an elevation of 6050 m a.s.l. on Volcán Socompa. Fumarolic soils had significantly higher OTU richness for both 16S and 18S than nearby dry tephra. Fumaroles contained on average 803 16S OTUs, whereas the non-fumarole site had only an average of 61 16S OTUs. For 18S communities, fumarole sites contained 319 OTUs, whereas the non-fumarole site had only 83 18S OTUs at 6050 m a.s.l. These data show that the fumarolic sites contained OTU richness comparable to a temperate forest soil (cf. Ferrenberg et al. 2013), whereas the dry sites at the same elevation supported among the lowest OTU richness yet reported for a terrestrial habitat in the high-throughput sequencing literature (cf. Crits-Christoph et al. 2013; Goordial et al. 2016; Solon et al. 2018). Thus, the presence of fumaroles allowed the proliferation of almost "normal" soil communities, whereas dry sites contained the simplest microbial communities yet described. Fumaroles also supported a diverse array of functional groups, including phototrophs and predators,

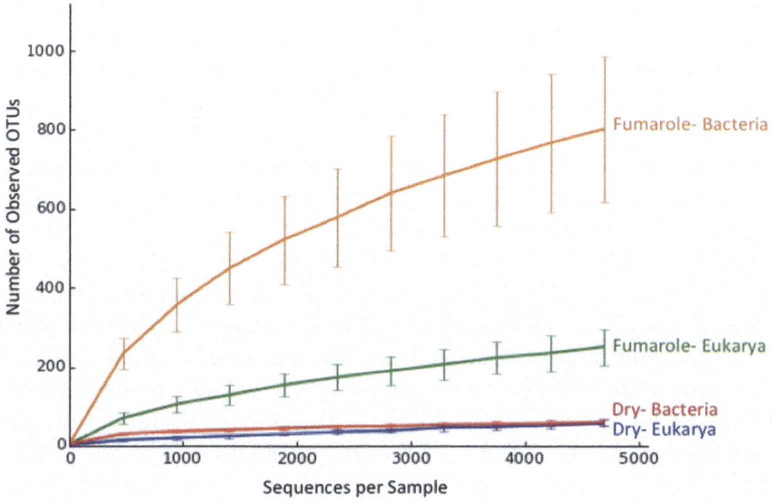

Fig. 7.4 Species (OTU) richness or alpha diversity for soils (tephra) affected by fumarolic activity at 6050 m a.s.l. compared to bare tephra at the same elevation on Volcán Socompa. Note that fumarolic soils contained bacterial diversity levels very similar to those found in vegetated soils from low elevation ecosystems, whereas the barren tephra contained among the lowest OTU richness yet recorded for a terrestrial ecosystem. (Data are redrawn from Solon et al. (2018) with permission)

indicating that fumarolic soils had functioning food webs and complex nutrient cycles, whereas the barren tephra sites lacked known phototrophs and trophic complexity (Solon et al. 2018). Bryozoans, or animals that associate with moss such as tardigrades, were also found primarily in samples from fumaroles, suggesting they may be active, if at low density, at those sites. It is also noteworthy that there were slightly more 18S OTUs (83) than 16S OTUs (61) at the barren tephra sites. To our knowledge this is the first soil ecosystem containing more eukaryotic diversity than bacterial diversity, perhaps indicating that eukaryotes may be better adapted to this austere system than are bacteria. However, more in depth, comparative work is needed to better understand the trophic structure of fumarolic communities and to compare them to fumarolic communities in other extreme landscapes (Benson et al. 2011; Herbold et al. 2014; Tebo et al. 2015).

7.5 Microbial Communities of Soils Associated with "Nieves Penitentes" (Penitentes)

Penitentes are ice formations found at high elevations throughout the Southern Andes. They consist of parallel rows of ice spires (Fig. 7.5) that form through the interaction of wind and solar radiation in areas of snow accumulation such as the leeward slopes of high-elevation ridges (Claudin et al. 2015; Lliboitry 1954). During expeditions in 2009 and 2016, we sampled damp soils between the rows of penitentes at elevations of 5825 and 5277 m a.s.l. on Socompa and Llullaillaco, respectively. We hypothesized that moisture from melting penitentes would support more microbial diversity than barren soils. To our surprise, soils in penitente fields did not have significantly higher OTU richness than nearby dry tephra soils on either Socompa or Llullaillaco (Solon et al. 2018). However, beta diversity of the eukaryotic (18S) microbial community was significantly affected by the presence of penitentes on Llullaillaco (Fig. 7.6). This difference was driven largely by the presence of algae in the penitente soil samples, whereas algae were not detected in any samples from dry sites on Llullaillaco (Solon et al. 2018). The dominant alga detected in the penitente soils is a phylotype related to the snow alga *Chloromonas brevispina,* and the dominant alga associated with penitentes on Socompa is related to an Antarctic snow alga (NCBI accession no. AB903007). These findings suggest that snow algae may be functioning in penitente fields, a hypothesis further supported by our observation of patches of red snow in the penitente field at 5277 m a.s.l. on Llullaillaco (Fig. 7.7) but not in penitentes at 5825 m a.s.l. on Socompa (Solon et al. 2018). Nonetheless, the findings of Solon et al. (2018) are the first evidence of algal communities on these high-elevation volcanoes, but more work is needed to confirm that these are true snow algae. A descriptiom of the microbes living in penitente ice on Llullaillaco was published as this chapter went to press (Vimercati et al. 2019).

In contrast to our detection of algae in penitente fields, cyanobacteria were not detected in penitente soils on either Llullaillaco or Socompa (Solon et al. 2018).

Fig. 7.5 Jack Darcy and Lara Vimercati standing on the edge of the field of penitentes at 5277 m a.s.l. on Volcán Llullaillaco. Basecamp (tiny yellow and red tents) can be seen in the lower left of the photo. (Photo credit S.K. Schmidt)

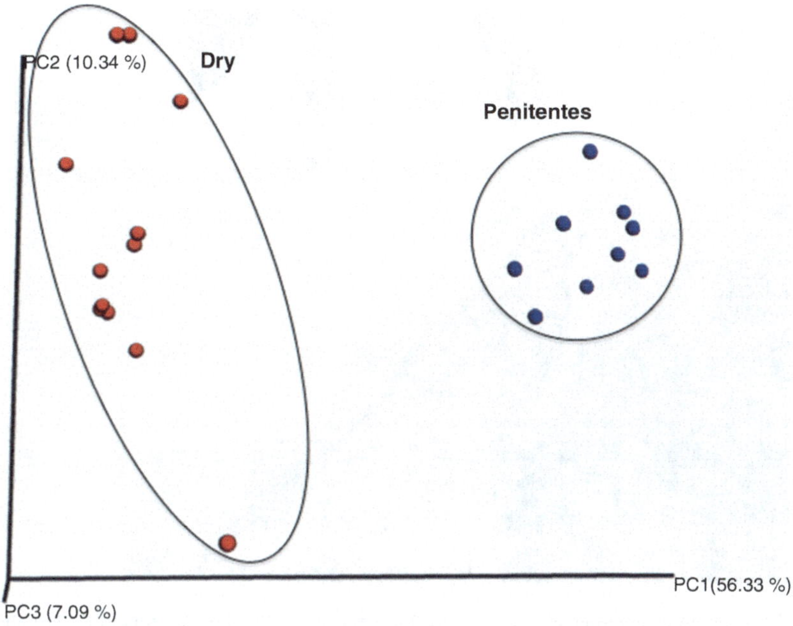

Fig. 7.6 Beta diversity plot demonstrating that eukaryotic microbial communities associated with the penitente field (blue dots) shown in Fig. 7.5 differed significantly ($P < 0.001$) from non-penitente soils sampled across a range of elevations (red dots) on Volcán Llullaillaco. (Redrawn from data originally presented in Solon et al. (2018))

This finding was surprising given the importance of cyanobacteria in most high-elevation and high-latitude ecosystems (Freeman et al. 2009; Nemergut et al. 2007; Schmidt et al. 2011; Vincent 2000). We did however detect putatively photosynthetic bacteria in the Rhodospirillales group of the Alphaproteobacteria in both penitente and dry soils on Volcán Llullaillaco (King et al. 2010; Solon et al. 2018). Many members of the Rhodospirillales can photosynthesize using longer wavelengths of light (e.g., 875 nm) than cyanobacteria or algae and therefore may be able to grow at deeper soil depths where they would be more protected from UV radiation. However, there are also non-photosynthetic species of Rhodospirillales, so more work is needed to determine the ecological roles of this ubiquitous group in high-elevation soils and sediments (Demergasso et al. 2010; King et al. 2010; Rasuk et al. 2016).

Despite the greater photosynthetic potential of penitente fields suggested by putative algae and Rhodospirillales, the food web in these soils appears to remain strikingly simple. Soil invertebrates such as rotifers and nematodes that are even widespread in the soils of the McMurdo Dry Valleys, Antarctica (Gooseff et al. 2003), remain as virtually undetectable in the penitente field as in the drier tephra surrounding it. Experimental water additions to MDV soils, however, do not increase the activity of invertebrates there (Buelow et al. 2016), and the other stressors of the high Atacama may still preclude their active growth.

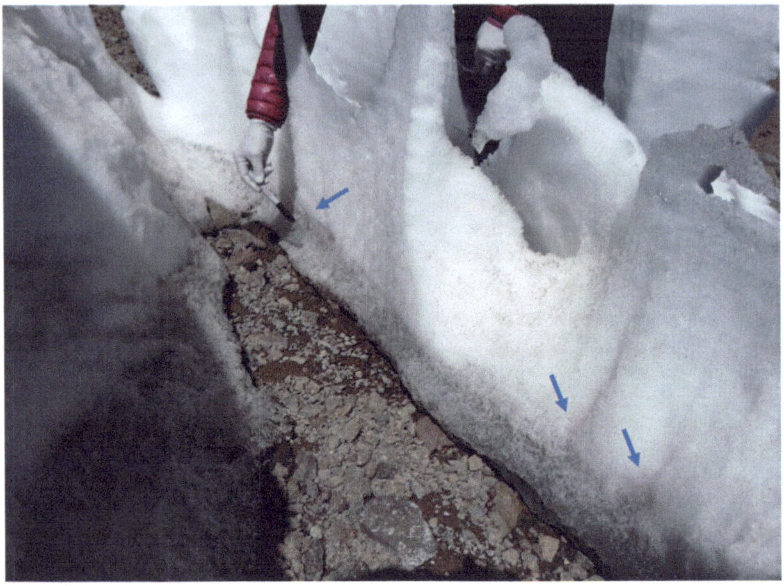

Fig. 7.7 Close up of penitentes at 5277 m a.s.l. on Volcán Llullaillaco with Dorota Porazinska's arm for scale. Blue arrows point to areas of red snow indicative of snow algae communities. Later examination of snow scraped from these red patches yielded abundant cells that resemble cysts of *Chloromonas* species. Molecular analyses of these patches will be reported in an upcoming publication. Note also the patches of damp soil between the penitentes. (Photo credit S.K. Schmidt)

7.6 Conclusion

Our initial expeditions to the high volcanoes of the Atacama region were inspired by reports of perfectly preserved mummies found buried on the top of Volcán Llullaillaco. To us as microbiologists, this finding indicated an environment that completely inhibited the growth and activity of microorganisms and we were intrigued as to how this could occur in a world dominated by microbes. As discussed above, this paradox is explained by the rare combination of extreme dryness combined with near *constant* below-freezing temperatures found beneath the soil surface at extreme elevations. Indeed, the answer to this dilemma was already known to archaeologists who have strived to mimic these conditions at the Museo de Arqueología de Alta Montaña in Salta Argentina (Previgliano et al. 2005).

The second big question we hoped to address was if and how microbes function on the exposed slopes of these majestic but harsh volcanoes. Unlike deeper soil layers, the near surface tephra layer is subjected to extreme fluctuations of radiation, temperature, and moisture. Indeed, the extremes of temperature and UV radiation are the most intense yet measured on the surface of our planet. In addition, the surface tephra at elevations over 6000 m a.s.l. has among the lowest levels of total carbon and nitrogen of any ecosystem yet described. This rare combination of extremes, combined with a thin atmosphere makes these high-elevation slopes arguably the best analogs for Mars on Earth (Lynch et al. 2012; Pulschen et al. 2015; Vimercati et al. 2016).

Obviously, we have only just scratched the surface in our understanding of how life functions under these extreme conditions, but the preliminary work discussed above indicates that some inhabitants (e.g., novel Chloroflexi and *Pseudonocardia*) of this extreme zone can utilize trace levels of atmospheric gases to aid in their subsistence. Other microbes such as the fungus *N. friedmannii* are likely dormant most of the time but are capable of rapid growth during rare periods of water and nutrient availability (Schmidt et al. 2017a, b). This organism is also the first organism known to be able to grow during extreme freeze–thaw cycles (Vimercati et al. 2016) and to withstand the intense UV radiation (Pulschen et al. 2015) that occur on a year-round basis at high elevations on these high volcanoes. Perhaps this explains why it is the dominant eukaryotic organism found at elevations above 6000 m a.s.l. on Llullaillaco and Socompa.

Finally, our explorations led us to look for oases of life in this vast landscape of barren tephra. Perhaps most surprising was our finding that soils formed as a result of fumarolic activity have levels of microbial diversity and activity rivaling those of temperate forest soils, even at elevations of over 6000 m a.s.l. (Costello et al. 2009; Solon et al. 2018). In contrast, tephra wetted by intermittent melt water from *nieves penitentes* did not show higher diversity levels than dry tephra, but did yield the first evidence for functioning snow algae communities in soils associated with penitentes. These potential oases of life surrounded by one of the most forbidding landscapes on Earth provide a perfect outdoor laboratory in which to carry out natural experiments on the cold–dry limits to life on Earth, and perhaps elsewhere in the universe (Schmidt et al. 2018; Vimercati et al. 2019).

Acknowledgments We thank M.E. Farías, C. Dorador, P. Aráns, C. Vitry, P. Maciel, G. Zimmerman, T. Bowen, G. Jesperson, and T. Harris for help in the field. Funding was provided by the National Science Foundation and the National Geographic Society Committee for Research and Exploration.

References

Allmendinger RW, Jordan TE, Kay SM, Isacks BL (1997) The evolution of the Altiplano-Puna plateau of the Central Andes. Annu Rev Earth Planet Sci 25:139–174

Arriaza BT, Càrdenas-Arroyo F, Kleiss E, Verano JE (1998) South American mummies: culture and disease. In: Cockburn A, Cockburn E, Reyman TA (eds) Mummies, disease and ancient cultures. Cambridge University Press, Cambridge, pp 336–350

Bastien JW (1978) Mountain of the Condor: metaphor and ritual in an Andean Ayllu. West Publishing Co., St. Paul

Benson CA, Bizzoco RW, Lipson DA, Kelley ST (2011) Microbial diversity in nonsulfur, sulfur and iron geothermal steam vents. FEMS Microbiol Ecol 76:74–88

Besom T (2009) Of summits and sacrifice: an ethno-historic study of Inka religious practices. University of Texas Press, Austin

Buelow HN, Winter AS, Van Horn DJ, Barrett JE, Gooseff MN, Schwartz E, Takacs-Vesbach CD (2016) Microbial community responses to increased water and organic matter in the arid soils of the McMurdo Dry Valleys, Antarctica. Front Microbiol 7:1040

Cabrol NA, Feister U, Häder D, Piazena H, Grin EA, Klein A (2014) Record solar UV irradiance in the tropical Andes. Front Environ Sci 2:19. https://doi.org/10.3389/fenvs.2014.00019

Cieza de Leon P (1864 [1553]) Travels of Pedro de Cieza de Leon A.D. 1532–1550: first part of his chronicle of Peru. Markham CR, translator/editor

Claudin P, Jarry H, Vignoles G, Plapp M, Andreotti B (2015) Physical processes causing the formation of penitentes. Phys Rev E 92:033015

Costello EK, Halloy SRP, Reed SC, Sowell P, Schmidt SK (2009) Fumarole-supported islands of biodiversity within a hyperarid, high-elevation landscape on Socompa Volcano, Puna de Atacama, Andes. Appl Environ Microbiol 75:735–747

Costello EK, Schmidt SK (2006) Microbial diversity in alpine tundra wet meadow soil: novel Chloroflexi from a cold, water-saturated environment. Environ Microbiol 8:1471–1486

Crits-Christoph A et al (2013) Colonization patterns of soil microbial communities in the Atacama Desert. Microbiome 1:28. https://doi.org/10.1186/2049-2618-1-28

de Betanzos J (1996) Narrative of the Incas 1557. Translated and edited by R. Hamilton and D. Buchanan. University of Texas Press, Austin

D'Altroy TN (2002) The Incas. Blackwell Publishing, Malden, MA

Demergasso C, Dorador C, Meneses D, Blamey J, Cabrol N, Escudero L, Chong G (2010) Prokaryotic diversity patterns in highaltitude ecosystems of the Chilean Altiplano. J Geophys Res 115:G00D09

Ferrenberg S, O'Neill SP, Knelman JE, Todd B, Duggan S, Bradley D, Robinson T, Schmidt SK, Townsend AR, Williams MW, Cleveland CC, Nemergut DR (2013) Changes in assembly processes in soil bacterial communities following a wildfire disturbance. ISME J 7:1102–1111

Freeman KR, Pescador M, Reed SC, Costello EK, Robeson MS, Schmidt SK (2009) Soil CO_2 uptake and photoautotrophic community composition in high-elevation, "barren" soils. Environ Microbiol 11:674–686

Friedmann EI (1982) Endolithic microorganisms in the Antarctic cold desert. Science 215:1045–1053

Garreaud RD (2009) The Andes climate and weather. Adv Geosci 22:3–11

Goordial J, Davila A et al (2016) Nearing the cold-arid limits of microbial life in permafrost of an upper dry valley, Antarctica. ISME J 10:1613–1624

Gooseff MN, Barrett JE, Doran PT, Fountain AG, Lyons WB, Parsons AN, Porazinska DL, Virginia RA, Wall DH (2003) Snow patch influence on soil biogeochemical processes and invertebrate distribution in the McMurdo Dry Valleys, Antarctica. Arct Antarct Alp Res 35:91–99

Halloy SRP (1991) Islands of life at 6000 m altitude: the environment of the highest autotrophic communities on Earth (Socompa Volcano, Andes). Arct Alp Res 23:247–262

Herbold CW, Lee CK, McDonald IR, Cary SC (2014) Evidence of global-scale Aeolian dispersal and endemism in isolated geothermal microbial communities of Antarctica. Nat Commun 5:3875

King AJ, Freeman KR, McCormick KF, Lynch RC, Lozupone C, Knight R, Schmidt SK (2010) Biogeography and habitat modeling of high-alpine bacteria. Nat Commun 1:53

King CE, King GM (2014) Description of Thermogemmatispora carboxidivorans sp. nov., a novel carbon-monoxide-oxidizing member of the class Ktedonobacteria isolated from a geothermally-heated biofilm, and analysis of carbon monoxide oxidation by members of the class Ktedonobacteria. Int J Syst Evol Microbiol 64:1244–1251

Lliboitry L (1954) The origin of penitentes. J Glaciol 2:331–338

Lynch R, Darcy JL, Kane NC, Nemergut DR, Schmidt SK (2014) Metagenomic evidence for metabolism of trace atmospheric gases by high-elevation desert Actinobacteria. Front Microbiol 5:698

Lynch R, King AJ, Farías ME, Sowell P, Vitry C, Schmidt SK (2012) The potential for microbial life in the highest-elevation (>6000 m a.s.l.) mineral soils of the Atacama region. J Geophys Res 117:G02028

Lynnerup N (2007) Mummies. Yearb Phys Anthropol 50:162–190

Marquet PA, Santoro CM, Latorre C, Standen VG, Abades SR, Rivadeneira MM, Arriaza B, Hochberg ME (2012) Emergence of social complexity among coastal hunter-gatherers in the Atacama Desert of northern Chile. Proc Natl Acad Sci 109(37):14754–14760

Medina Rojos A (1958) Hallazgos arqueológicos en el Cerro El Plomo. In: Arqueología Chilena. Centro de Estudios Anthroploogicos, Universidad de Chile, Santiago

Micozzi MS (1991) Postmortem change in human and animal remains. A systematic approach. Charles C. Thomas, Springfield, IL

Nemergut DR et al (2007) Microbial community succession in unvegetated, recently-deglaciated soils. Microb Ecol 53:110–122

Previgliano CH, Ceruti C, Aráoz FA, Diez JG, Reinhard J (2005) Radiología en estudios arqueológicos de momias Incas. Rev Argent Radiol 68:199–210

Pulschen AA, Rodrigues F, Duarte RTD, Araujo GG, Santiago IF, Paulino-Lima IG, Rosa CA, Kato MJ, Pellizari VH, Galante D (2015) UV-resistant yeasts isolated from a high-altitude volcanic area in the Atacama Desert as eukaryotic models for astrobiology. Microbiologyopen 4:574–588

Ramos VA, Aleman A (2000) Tectonic evolution of the Andes. In: Cordani UG, Milani EJ, Thomaz Filho A, Campos Neto MC (eds) Tectonic evolution of South America: 31st International Geological Congress. Institut de Recherche pour le Développement, Rio de Janeiro, pp 635–685

Rasuk MC, Fernández AB, Kurth D, Contreras M, Novoa F, Poiré D, Farías ME (2016) Bacterial diversity in microbial mats and sediments from the Atacama Desert. Microb Ecol 71:44–56

Reinhard J (2014) Andes: high-altitude archaeological sites as cultural heritage. In: Encyclopedia of global archaeology. Springer, New York

Reinhard J, Ceruti MC (2010) Inca Rituals and sacred mountains. The Cotsen Institute of Archaeology Press, Los Angeles, CA

Schmidt D (1999) Das Extremklima der nordchilenischen Hochatacama unter besonderer Berücksichtigung der Höhengradienten. Dresdener Geogr Beiträge 4:1–122

Schmidt SK, Darcy JL, Sommers P, Gunawan E, Knelman JE, Jager K (2017a) Freeze–thaw revival of rotifers and algae in a desiccated, high elevation (5500 meters) microbial mat, high Andes, Perú. Extremophiles 21:573–580

Schmidt SK, Lynch R, King AJ, Karki D, Robeson MS, Nagy L, Williams M, Mitter M, Freeman KR (2011) Phylogeography of microbial phototrophs in the dry valleys of the high Himalayas and Antarctica. Proc R Soc B 278:702–708

Schmidt SK, Gendron EMS, Vincent K, Solon AJ, Sommers P, Schubert ZR, Vimercati L, Porazinska DL, Darcy JL, Sowell P (2018) Life at extreme elevations on Atacama volcanoes: the closest thing to Mars on Earth?. Antonie van Leeuwenhoek 111(8):1389–1401

Schmidt SK, Vimercati L, Darcy JL, Arán P, Gendron E, Solon A, Porazinska D, Dorador C (2017b) A Naganishia in high places: functioning populations or dormant cells from the atmosphere? Mycology 8:153. https://doi.org/10.1080/21501203.2017.1344154

Schmidt SK, Wilson KL, Monson RK, Lipson DA (2009) Exponential growth of snow molds at sub-zero temperatures: an explanation for high beneath-snow respiration rates and Q10 values. Biogeochemistry 95:13–21

Schubert ZR (2014) Dew formation and water availability at high elevation in the Atacama Desert, Chile. Undergraduate honors thesis. Paper 192. University of Colorado, Boulder

Solon AJ, Vimercati L, Darcy JL, Arán P, Porazinska D, Dorador C, Farias ME, Schmidt SK (2018) Microbial communities of high-elevation fumaroles, penitentes and dry tephra "soils" of the Puna de Atacama Volcanic Zone. Microb Ecol 76:340

Tebo BM, Davis RE, Anitori RP, Connell LB, Schiffman P, Staudigel H (2015) Microbial communities in dark oligotrophic volcanic ice cave ecosystems of Mt. Erebus, Antarctica. Front Microbiol 6:179. https://doi.org/10.3389/fmicb.2015.00179

Vimercati L, Hamsher S, Schubert Z, Schmidt SK (2016) Growth of a high-elevation Cryptococcus sp. during extreme freeze-thaw cycles. Extremophiles 20:579–588

Vimercati L, Solon AJ, Krinsky A, Arán P, Porazinska DL, Darcy JL, Dorador C, Schmidt SK (2019) Nieves penitentes are a new habitat for snow algae in one of the most extreme high-elevation environments on Earth. Arct Antarct Alp Res 51(1):190–200

Vincent WF (2000) Cyanobacterial dominance in the polar regions. In: Whitton BA, Potts M (eds) The ecology of cyanobacteria. Kluwer Academic Publishers, Dordrecht, pp 321–340
Vishniac HS (1985) Cryptococcus friedmannii, a new species of yeast from the Antarctic. Mycologia 77:149–153
Weber CF, King GM (2010) Distribution and diversity of carbon monoxide-oxidizing bacteria and bulk bacterial communities across a successional gradient on a Hawaiian volcanic deposit. Environ Microbiol 12:1855–1867
Wilson AS, Brown EL, Villa C, Lynnerup N, Healey A, Ceruti MC, Reinhard J, Previgliano CH, Araoz FA, Gonzalez Diez J, Taylor T (2013) Archaeological, radiological, and biological evidence offer insight into Inca child sacrifice. Proc Natl Acad Sci U S A 110:13322–13327

Chapter 8
A Unique Natural Laboratory to Study Polyextremophile Microorganisms: Diamante Lake as a Window to the Origin of Life

Tatiana Stepanenko, Mariana N. Soria, Luis Alberto Saona Acuña, Agustina I. Lencina, and María Eugenia Farías

8.1 Diamante Lake: A Unique Environment on Earth

Diamante Lake is located in the caldera of the Galán Volcano at a height of 4589 m a.s.l. (coordinates: 26°00′51.04″S, 67°01′46.42″O; Fig. 8.1a, b). The Galán, located in the province of Catamarca in the Puna Argentina, is cataloged as a supervolcano and multiple eruptions between the upper Miocene to the Pleistocene are recorded (Ruggieri et al. 2011; Soler et al. 2007). The caldera is elliptical in shape, 40 km by 24 km in size, with its longest axis extending in a north–south direction. The floor of the caldera is at an altitude of about 4.5 km; the highest points on the caldera wall reach about 6 km and the resurgent centre reaches a similar height. Doming and uplift in this area presumably preceded the eruption of ignimbrites from ring fractures and subsequent collapse of the caldera floor along the ring fractures. Several other major volcanic structures are known in the Central Andes; however, the caldera of the Galán Volcano seems to be unique at present in possessing well defined ring fractures and a resurgent centre and in its association with rift faulting (Francis et al. 1978).

After the Galán Caldera eruption, a lake formed in the caldera on top of ignimbrites deposits. Talus apron and fluvial fans formed, sloping into the depression as the caldera walls eroded (Sparks et al. 1985).

Diamante Lake is ~7 km long and ~3 km wide at the widest part, with the longest axis oriented NW–SE. A larger lake probably existed before resurgence took place (Francis et al. 1978).

T. Stepanenko · M. N. Soria · L. A. S. Acuña · A. I. Lencina
Laboratorio de Investigaciones Microbiológicas de Lagunas Andinas (LIMLA), Planta Piloto de Procesos Industriales Microbiológicos (PROIMI-CONICET), Tucumán, Argentina

M. E. Farías (✉)
LIMLA-PROIMI-CONICET, Tucuman, Argentina

M. E. Farías (ed.), *Microbial Ecosystems in Central Andes Extreme Environments*, https://doi.org/10.1007/978-3-030-36192-1_8

Fig. 8.1 (**a**) Map of Argentina showing location of Diamante Lake. (**b**) Diamante Lake is located in the crater of the Galán Volcano. (**c**) Panoramic photography of Diamante Lake carbonates rocks and red biofilm. (**d**, **e**) Carbonates rocks showing red biofilm coating and gaylussite crystals (**e**—sparkling glow)

Diamante Lake receives freshwater inputs from snowmelt, a peatland (named *vega andina*) located at the south and a temporary river of mixing water at the southeast. These differences in the water inputs of the lake give rise to a hydrological mixing zone at the east and south margin, where carbonate rocks are located.

Diamante Lake presents characteristics that make it a place not only of tourist interest but also a key site for scientific research. It is a natural laboratory—still unexplored—where we can study, among other things, the adaptation, survival and evolution of microbial ecosystems under extreme conditions. The high UV radiation (84 W m^{-2} of UVA-B at noon), the high salinity (270 g l^{-1}, 217 mS cm^{-1}), alkaline pH (9–11), the low oxygen pressure, the high heavy metal concentration from volcanic action and high day–night thermal fluctuation (from −20 to 20 °C) are some of the conditions found in Diamante Lake. In the midst of these extreme natural conditions, Andean microbial ecosystems (AME) forming a red biofilm have been found and studied (Rascovan et al. 2016; Ordoñez et al. 2018), and they are associated with gaylussite crystals on the bottom of the calcareous rocks present in the lake (Fig. 8.1c, d).

Carbonate rocks are grainstones with a coating of laminated crust. Grainstone consists of clastic grains (quartz and plagioclase), volcanic intraclast and micrite cement. Laminated crust consists of calcite laminae that occur as isopachous layers produced by regular alternation of sparite and micrite. In addition, isopachous laminae infilled interparticle porosities (Lencina et al. unpublished data; Fig. 8.2).

This kind of ecosystem has already been reported in other lakes with similar physicochemical conditions, such as the Mono Lake located in California, USA. The

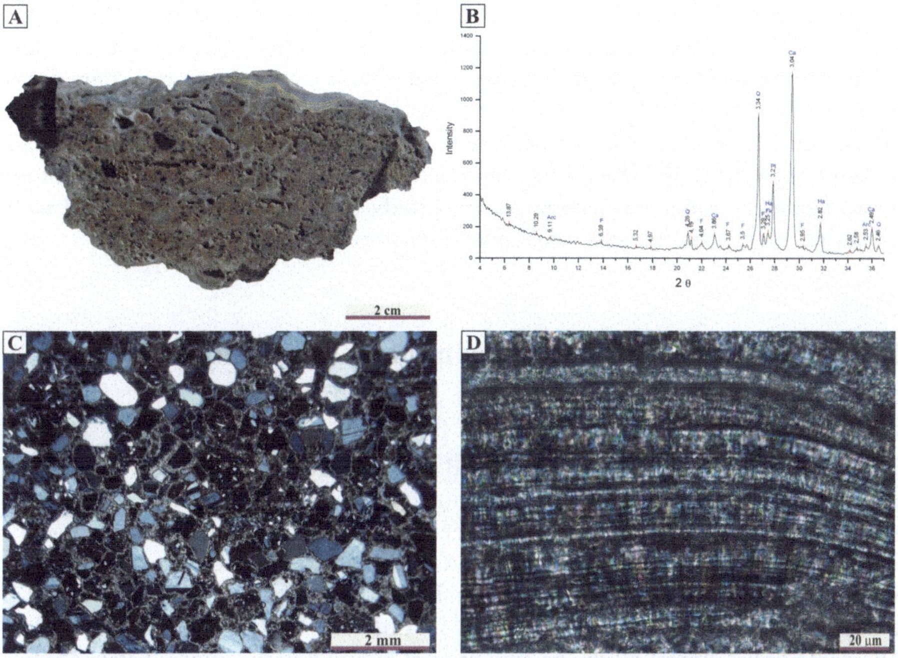

Fig. 8.2 (**a**) Cross section of the carbonatic rock showing the internal structure and laminated crust (yellow dotted line). (**b**) XRD analyses showing the peaks of calcite (calcium carbonate). (**c**) Petrographic microscope image of the grainstone in which most of the grains are quartz and plagioclase. (**d**) Petrographic microscope image of the laminated crust showing isopachous laminae produced by regular alternation of bright (sparry) and dark (micritic) laminae (unpublished results)

Mono Lake is characterized—like the Diamante Lake—as an alkaline lagoon, and they are called Soda Lakes (Kulp et al. 2008). An alkaline lake is a lake rich in dissolved sodium salts, especially sodium carbonate, sodium chloride and sodium sulfate, as is the case of Mono Lake and Magadi Lake at Kenia, Africa, crater lake at Malha, northern Sudan, and Baer Soda Lake at Inner-Mongolia (Jones and Grant 1999; Pecoraino et al. 2015; Ma et al. 2004).

The precipitation of minerals such as gaylussite crystals—to which the biofilm of Diamante Lake is attached—has been reported in different alkaline lakes. This mineral is a pentahydrate sodium–calcium carbonate of formula $Na_2Ca(CO_3)_2{\cdot}5H_2O$, which is formed as an alkaline water evaporite with the appearance of a hard substrate at the bottom of rocks, submerged under water. It is an unstable mineral that dehydrates in dry air and decomposes in water (Bischoff et al. 1991). Particularly in Diamante during the winter, when there is more water in the lake, the gaylussite crystals are not very abundant and even disappear from the rocks.

During the past 10 years, the microbial ecosystems of Puna's lakes have been considered exceptional environments for studying the extreme microbial diversity (Farias et al. 2019). This diversity displays particular mechanisms that allow them to adapt to the above-mentioned extreme conditions, especially strong UV-radiation and hypersalinity (Albarracín et al. 2016). However, a particular feature of Diamante Lake is the very high concentration of arsenic (As)—175 mg l^{-1}—(Rascovan et al. 2016). The arsenic content present in the lake makes it a unique environment to study its relationship with biological systems. The Mono Lake was characterized as an alkaline lake where the arsenic concentration was extremely high (around 10 mg l^{-1}) (Kulp et al. 2008); however, Diamante Lake has almost 20 times more. Thus, Diamante Lake is a unique place to study the metabolisms associated with obtaining energy through electrochemical transformations of arsenic.

8.2 Biological Role of Arsenic in the Central Andean Wetlands

Arsenic is an element of high toxicity and affects all forms of life we know (Parke 2013). This metalloid is distributed throughout the Earth's crust and represents a serious health problem due to its carcinogenic effect. In its inorganic form it occurs in nature as arsenate (AsV) and arsenite (AsIII). Arsenite has greater toxicity affecting the actions of enzymes within the cell (Sukhvinder et al. 2011); however, it oxidizes rapidly, and its concentration in different environments is minor. In Diamante Lake, almost all the arsenic present is found in its oxidized state (AsV). Arsenate, in spite of being less toxic than arsenite, interacts with cells due to its chemical similarity with phosphate, and it enters the cell through phosphate transporters (Pit and Pst) and takes the place of phosphate in different proteins. Thus, despite being such a toxic element, the interaction of arsenic with biological systems is very widespread, especially in Diamante Lake.

The relationship of arsenic with living organisms is not new; for several years different topics have been investigated: (1) Obtaining energy through electrochemical transformations of arsenate/arsenite (van Lis et al. 2013), (2) phylogeny and evolution of enzymes involved in the arsenic bioenergetic metabolism (Rascovan et al. 2016; van Lis et al. 2013), (3) phosphate–arsenic relationship (Mukhopadhyay et al. 2002) and (4) stability of key arsenic-based molecules for life, among others. All these topics can be studied in the natural laboratory represented by Diamante Lake, which is a unique environment for the study of the origin of life (Fig. 8.3).

8.3 Microbial Diversity in Diamante Lake

To evaluate the taxonomic composition and the metabolic potential of the Diamante Lake Red Biofilms (DLRB), we performed metagenomic shotgun sequencing of the extracted total DNA, then prepared a Metagenomic Shotgun Library. Metagenomic analysis showed that the DLRB were constituted by 94% *Archaea*

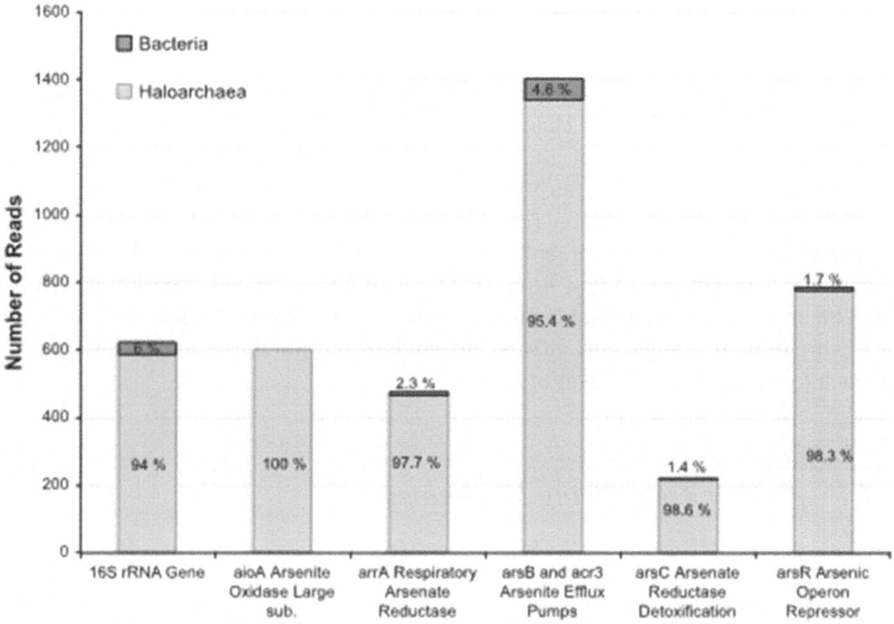

Fig. 8.3 Abundance (number of reads) of genes related to the arsenic metabolism in Diamante Lake. Read counts for 16S rRNA is shown as a reference gene that is present in all organisms in the biofilm. (Reprinted from "Metagenomic study of red biofilms from Diamante Lake reveals ancient arsenic bioenergetics in haloarchaea" (from Rascovan et al. 2016))

and 6% *Bacteria* (Rascovan et al. 2016). This low diversity found in the DLRB can be explained by the previously mentioned physicochemical conditions in the environment. It is known that hypersaline environments have a low diversity, where halophilic microorganisms are able to survive in these extreme conditions because they have specific strategies to balance the osmotic pressure (Ventosa et al. 2014). The presence of *Archaea* suggests they are responsible for the color of the biofilm, due to the presence of bacterioruberin, a red carotenoid pigment present in this microorganism (Oren 1994). *Euryarchaeota* (96%) was the most abundant, and other phyla present in the biofilm included *Proteobacteria* (3%), *Archaeplatida* and *Firmicutes* in smaller percentages. The results also showed that DLRB were composed of two genera of haloarchaea: *Halorubrum* (55%) and *Natronomonas* (14%) (Rascovan et al. 2016). With respect to 7% of bacteria, this was mainly composed of *Firmicutes* of the class *Clostridia* and *Gammaproteobacteria*. It is interesting to note that all the bacteria found belong to anaerobic taxa; therefore, this indicates that the microenvironment from biofilms is anaerobic or microaerobic (Rascovan et al. 2016).

8.4 Arsenic Metabolism in Diamante's Biofilms

Given the high As content, arsenic metabolism was studied by metagenomics analysis of the Diamante Lake Red Haloarchaea Biofilms (DLRB). Metagenomic analyses revealed the presence of the genes necessary for the arsenic detoxifying mechanism (*arsABC*) and a surprisingly high abundance of arsenate respiration (*arrBA*) and arsenite oxidation (*aioBA*). In addition, in this study, it was also observed that the genomic organization of the haloarchaea arsenic bioenergetic genes was very similar to that found in the well-characterized arsenite-oxidizing bacteria and arsenate-respiring bacterial models (Rascovan et al. 2016).

Owing to these results, currently we are studying communities and isolated haloarchaea to learn the metabolism of DLBR.

8.4.1 Arsenic Detoxifying Mechanism

One of the resistance mechanisms widely used by *Bacteria* and *Archaea* consists of the *ars* operon expression encoding for the cytoplasmic arsenate reductase ArsC that catalyzes the transformation of arsenate to arsenite, and the posterior arsenite extrusion from the cell interior by ArsB/ArsA efflux pumps. All *ars* operon genes are present in high abundance in the Diamante Lake metagenome, suggesting that this mechanism of detoxification is used by the haloarchaea from DLRB to limit the intracellular concentration of arsenic (Rascovan et al. 2016).

Arsenate can use phosphate transporters to enter the cell cytoplasm, due to the structural similarity of both molecules. For this reason, another resistance mechanism proposed for the Diamante Lake ecosystem is the differential expression of phosphate transporters, with a strong preference for those with greater affinity for phosphate, such as the Pts system, thus preventing the entry of arsenate into the cell interior.

8.4.2 Arsenotrophy

Most of the organisms that live in high arsenic concentration conditions not only have developed detoxification mechanisms but can also obtain energy catalyzing the dissimilatory transformation of arsenic (Muller et al. 2003).

The *aioA* and *aioB* genes encode the heterodimeric enzyme arsenite oxidase, which participates both in the detoxification of As in heterotrophic microorganisms and in the generation of energy in chemoheterotrophic and quimiolitoautotrophic microorganisms, at the expense of arsenite, which is a source of electrons (Oremland et al. 2002). On the other hand, the ArrA enzyme is responsible for the arsenate reduction process, using it as the final acceptor of the electron chain (anaerobic arsenic respiration). Both enzymes are members of the DMSO reductase family.

To verify the importance of arsenic as a bioenergetic component, physiological and molecular assays were carried out in haloarchaea isolates from DRLB (Ordoñez et al. 2018). RT-PCR results revealed that arsenic metabolism genes were upregulated when As is supplemented to the growth medium and the isolated DM2 (from DRLB) was able to oxidize arsenite in minimal medium (Ordoñez et al. 2018). Moreover, it was demonstrated that arsenate promoted the growth of DRLB isolates in anaerobic conditions, suggesting that this component participates in the metabolic process of respiration as an electron acceptor and bioenergetic substrate (unpublished data). These findings support the idea that the microbial biotransformation plays an important role in the biogeochemical cycle and bioavailability of this metalloid.

8.5 Conclusions

Placed inside Galán Volcano, Diamante Lake hosts one of the most extreme microbial ecosystems of the Puna, not only because of the harsh conditions that have developed in the volcano caldera at 4560 m a.s.l. but also because of the mineralogy (Gaylussite) that precipitates in its extremely alkaline, saline and arsenic waters. These microbial communities, dominated by *Archaea*, associated with carbonated mineral, have proven to function by breathing arsenic as has been demonstrated by metagenomes analyses, genes expression and physiology of the haloarchaea isolated from these biofilms. Ongoing metatranscriptomic studies are also supporting these results. These results add value to the extreme environments of the Andean highlands of Argentina and Chile as excellent models of the early Earth.

References

Albarracín VH, Gärtner W, Farias ME (2016) Forged under the Sun: life and art of extremophiles from Andean lakes. Photochem Photobiol 92:14–28. https://doi.org/10.1111/php.12555

Bischoff JL, Herbst DB, Rosenbauer RJ (1991) Gaylussite formation at Mono Lake, California. (1991. Geochim Cosmochim Acta 55(6):1743–1747

Francis PW, Hammill M, Kretzschmar G, Thorpe RS (1978) The Cerro Galan Caldera, North-west Argentina and its tectonic setting. Nature 274:749

Jones BE, Grant WD (1999) Microbial diversity and ecology of the Soda Lakes of East Africa. In: Proceedings of the 8th international symposium on microbial ecology. http://Plato.acadiu.ca/isme/Symposium22/jones.

Kulp TR, Hoeft SE, Asao M, Madigan MT, Hollibaugh JT, Fisher JC et al (2008) Arsenic(III) fuels anoxygenic photosynthesis in hot spring biofilms from Mono Lake, California. Science 321:967–970

van Lis R, Nitschke W, Duval S, Schoepp-Cothenet B (2013) Arsenics as bioenergetic substrates. Biochim Biophys Acta 1827:176–188

Ma Y, Zhang W, Xue Y, Zhou P, Ventosa A, Grant WD (2004) Bacterial diversity of the inner Mongolian Baer soda lake as revealed by 16S rRNA gene sequence analyses. Extremophiles 8:45–51

Mukhopadhyay R, Rosen BP, Phung L, Silver S (2002) Microbial arsenic: from geocycles to genes and enzymes. FEMS Microbiol Rev 26(3):31–324. https://doi.org/10.1111/j.1574-6976.2002.tb00617.x

Muller D, Lièvremont D, Simeonova D, Hubert JC, Lett MC (2003) Arsenite oxidase aox genes from a metal-resistant-β proteobacterium. J Bacteriol 185(1):135–141

Ordoñez OF, Rasuk MC, Soria MN, Contreras M, Farías ME (2018) Haloarchaea from the Andean Puna: biological role in the energy metabolism of arsenic. Microb Ecol 76:1–11. https://doi.org/10.1007/s00248-018-1159-3.

Oremland RS, Hoeft SE, Santini JM, Bano N, Hollibaugh RA, Hollibaugh JT (2002) Anaerobic oxidation of arsenite in Mono Lake water and by a facultative, arsenite-oxidizing chemoautotroph, strain MLHE-1. Appl Environ Microbiol 68(10):4795–4802

Oren A (1994) The ecology of extremely halophilic archaea. FEMS Microbiol Rev 13:415–440

Parke EC (2013) What could arsenic bacteria teach us about life? Biol Philos 28(2):205–218

Pecoraino G, D'Alessandro W, Inguaggiato S (2015) The other side of the coin: geochemistry of alkaline lakes in volcanic areas. In: Volcanic lakes, Advances in volcanology. Springer-Verlag, Berlin, Heidelberg. https://doi.org/10.1007/978-3-642-36833-2_9

Rascovan N, Maldonado J, Vazquez MP, Farías ME (2016) Metagenomic study of red biofilms from Diamante Lake reveals ancient arsenic bioenergetics in haloarchaea. ISME J 10:299–309. https://doi.org/10.1038/ismej.2015.109

Ruggieri F, Fernandez-Turiel JL, Saavedra J, Gimeno D, Polanco E, Naranjo JA (2011) Environmental geochemistry of recent volcanic ashes from Southern Andes. Environ Chem 8:236–247

Soler M, Caffe P, Coira B, Onoe A, Kay S (2007) Geology of the Vilama caldera: a new interpretation of a large-scale explosive event in the Central Andean plateau during the Upper Miocene. J Volcanol Geotherm Res 164:27–53

Sparks R, Francis P, Hamer R, Pankhurst R, O'callaghan L, Thorpe R, Page R (1985) Ignimbrites of the Cerro Galan Caldera, NW Argentina. J Volcanol Geotherm Res 24(3–4):205–248

Sukhvinder K, Majid K, Arif A (2011) Role of arsenic and its resistance in nature. Can J Microbiol 57(10):769–774

Ventosa A, Fernández AB, León MJ, Sánchez-Porro C, Rodriguez-Valera F (2014) The Santa Pola saltern as a model for studying the microbiota of hypersaline environments. Extremophiles 18(5):811

Part IV
Evaporitic Microbial Ecosystems at Salars: Living in the Salt

Chapter 9
Microbial Communities Composition, Activity, and Dynamics at Salar de Huasco: A Polyextreme Environment in the Chilean Altiplano

Cristina Dorador, Verónica Molina, Martha Hengst, Yoanna Eissler, Marcela Cornejo, Camila Fernández, and Vilma Pérez

9.1 Introduction

Salar de Huasco (20°17.7′S, 68°50.7′W) is located at 3800 m above sea level in the Tarapaca Region of northern Chile (Fig. 9.1). This ecosystem can be considered as broadly representative of ca. 52 active salt flats and wetlands of the high-altitude plateau of the Chilean Altiplano (Risacher et al. 2003) located at the western Cordillera. The majority of aquatic systems currently present in the Altiplano are derived from ancient paleolakes. During the late Holocene (since 2000 YBP), arid conditions prevailed, and these paleolakes became gradually limited forming the current evaporitic closed basins. Salar de Huasco was formed during the Pleistocene, and its current form reflects a series of processes of evaporation and erosion (Chong Dıaz 1984).

C. Dorador (✉)
Laboratorio de Complejidad Microbiana y Ecología Funcional, Departamento de Biotecnología, Facultad de Ciencias del Mar y Recursos Biológicos, Universidad de Antofagasta, Antofagasta, Chile

Centre for Biotechnology and Bioengineering, Santiago, Chile
e-mail: cristina.dorador@uantof.cl

V. Molina
Programa de Biodiversidad and Departamento de Biología, Facultad de Ciencias Naturales y Exactas, Observatorio de Ecología Microbiana, Universidad de Playa Ancha, Valparaíso, Chile

M. Hengst · V. Pérez
Centre for Biotechnology and Bioengineering, Santiago, Chile

Departamento de Ciencias Farmacéuticas, Facultad de Ciencias, Universidad Católica del Norte, Antofagasta, Chile

M. E. Farías (ed.), *Microbial Ecosystems in Central Andes Extreme Environments*, https://doi.org/10.1007/978-3-030-36192-1_9

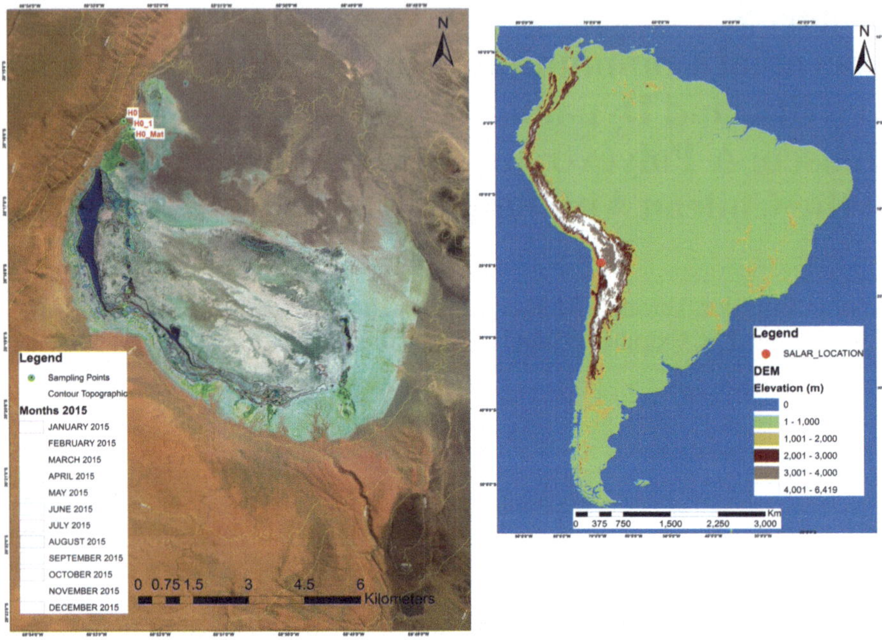

Fig. 9.1 Satellite images of Salar de Huasco. Left panel: evaporitic basin of Salar de Huasco, sampling points are indicated with green points. Right panel: the high plateau of the Andes, the location of Salar de Huasco is indicated by the red point

The Collacagua River is the main inflow of the Salar (Risacher et al. 1999), but a number of other springs and streams contribute to system recharge. Salar de Huasco is characterized by spatial heterogeneity, and includes peatlands (locally referred as 'bofedales'), permanent and non-permanent lakes, and salt crusts. The catchment has a total area of 1572 km^2, and the salar has a surface area of ca. 50 km^2, with open water on average extending to ca. 2.5 km^2 (Risacher et al. 2003). Satellite images analysis carried out from our group indicates that the main shallow lagoon area can vary between 1.75 and 4.64 km^2 (see Fig. 9.1, provided by C. Romero).

Y. Eissler
Instituto de Química y Bioquímica, Facultad de Ciencias, Universidad de Valparaíso, Valparaíso, Chile

M. Cornejo
Escuela de Ciencias del Mar e Instituto Milenio de Oceanografía, Pontificia Universidad Católica de Valparaíso, Valparaíso, Chile

C. Fernández
Sorbonne Université, UPMC Univ Paris 06, Laboratoire d'Océanographie Microbienne (LOMIC), Observatoire Océanologique de Banyuls/Mer, Banyuls/Mer, France

COPAS SUR-AUSTRAL Program, Universidad de Concepción, Concepción, Chile

Centro Fondap de Investigación Dinámica de Ecosistemas Marinos de Altas Latitudes (IDEAL), Valdivia, Chile

Life in this high-altitude ecosystem is subject to a wide range of extreme conditions, including: daily temperature changes between −15 and 20 °C, high irradiance (up to 1000 W m^{-2}) including high UV radiation (>70 W m^{-2}), turbulence and mixing associated with high wind velocities (up to 17 ms^{-1}) and shifts in direction, as well as the low gas partial pressures (including O_2) associated with altitude. Salar de Huasco includes a diverse patchwork of aquatic habitats, including freshwater springs and streams that vary greatly in conductivity from <800 μS cm^{-1} through to hypersaline (>50,000 μS cm^{-1}) conditions driven by extreme evaporation rates in shallow ponds and evaporitic lagoons. Rain and snow melt can also be important sources of water, with rain usually falling during the summer monsoon (known locally as the Bolivian or Altiplanic winter), which also shows marked interannual variability due to ENSO.

9.2 Microbial Diversity in Salar de Huasco

9.2.1 Bacterial Diversity

A series of studies of microbial diversity of Salar de Huasco have revealed a system where bacterial communities generally dominate, but this depends on the methodological approach followed and the biome analyzed (water, sediment or microbial mat). Using qPCR, it was shown that although Bacteria dominated water samples, Archaea dominated saline sediments (Dorador et al. 2010). The first descriptions of microbial assemblages were made using PCR-based techniques (e.g., DGGE and clone libraries) and revealed high bacterial diversity in water, sediment, and microbial mats characterized by the dominance of Proteobacteria, followed by Bacteroidetes, Cyanobacteria, and Actinobacteria (Dorador et al. 2013). These initial results have been corroborated with the utilization of next generation techniques using tag-sequencing of the 16S rRNA gene (e.g., pyrosequencing and Illumina) and shotgun sequencing (metagenomes).

The phylum Proteobacteria is clearly abundant in water samples. Generally, its relative abundance reached up to 76% in some sites, where the presence of Alpha, Beta, Gamma, and Deltaproteobacteria was notorious (e.g., Aguilar et al. 2016; Molina et al. 2018b; Eissler et al. 2019). Epsilonproteobacteria were reported as being rare using clone libraries (Aguilar 2012), but pyrosequencing has shown that it can reach up to 9.5% of relative abundance, e.g., in water samples collected at site H6 (Fig. 9.2) (Eissler et al. 2019). Alphaproteobacteria is the most frequent group recorded from microbial mats (Dorador et al. 2013). The class Rhodobacterales has been widely described in different surveys as one of the most frequent groups in microbial mats and water samples. Also, members of the *Roseobacter* clade are frequently isolated using culture media such as marine agar (Aguilar 2012). The strain *Rhodobacter* sp. Rb3 has been used as a reference strain to study adaptations to high solar radiation (Pérez et al. 2017), and their whole genome has been sequenced (Pérez et al. 2018). The class Gammaproteobacteria is frequent in saline

Fig. 9.2 Spatial heterogeneity characterizes Salar de Huasco. (**a**) Freshwater stream and "bofedales", site H0; (**b**) Microbial mats and *Nostoc* colonies at site H0; (**c**) Hypersaline ponds at site H6; (**d**) Hypersaline lake "playa" type, site H4

waters, where the cultivation of members of the family Halomonadaceae is frequent (Dorador 2007; Aguilar 2012). Other gammaproteobacterial taxa detected in Salar de Huasco are typical psychrophilic, halophilic, and methanotrophic bacteria, such as *Psychrobacter*, *Halochromatium*, and *Methylococcales* (Dorador et al. 2013; Molina et al. 2018b). Betaproteobacteria typically detected in Salar de Huasco are *Hydrogenophaga* (hydrogen oxidizing bacteria) (Aguilar et al. 2016) and *Rhodoferax antarcticus* (Dorador et al. 2013). Strains such as *Alcaligenes* have been isolated from sediment samples (Pérez 2014). Other proteobacteria classes are detected from different biomes, including Deltaproteobacteria, characterized mostly by the presence of sulfate reducing bacteria in sediment samples (Dorador et al. 2013; Aguilar et al. 2016).

Bacteroidetes represent ca. 30% of the typical bacterial composition in water and ca. 3% in sediment (Dorador et al. 2013). This is a group that has a broad distribution in high-altitude wetlands as well as aquatic ecosystems from the Atacama Desert (Dorador et al. 2009). In particular, Flavobacteriaceae and Cytophagaceae are the dominant families representing this phyla (Dorador et al. 2013; Eissler et al. 2019), including genera such as *Psychroflexus*, *Flavobacterium*, and *Algoriphagus* (Dorador et al. 2013; Aguilar et al. 2016; Cruz 2019).

Cyanobacteria are commonly visible as a green layer within microbial mats and also as *Nostoc* colonies above the sediments (Fig. 9.2b). Using microscopic observation of water samples, Cyanobacteria occupied <30% of the phytoplankton community that was dominated by diatoms (Dorador et al. 2008b; Hernández et al. 2016). *Oscillatoria* sp. and *Anabaena* sp. are frequently described (Dorador et al. 2008b; Hernández et al. 2016). Furthermore, one species from the genus *Leptolyngbya* has been isolated (Huanca 2015). Early studies using cyanobacterial-specific primers for the 16S rRNA gene described community structures that differed regarding site and sample type. Importantly, a characteristic feature of Salar de Huasco has been that many sequences have low similarity with those present in databases, something found using either clone libraries or massive sequencing (Dorador et al. 2008b; Aguilar et al. 2016; Molina et al. 2016b; Eissler et al. 2019). *Arthrospira* reached up to 15% of relative abundance in a pond with high dissolved organic carbon content at site H3, followed by *Leptolyngbya*, *Oscillatoria*, and *Nodularia* (Aguilar et al. 2016). In general, Cyanobacteria are potentially actively involved in nitrogen and carbon cycles linked with photosynthesis and nitrogen fixation, including a potential sink of N_2O.

An important component of bacterial communities from the Salar de Huasco is the phylum Actinobacteria. This group has been subject to considerable attention in under-explored environments, such as the Altiplano and the Atacama Desert, due to their capacity to produce bioactive compounds (Bull et al. 2016). In dry soils surrounding the main water body of Salar de Huasco, 65 different actinobacterial strains (mostly *Streptomyces*) have been isolated. These exhibited elevated bioactivity, especially against methicillin-resistant *Staphylococcus aureus* (MRSA) and other antibacterial, antifungal, and cytotoxic activities (against cell lines HepG2 and NIH-3T3 for carcinoma and fibroblast) (Cortés-Albayay et al. 2019a). Several of these *Streptomyces* strains showed low 16S rRNA sequence similarity with other species, and two new species have been recently described from Salar de Huasco: *Streptomyces huasconensis* sp. nov (Cortés-Albayay et al. 2019b) and *Streptomyces altiplanonensis* sp. nov (Cortés-Albayay et al. 2019c).

Firmicutes have been reported both through the use of clone libraries and next generation sequencing based on the amplification of the 16S rRNA gene: typically their relative abundance is lower than 10% (Aguilar et al. 2016). Sequences associated with the *Halanaerobium* have been reported for Salar de Huasco (Dorador et al. 2013). Also, different strains of *Bacillus* have been isolated: specifically, strain H4.8/5 which can oxidize Mn(II) including the presence of the *mnxG* gene (coding for a multicopper oxidase) and tolerate up to 1 mM of Cu(II) (Sathiyanarayanan et al. 2016). *Exiguobacterium* have been frequently isolated from different environments from the Altiplano, and their genomes have been sequenced. *Exiguobacterium* strain SH31 showed a broad range of genes conferring resistance to different heavy metals, including cadmium, copper, mercury, tellurium, chromium, and arsenic (Castro-Severyn et al. 2017). Strain *Exiguobacterium* PN47 also encoded genes related to hot and cold temperatures (Strahsburger et al. 2018).

In total, more than 45 bacterial phyla have been described from Salar de Huasco using the 16S rRNA gene as a molecular marker. The estimated number has increased following the application of different sequencing technologies as well as the increased information available in databases. For example, ca. 10 bacterial phyla were detected using clone libraries (Dorador et al. 2013), between 23 and 42 using pyrosequencing (Aguilar et al. 2016; Molina et al. 2016b; Eissler et al. 2019), and up to 45 with Illumina sequencing (Molina et al. 2018b). An important contribution to the understanding of microbial diversity in Salar de Huasco was the report of new bacterial phyla by Rinke et al. (2013) and new subdomains of Bacteria and Archaea using a combination of single cell genomics and metagenome-assembled genomes (MAGs) (Brown et al. 2015), because this new information aided the identification of typically "unclassified sequences". Notably, the majority of the bacterial phyla currently described from Salar de Huasco are members of the 'rare biosphere' (Sogin et al. 2006), i.e., those taxa present in low relative abundance (<0.1%) and mostly uncultured. Generally, between 25 and 35 phyla are members of the rare biosphere, with most belonging to Candidate Phyla Radiation (CPR) groups (Brown et al. 2015). These taxa were characterized as active based on 16S RNA versus 16S RNA genes. Furthermore, they exhibited different temporal and spatial dynamics (Eissler et al. 2019), responding to changes in their physical and chemical conditions, such as during in situ manipulative experiments (Molina et al. 2016b). Considering recent changes to the tree of life and bacterial phyla classification (Brown et al. 2015; Parks et al. 2018), ca. 50% of all bacterial phyla described globally to date can be found at Salar de Huasco.

9.2.2 Archaeal Diversity

Archaeal communities from Salar de Huasco were initially described using the 16S rRNA gene as phylogenetic and functional molecular markers, the latter based on the detection of ammonia oxidizing archaea through *amoA* gene encoding from subunit A of Ammonia Monooxygenase (Dorador et al. 2010). A large number of sequences exhibited low sequence similarity with those included in public repositories at the time of the study, and two major phyla in the archaeal domain were detected in samples collected during summer (rainy season). In general, Euryarchaeota dominate in water and sediment samples and Crenarchaeota (14% of the analyzed sequences) was present in water samples from site H0 (main springwater site) (Fig. 9.2). Several uncultured euryarchaeotal clusters have been identified, including: MBGD, Marine Benthic Group D; MBGB, Marine Benthic Group B (Vetriani et al. 1999); TMEG, Terrestrial Miscellaneous Euryarchaeotal Group (Takai et al. 2001); MG-I, Marine Group I (DeLong 1998); Euryarchaeota (E) and Crenarchaeota (C); MCG, Miscellaneous Crenarchaeota Group (Teske 2006); and five clusters of unidentified Euryarchaeota exclusively containing sequences from Salar de Huasco. However, during another study conducted in winter (dry season), a high number of sequences related to the family Halobacteriaceae, including the genera *Halorubrum*, *Halomicrobium*, *Natronomonas*, *Natronorubrum*, and *Halalkalicoccus*, were detected (Dorador et al. 2013). Also, in the same study different methanogens were described related to *Methanosarcinales* and *Methanosaeta*.

Archaeal phylogeny has recently undergone marked changes, largely reflecting the improved accessibility of data from sequences obtained using high throughput sequencing, metagenomics, and single cell genomics techniques (e.g., Adam et al. 2017). Currently, the archaeal domain includes 9 phyla and 3 superphylum: Asgaard, TACK/Proteoarchaeota, and DPANN. Considering the initial classification of archaeal sequences in Salar de Huasco, four archaeal domains were detected: Euryarchaeota, Thaumarchaeota, Bathyarchaeota, Thorarchaeota; and two superphylum: Asgard and TACK/Proteoarchaeota. Functional approaches (e.g., metatranscriptomics and metagenomics) to the analyses of these new groups has enabled us to infer the presence of different metabolic pathways, with important implications for our understanding of how Archaea contribute to ecosystem function in polyextreme environments such as Salar de Huasco.

For example, in Salar de Huasco, MBG-D represents ca. 20% of sequences from water and sediment samples. MBG-D has been classified as Thermoprofundales, a new order in the class Thermoplasmata (Euryarchaeota) that utilize mixotrophic metabolisms. Their capacity to ferment peptides forming acetate and ethanol has been demonstrated, and their genomes include two autotrophic pathways: Wood–Ljundahl (WL) and incomplete dicarboxylate/4-hydroxybutyrate cycle, revealing the importance of this group in sedimentary environments (Zhou et al. 2019).

A recent study using specific archaeal primers for 16S rRNA (Illumina sequencing) demonstrated the presence of three active archaeal phylum (Parvarchaeota, Euryarchaeota, and Crenarchaeota) in microbial mats, water, and sediment samples of Salar de Huasco (Molina et al. 2018b). Parvarchaeota is a member of the DPANN superphyla (Rinke et al. 2013), characterized by both small cell and genome size. Originally, this phyla was recognized as the ARMAN group (Baker et al. 2010), described from acid mine drainage environments. Undoubtedly, the archaeal diversity in Salar de Huasco and other high-altitude salares has been underestimated, largely because of the differential use of specific 16S rRNA primers, and the general low abundance of Archaea compared to Bacteria.

9.2.3 Viral Abundance and Diversity

Viruses are the most abundant entities on planet Earth, and Salar de Huasco is no exception. A survey of water column viruses conducted across a series of sites characteristic of Salar de Huasco, including the main lagoon, springs, and ponds, revealed viral abundances ranging from 8.44×10^5 to 4.78×10^8 virus-like particles (VLP) mL^{-1} during dry (February) and wet (July) season, respectively. VLP abundance increased with salinity from freshwater (<800 $\mu S\ cm^{-1}$, spring sites) to mixosaline (1400–51,000 $\mu S\ cm^{-1}$, i.e., ponds and lagoon) and high picoplankton abundances (ranged from 1.07×10^5 to 1.83×10^7 cells mL^{-1}). The viral to picoplankton ratio (VPR), which indicates the level of host–viral interactions, ranged from 2 to 351, also showed greater values at higher salinity locations such as ponds compared to spring sites. Viral abundance was not only related to physicochemical properties and picoplankton abundance in the water column but also with microbial diversity and complexity, which were higher in the presence of microbial mats located in ponds compared with springs or the lagoon (Shannon index *H′* 2.6–3.9 vs <2.0). Consequently, as expected for microbial mats that harbor high host abundance, richness, and diversity, viral abundance detected was high, between 9.6×10^8 and 1.2×10^9 VLP g^{-1} (November 2014) (Eissler et al. 2019).

Metagenomic analyses have allowed host–viral interactions in Salar de Huasco to be inferred, showing the high presence of viruses that belong to the Caudovirales order, which includes bacterial and archaeal viruses. Both hosts have common viral families within the Caudovirales, such as Myoviridae, Podoviridae, and Siphoviridade, whose sequences dominated the metagenomes. Present, but less well represented were viral families such as Phycodnaviridae, which infect microalgae, bacteriophages such as Microviridae, Iridoviridae that infect invertebrates and some vertebrates such as fish and amphibia, and Caulimoviridae that infect plants.

Therefore, viruses may be playing a major role in Salar de Huasco driving microbial community diversification through top-down control, and certainly are influencing biogeochemical cycles enhancing microbial activity such as increasing the availability of organic matter degradation (Eissler et al. 2019; Molina et al. 2018a).

9.2.4 Eukaryotic Microorganism Diversity

To date, eukaryotic communities have been poorly characterized in Salar de Huasco. Phytoplankton communities have been the most studied, mainly through morphological characterization, and as such, it is likely that their diversity has been underestimated. The abundance of Picoeukaryotes and Nanoflagellates has been reported, fluctuating between 4.0×10^5 and 3.3×10^6 and 1.05×10^4 and 1.47×10^5 cells/L, respectively (Hernández et al. 2016).

The abundance of microalgae in distinct water samples of Salar de Huasco ranged between 1.7×10^5 and 3.8×10^6 cells/L (Hernández et al. 2016). A series of different studies of microalgae have described the high abundance of Bacillariophyta (Diatoms) in both water and sediments of Salar de Huasco (Dorador et al. 2008a, b; Marquez-Garcia et al. 2009; Hernández et al. 2016). *Frankophila sudamericana* is a new species of Bacillariophyta described from the Salar de Huasco (Rivera and Cruces 2015). Also, *Nitzschia epithemioides* and *Nitzschia* sp. have been isolated from the salar (Ramírez et al. 2015). The cosmopolitan species *Achnanthidium exiguum* has recently been shown to be present in Salar de Huasco and other localities of the Chilean Altiplano such as Salar de Aguas Calientes 2 (Rivera and Cruces 2018).

Heterotrophic protists have recently been studied in Salar de Huasco. In general, protists play a key role in microbial food webs, acting as grazers of picoplankton—transferring carbon to higher trophic levels (Arndt et al. 2000). *Salpingoeca huasca* is a new species of Choanoflagellate, whose type locality is Salar de Huasco. This species is highly adapted to saline environments, with a salt tolerance ranging between 5 and 150 PSU (Schiwitza et al. 2018).

Yeasts have been described from microbial mats sampled from the Salar de Huasco: three different strains were isolated and identified as *Cryptococcus*, *Dioszegia*, and *Rhodotorula*. Interestingly, *Cryptococcus* and *Rhodotorula* exhibited high tolerance to NaCl, reaching up to 10% (Fuentes 2014).

9.3 Adaptations of Microorganisms to Polyextreme Conditions

The harsh environmental conditions found in high-altitude aquatic ecosystems from the Atacama Desert are associated with diverse and abundant microbial life. In order to survive and even thrive, microbial communities have evolved a plethora of adaptive mechanisms to cope with a wide range of stressors, including salinity, temperature, pH, high UV radiation, desiccation, among others, known collectively as the environmental resistome (Pérez et al. 2018). A well-studied microbial adaptation to environmental pressures is spore forming: this mechanism allows resistance to desiccation, temperature or starvation and is reported for Firmicutes and Gammaproteobacteria from geothermal environments of the Chilean Altiplano (Filippidou et al. 2016, 2019). These authors propose that sporulation or the resistant resting state of cells could compensate for the energy costs of living in the environmental conditions encountered in geothermal ecosystems of the Atacama Desert.

The Atacama Desert receives extremely high levels of UV radiation and is considered among the most exposed environments globally (Cordero et al. 2016; Hernández et al. 2016; Molina et al. 2016b). The UV-resistome has been analyzed in bacteria isolated from different ecosystems using metagenomic, proteomic, and genome sequencing approaches (Kurth et al. 2015; Pérez et al. 2017, 2018). Genome sequencing of *Acinetobacter* sp. Ver3 (a Gammaproteobacteria isolated from Laguna Verde, Argentina) together with proteomic analysis of cultures exposed to UVB radiation under controlled laboratory conditions demonstrated the occurrence of outstanding capabilities to withstand UV radiation. The genome of *Acinetobacter* sp. ver3 highlighted a number of unique genes compared with other *Acinetobacte*r strains, such as a novel cryptochrome. Furthermore, UVB exposure experiments demonstrated that *Acinetobacter* sp. ver3 showed a higher growth compared to congenerics, also showing increased upregulation of proteins belonging to specific cytoplasmic catalases and proteins associated with amino acid and protein synthesis (Kurth et al. 2015).

Pérez et al. (2018) similarly examined the genomic capabilities of *Rhodobacter* sp. Rb3 to thrive under polyextreme environmental conditions. *Rhodobacter* sp. Rb3 is an Alphaproteobacteria (Rhodobacterales), and as mentioned above, makes one of the largest contributions to microbial communities in Salar de Huasco. This group has been used as a model species to explore multi-stress responses based on genomic and proteomic in situ experiments, and in particular to have a deeper understanding of UV adaptations. Using in situ experiments, Molina et al. (2016b) demonstrated that bacteria belonging to the order Rhodobacterales decreased their metabolic activity when exposed to all radiation spectra. This activity was shown to recover after UVA irradiation, which suggests the existence of a probable photorepair mechanism in this group. The *Rhodobacter* sp. Rb3 resistome includes genes to repair and protect DNA integrity (e.g., photolyases, helicases, ligases, endonucleases, histone-like proteins, and glycosylases). Additionally, this bacteria has genes encoding enzymes from the base and nucleotide excision repair systems, a light-independent mechanism that could improve survival in this high-UV environment (Pérez et al. 2018). This kind of enzyme has also been detected in bacteria with the highest gamma radiation tolerance shown so far, *Deinococcus radiodurans* (Slade and Radman 2011). Furthermore, *Rhodobacter* sp. Rb3 possess many other molecular mechanisms identified as potential survival mechanisms to cope with polyextreme conditions, including antioxidants, motility, chemotaxis, biofilm synthesis, compatible solutes, metal resistance, antimicrobial resistance, sporulation, cold and heat shock stress resistance, toxin–antitoxin system, and genetic transference elements (Pérez et al. 2018).

A comparative proteomics study using different sources of radiation (in situ sunlight and UVB lamps) demonstrated that the cellular adaptation of *Rhodobacter* sp. Rb3 to damaging UV radiation is supported by the use of compatible solutes as efficient reactive scavengers to prevent oxidative stress. Indeed, proteins involved in the synthesis and transport of osmolytes, such as glycine betaine and inositol, were upregulated. Proteomic analysis also revealed down regulation of RecA, upregulation of proteins involved in physical DNA protection such as a histone-like protein, and a tightly regulated network of proteins involving proteases and chaperones that would selectively degrade misfolded and/or damaged proteins (Pérez et al. 2017).

9.3.1 Effects of High Solar Radiation on the Microbial Community Composition of Salar de Huasco

The microbial communities inhabiting Salar de Huasco are exposed to the highest solar irradiance found at the Earth's surface, a characteristic of the arid ecosystems (Cordero et al. 2016). Solar radiation could negatively influence microbial metabolism, generating photoinhibition of light-harvesting organisms. This could also extend to other taxa such as chemoautotrophs, e.g., nitrifiers, with subsequent impacts on nutrient recycling. It may also result in positive effects for ecosystem function such as photodegradation of organic matter, stimulating heterotrophic activities, and switching other ecological responses (see Ruiz-Gonzalez et al. 2013). In situ experiments conducted in Salar de Huasco indicate that microbial communities show differential responses to radiation intensity and quality, and that UV radiation could play a significant role (Hernández et al. 2016; Molina et al. 2016b). Moreover, a potential selection or acclimation mechanism associated with the photo-history of the microbial community (Hernández et al. 2016) has been proposed as an explanation for the high heterotrophic activity observed for treatments exposed to solar radiation in isolated ponds compared with their counterparts from spring areas. Moreover, in situ mesocosm experiments have shown that the composition and community structure of active bacteria from the mainspring wetland responded differentially to the intensity of solar radiation as well as its quality during the day (Molina et al. 2016b). Here, evidence was shown for photoinhibition in Cyanobacteria based on the decrease in the contribution of 16S rRNA reads to the total community at noon, compared with the afternoon during a daily cycle study. Along with the decrease in Cyanobacteria, an increase in infrequent and rare bacteria taxa (Burkholderiales, Clostridiales, Opitutales, Rhodobacterales) has been reported from the initial samples that were seen to positively respond to UVR compared to full solar radiation. In addition, in a similar study, ammonia oxidizing bacteria were also shown to decrease their potential activity based on *amoA* copy numbers and nutrient recycling compared with initial values (Molina et al. 2016a).

9.4 Biogeochemical Cycles and Greenhouse Production

Microbial basal support processes such as primary and secondary productivity are likely important contributors to the food web of Salar de Huasco, supporting aquatic life including large flamingo colonies as well as other animals living on planktonic grazers of picoplankton. A summary of the process rates measured and net greenhouse gas budgets (CO_2, N_2O, and CH_4) for Salar de Huasco are shown in Table 9.1. In general, microbial communities are highly active in the salar, with rates of carbon productivity reaching magnitudes similar to those from highly productive marine ecosystems such as the Pacific Ocean (see Troncoso et al. 2003).

High primary productivity was reported from sediments of the main lagoon of Salar de Huasco, potentially related with oxygenic photosynthetic organisms such as diatoms and Cyanobacteria, as discussed above. However, anoxygenic photosynthesis is likely playing an important role in the local carbon cycle due to the constant presence and abundance of related taxa (Dorador et al. 2013). Studies using functional genes related to photosynthetic unit form (*pufM* and L genes) and the subunit Y of the chlorophyllide reductase (*bchY*) have reported a high diversity of aerobic and anaerobic phototrophic bacteria in microbial mats (Cruz 2019; Vejar 2019). Also, different isolates have been obtained affiliated with the genera: *Rhodobacter*, *Thiocapsa*, *Rhodovulum*, and *Ectothiorhodospira* (Dorador 2007).

Despite the high level of adaptation displayed by microbial communities to the extreme environmental conditions found in Salar de Huasco, physical and chemical changes in the environment are so intense they play a regulatory role in biogeochemical processes. This has been shown by changes in heterotrophic and chemoautotrophic activities linked to carbon and nitrogen cycles at the different aquatic sites in association with variables such as conductivity, solar radiation, nutrients, and substrate availability (see Hernández et al. 2016; Molina et al. 2018a).

Previous work has revealed a high demand for nitrogen in Salar de Huasco by those microbial communities related to heterotrophic and chemolithoautotrophic lifestyles. This includes microbial groups dependent on ammonia and nitrite as electron donors, e.g., those associated with the nitrification process (Table 9.1).

Table 9.1 Budgets of biogeochemical processes and greenhouse gases analyzed at Salar de Huasco

Process	Biome at wetland	Range in rates recorded	Reference
Primary productivity	Sediments main lake (oxygen microprofiling)	0.6–1.53 g C m^{-2} day^{-1} [a]3,065 kg C day^{-1}	De la Fuente (2014)
Secondary productivity	Water spring—evaporitic ponds (H3-Leucine incorporation)	1.2–264 μg C L^{-1} day^{-1} [a]0.48–105 kg C day^{-1}	Hernández et al. (2016)
Ammonium uptake	Water spring—evaporitic ponds	80–6800 nM day^{-1} 0.48–40.8 kg N day^{-1}	Molina et al. (2018a)
Ammonia and nitrite oxidation	Water spring—evaporitic ponds	12–85 nM day^{-1} 0.07–0.5 kg N day^{-1}	
N_2O fluxes	Rate determined as fluxes per area based on discrete measurements, a positive value indicates an efflux, whereas a negative value an influx between the water and atmosphere	−0.5 to −1.56 μmol m^{-2} day^{-1} [b]−0.28 to −87.4 g N day^{-1}	Molina et al. (2018b)
CH_4 fluxes		4.2–114 μmol m^{-2} day^{-1} [b]101–2.743 g C day^{-1}	
CO_2 fluxes		54–75 mmol m^{-2} day^{-1} [b]1296–1800 g C day^{-1}	

[a]Estimations based on volume (entire lagoon = average of area of 2 km^2 and 0.2 m depth for volume estimations)
[b]Estimation based on area

An active and diverse nitrifying assemblage has been described from Salar de Huasco, including ammonia oxidizing bacteria and archaea, as well as nitrite oxidizers (Molina et al. 2018a). Moreover, anammox bacteria were also detected in Salar de Huasco, likely associated with particles, suggesting an important role of aerobic and anaerobic chemolithoautotrophic communities (Dorador et al. 2008a; Molina et al. 2018a).

Greenhouse gas budgets (Table 9.1) indicate that CO_2 and CH_4 usually accumulate in the water column and result in positive effluxes to the atmosphere. CH_4 supersaturation of the water column was seen across a diversity of aquatic sites studied at the Salar de Huasco, including springs, streams, ponds, and the main lagoon. This suggests an active contribution by methanogens, particularly associated with the sediments where bubbles containing huge amounts of methane sometimes get trapped beneath microbial mats in shallow water habitats. In contrast to CO_2 or CH_4, negative fluxes of N_2O were determined in the Salar de Huasco, suggesting the intense utilization of this gas by microbial communities (Molina et al. 2018b). However, observations of significant variation in GHG concentrations suggest that the magnitudes and direction of fluxes were determined by abiotic factors such as temperature and solar radiation: negative N_2O fluxes were detected during the day compared with positive fluxes during the night (Molina et al. 2018b).

9.5 Microbial Conservation

Salar de Huasco is located in the heart of the Chilean copper mining industry, and is surrounded by copper-mining activities. Copper mining and subsequent processing of ores requires large volumes of water. Given its location in an arid area, this requires extraction of water from different altiplanic aquifers, as well as direct extraction from closed evaporitic basins to support these processes. This has repeatedly resulted in severe environmental damage to salares and bofedales in the region (see Scheihing and Tröger 2018). Also, the presence of large mine tailings could be considered as harmful because of the presence of toxic elements that could be dispersed by the wind and birds (flamingos). In this context, it is urgent to establish conservation strategies for Salar de Huasco and other similar evaporitic basins in the Altiplano that extend beyond the current situation, which focuses on certain taxa such as flamingos and endemic fish such as *Orestias agassi* (Riveros et al. 2012). Microbial ecosystems are complex units, where even small changes could have consequences at taxonomical and functional levels. Water extraction not only causes the desiccation of superficial waters, it also affects subterranean water that is the source of freshwater in Salar de Huasco and supports different active microbial communities (see Molina et al. 2016b). Climate change predictions in Salar de Huasco (reduced precipitation which will fall in intense, short events) suggest that the aquifer will have the capacity to buffer the external variations; nevertheless, any human intervention will have an effect on the discharge flows to bofedales impacting the natural conditions of the basin (Blin 2018).

The recent period has been particularly fruitful in microbial ecology: the development of new sequencing techniques, the possibility to discover the 'dark matter of life' (Rinke et al. 2013) and to understand the function and metabolic capacities of microorganisms is still intriguing. Extreme environments such as Salar de Huasco offer a window of opportunities to study microbial ecology in action, including adaptations and biological interactions (e.g., simbiosis; viral–host interactions). To understand ecological and trophic interactions between different components of aquatic systems is without a doubt a major challenge, for example, using complementary approaches, such as fatty acids and stable isotopes to understand the contribution of microbial communities living in different matrices (e.g., microbial mats, biofilms, water, sediment, macrophytes) to food webs (Dorador et al. 2018). Therefore, multidisciplinary research is crucial to combine knowledge and build an ecosystemic view of Salar de Huasco.

Microbial conservation necessarily requires the preservation of the ecosystem. Under this concept, ecosystem management should include microbial ecologists and an integral ecosystemic view. When an aquatic ecosystem such as the Salar de Huasco (or others in the Altiplano) dries out, not only is a keystone environmental lost but also thousands of years of evolution and unique biodiversity disappear forever.

Acknowledgements A large number of people and institutions have collaborated with us over the past decade, contributing with sampling, fruitful discussions, and good science. We would like to express our deep gratitude to Irma Vila, Karl-Paul Witzel, Johannes F. Imhoff, Wade Jeffrey, Pilar Junier, Brad Bebout, Joseline Tapia, Lasse Olsen, Klaudia Hernández, Chris Harrod, Felipe Docmac, Francisco Remonsellez, Mario Tapia, Pablo Aguilar, Daniela Meneses, Drina Vejar, Carolina Cubillos, Bernardita Valenzuela, Gonzalo Icaza, Diego Cornejo, Johanna Cortés, Jonathan García, Pablo Arán, Eduardo Acosta, Paulina Huanca, Gonzalo Fuentes, Viviana Araya, Yaneisi Vázquez, Vinko Zadjelovic, Jonathan Pérez, Giannina Maya, Claudia Muñoz, Claudia Elizondo, María Jesús Gálvez, Angel Rain, Nicolás Cheddri, Hermann Peña, Agustina Undabarrena, Carlos Romero, Gabriel Soto, and the local inhabitants of Salar de Huasco: Pedro and Margarita Luca. A special recognition to Jaime Guerrero, as without his help, all this would not have been possible.

Funding: Fondecyt grants N°11080228, 1110824, 1110953, 1140356, 1140179, 1171324, 1181773, 1150891; DAAD-CONICYT N°698/2011; PIA-CONICYT FB0001; LIA-MORFUN.

References

Adam PS, Borrel G, Brochier-Armanet C, Gribaldo S (2017) The growing tree of Archaea: new perspectives on their diversity, evolution and ecology. ISME J 11:2407–2425. https://doi.org/10.1038/ismej.2017.122

Aguilar P (2012) Diversidad filogenética y caracterización de bacterias halófilas y halotolerantes provenientes del Salar de Huasco. Undergraduate thesis, Biotechnology degree. Universidad de Antofagasta, Antofagasta, Chile, pp. 83

Aguilar P, Dorador C, Sommaruga R (2016) Large differences in bacterial community composition among three nearby extreme waterbodies of the high Andean plateau. Front Microbiol 7:976. https://doi.org/10.3389/fmicb.2016.00976

Arndt H, Dietrich D, Auer B, Cleven E-J, Gräfenhan T, Weitere M, Mylnikov AP (2000) Functional diversity of heterotrophic flagellates in aquatic ecosystems. In: Leadbeater BSC, Green JC (eds) The flagellates. Taylor and Francis Ltd, London, pp 240–268

Baker BJ, Comolli LR, Dick GJ, Hauser LJ, Hyatt D, Dill BD, Land ML, VerBerkmoes NC, Hettich RL, Banfield JF (2010) Enigmatic, ultrasmall, uncultivated Archaea. Proc Natl Acad Sci U S A 107(19):8806–8811. https://doi.org/10.1073/pnas.0914470107

Blin N (2018) Efectos del cambio climático en el acuífero del Salar del Huasco. Master thesis, Facultad de Ingeniería, Universidad Católica de Chile, Santiago, Chile, p 81

Brown CT, Hug LA, Thomas BC, Sharon I, Castelle CJ, Singh A, Wilkins MJ, Wrighton KC, Williams KH, Banfield JF (2015) Unusual biology across a group comprising more than 15% of domain Bacteria. Nature 523(7559):208–211. https://doi.org/10.1038/nature14486

Bull AT, Asenjo JA, Goodfellow M, Gomez-Silva B (2016) The Atacama Desert: technical resources and the growing importance of novel microbial diversity. Annu Rev Microbiol 70:215–234. https://doi.org/10.1146/annurev-micro-102215-095236

Castro-Severyn J, Remonsellez F, Salinas C, Valenzuela SL, Fortt J, Aguilar P, Pardo-Esté C, Aguayo D, Dorador C, Quatrini R, Molina F, Castro-Nallar E, Saavedra CP (2017) Comparative genomics analysis of a new Exiguobacterium strain from Salar de Huasco reveals a repertoire of stress-related genes and arsenic resistance. Front Microbiol 8:456. https://doi.org/10.3389/fmicb.2017.00456

Chong Dıaz G (1984) Die Salare in Nordchile: Geologie, Struktur und Geochemie. Geotekt Forsch 67:1–146

Cordero RR, Damiani A, Seckmeyer G, Jorquera J, Caballero M, Rowe P, Matus M (2016) The solar spectrum in the Atacama Desert. Sci Rep 6:22457. https://doi.org/10.1038/srep22457

Cortés-Albayay, C, Silber J, Imhoff JI, Asenjo JA, Andrews B, Nouioui I, Dorador C (2019a) The polyextreme ecosystem, Salar de Huasco at the Chilean Altiplano of the Atacama Desert houses diverse *Streptomyces* spp. with promising pharmaceutical potentials. Diversity 11(59):69

Cortés-Albayay C, Dorador C, Schumann P, Andrews B, Asenjo J, Nouioui I (2019b) *Streptomyces huasconensis* sp. nov., an haloalkalitolerant actinobacterium isolated from a high altitude saline wetland at the Chilean Altiplano. Int J Syst Evol Microbiol 69(8):2315–2322

Cortés-Albayay C, Dorador C, Schumann P, Schniete, JK, Herron P, Andrews B, Asenjo J, Nouioui I (2019c) *Streptomyces altiplanensis* sp. nov., an alkalitolerant species isolated from Chilean Altiplano soil, and emended description of *Streptomyces chryseus* (Krasil'nikov et al. 1965) Pridham 1970. Int J Syst Evol Microbiol 69(8):2498–2505

Cruz C (2019) Valoración de bacterias del Salar de Huasco desde un enfoque de provisión de compuestos bioactivos y potenciales servicios ecosistémicos. Master thesis, Facultad de Ciencias de la Salud, Universidad de Antofagasta, Antofagasta, Chile, p 96

de la Fuente A (2014) Heat and dissolved oxygen exchanges between the sediment and water column in a shallow salty lagoon. J Geophys Res Biogeo 119:596–613. https://doi.org/10.1002/2013JG002413

DeLong EF (1998) Everything in moderation: archaea as 'nonextremophiles'. Curr Opin Genet Dev 8:649–654. https://doi.org/10.1016/S0959-437X(98)80032-4

Dorador C (2007) Microbial diversity in high altitude wetlands of the Chilean altiplano: phylogeny, diversity and function. PhD Thesis, University of Kiel, Germany, Kiel

Dorador C, Busekow A, Vila I, Imhoff JF, Witzel K-P (2008a) Molecular analysis of enrichment cultures of ammonia oxidizers from the Salar de Huasco, a high altitude saline wetland in northern Chile. Extremophiles 12:405–414. https://doi.org/10.1007/s00792-008-0146-x

Dorador C, Vila I, Imhoff JF, Witzel K-P (2008b) Cyanobacterial diversity in Salar de Huasco, a high altitude saline wetland in northern Chile: an example of geographical dispersion? FEMS Microbiol Ecol 64:419–432. https://doi.org/10.1111/j.1574-6941.2008.00483.x

Dorador C, Fink P, Hengst M, Icaza G, Villalobos AS, Vejar D, Meneses D, Zadjelovic V, Burmann L, Moelzner J, Harrod C (2018) Microbial community composition and trophic role along a marked salinity gradient in Laguna Puilar, Salar de Atacama, Chile. Antonie Van Leeuwenhoek 111(8):1361–1374. https://doi.org/10.1007/s10482-018-1091-z

Dorador C, Meneses D, Urtuvia V, Demergasso C, Vila I, Witzel K-P, Imhoff JF (2009) Diversity of Bacteroidetes in high altitude saline evaporitic basins in northern Chile. J Geophys Res Biogeosci 114:G00D05. https://doi.org/10.1029/2008JG000837

Dorador C, Vila C, Remonsellez F, Imhoff JF, Witzel K-P (2010) Unique clusters for Archaea in Salar de Huasco, an athalassohaline evaporitic basin of the Chilean Altiplano. FEMS Microbiol Ecol 73:291–302. https://doi.org/10.1111/j.1574-6941.2010.00891.x

Dorador C, Vila I, Witzel K-P, Imhoff JF (2013) Bacterial and archaeal diversity in high altitude wetlands of the Chilean Altiplano. Fundam Appl Limnol 182:135–159. https://doi.org/10.1127/1863-9135/2013/0393

Eissler Y, Gálvez MJ, Dorador C, Hengst M, Molina V (2019) Active microbiome structure and its association with environmental factors and viruses at different aquatic sites of a high altitude wetland. Microbiologyopen 8(3):e00667. https://doi.org/10.1002/mbo3.667

Filippidou S, Junier T, Wunderlin T, Kooli WM, Palmieri I, Al-dourobi A, Molina V, Lienhard R, Spangenberg JE, Johnson SL, Chain PSG, Dorador C, Junier P (2019) Adaptive strategies in a poly-extreme environment: differentiation of vegetative cells in Serratia ureilytica and resistance to extreme conditions. Front Microbiol 10:102. https://doi.org/10.3389/fmicb.2019.00102

Filippidou S, Wunderlin T, Junier T, Jeanneret N, Dorador C, Molina V, Johnson D, Junier P (2016) A combination of extreme environmental conditions favor the prevalence of Endospore-forming Firmicutes in geothermal and mineral springs. Front Microbiol 7:1707. https://doi.org/10.3389/fmicb.2016.01707

Fuentes G (2014) Búsqueda de compuestos bioactivos desde levaduras halotolerantes del Salar de Huasco, Altiplano Chileno. Undergraduate thesis, Biotechnology degree. Universidad de Antofagasta, Antofagasta, Chile, p 105

Hernández KL, Yannicelli B, Olsen LM, Dorador C, Menschel EJ, Molina V, Remonsellez F, Hengst M, Jeffrey WH (2016) Microbial activity response to solar radiation across contrasting environmental conditions in Salar de Huasco, Northern Chilean Altiplano. Front Microbiol 7:1857. https://doi.org/10.3389/fmicb.2016.01857

Huanca P (2015) Detección de compuestos con actividad fotoprotectora aislados desde microorganismos fotótrofos oxigénicos del Salar de Huasco. Undergraduate thesis, Biotechnology degree. Universidad de Antofagasta, Antofagasta, Chile, 83p

Kurth D, Belfiore C, Gorriti MF, Cortez N, Farias ME, Albarracín VH (2015) Genomic and proteomic evidences unravel the UV-resistome of the poly-extremophile Acinetobacter sp. Ver3. Front Microbiol 6:328. https://doi.org/10.3389/fmicb.2015.00328

Marquez-Garcia M, Vila I, Hinojosa LF, Mendez MA, Carvajal JL, Sabando MC (2009) Distribution and seasonal fluctuations in the aquatic biodiversity of the southern Altiplano. Limnologica 39:314–318. https://doi.org/10.1016/j.limno.2009.06.007

Molina V, Dorador C, Fernandez C, Bristow L, Eissler Y, Hengst M, Hernández K, Olsen L, Harrod C, Marchant F, Anguita C (2018a) The activity of nitrifying microorganisms in a high altitude Andean wetland. FEMS Microbiol Ecol 94(6):fiy062. https://doi.org/10.1093/femsec/fiy062

Molina V, Eissler Y, Cornejo M, Galand PE, Dorador C, Hengst M, Fernandez C, Francois JP (2018b) Greenhouse gases distribution at hyperarid and arid areas of northern Chile and the contribution of high altitude wetland microbiome (Salar de Huasco, Chile). Antonie Van Leeuwenhoek 111(8):1421–1432. https://doi.org/10.1007/s10482-018-1078-9

Molina V, Hernández K, Dorador C, Eissler Y, Hengst M, Anguita C, Peña H (2016a) Respuesta de bacterias amonio-oxidantes naturales a distintos rangos de radiación en un humedal Altiplánico (Salar de Huasco). Revista Visiones Científicas 11(2):17–23

Molina V, Hernandez K, Dorador C, Hengst M, Pérez V (2016b) Daily cycle of bacterial active community structure and nutrient recycling under natural solar radiation in a high altitude wetland. Front Microbiol 7:1823. https://doi.org/10.3389/fmicb.2016.01823

Parks DH, Chuvochina M, Waite DW, Rinke C, Skarshewski A, Chaumeil P-A, Hugenholtz P (2018) A standardized bacterial taxonomy based on genome phylogeny substantially revises the tree of life. Nat Biotechnol 36:996–1004. https://doi.org/10.1038/nbt.4229

Pérez P (2014) Humedales altoandinos en el Norte de Chile: sitios clave para la búsqueda de nuevos compuestos bioactivos y su posible aplicación en la Acuicultura. Undergraduate thesis, Biotechnology degree. Universidad de Antofagasta, Antofagasta, Chile, 105pp

Pérez V, Dorador C, Molina V, Yáñez C, Hengst M (2018) Rhodobacter sp. Rb3, an aerobic anoxygenic phototroph which thrives in the polyextreme ecosystem of the Salar de Huasco, in the Chilean Altiplano. Antonie Van Leeuwenhoek 111(8):1449–1465. https://doi.org/10.1007/s10482-018-1067-z

Pérez V, Hengst MB, Kurte L, Dorador C, Jeffrey WH, Wattiez R, Molina V, Matallana-Surget S (2017) Life under extreme UV radiation: A comparative proteomics study of Rhodobacter sp., isolated from high altitude wetlands in Chile. Front Microbiol 8:1173. https://doi.org/10.3389/fmicb.2017.01173

Ramírez EE, González MA, Cifuentes AS, Inostroza I, Urrutia R (2015) Culture and growth of two benthic diatoms species isolated from the Salar del Huasco (north of Chile, 20°S) at different conditions of temperature, light and nutrient. Gayana Bot 72(2):165–176. https://doi.org/10.4067/S0717-66432015000200001

Rinke C, Schwientek P, Sczyrba A, Ivanova NN, Anderson IJ, Cheng JF, Darling A, Malfatti S, Swan BK, Gies EA, Dodsworth JA, Hedlund BP, Tsiamis G, Sievert SM, Liu WT, Eisen JA, Hallam SJ, Kyrpides NC, Stepanauskas R, Rubin EM, Hugenholtz P, Woyke T (2013) Insights into the phylogeny and coding potential of microbial dark matter. Nature 499(7459):431–437. https://doi.org/10.1038/nature12352

Risacher F, Alonso H, Salazar C (1999) Geoquímica de aguas en cuencas cerradas: I, II y III regiones—Chile. Ministerio de Obras Públicas, Dirección General de Aguas, Santiago, Chile

Risacher F, Alonso H, Salazar C (2003) The origin of brines and salts in Chilean salars: a hydrochemical review. Earth-Sci Rev 63:249–293. https://doi.org/10.1016/S0012-8252(03)00037-0

Rivera P, Cruces F (2015) Frankophila sudamericana sp. nov., a new diatom species (Bacillariophyta) found in Salar de Aguas Calientes and Salar de Huasco, high altitude Andean localities in northern Chile. Gayana Bot 72(2):373–376. https://doi.org/10.4067/S0717-66432015000200017

Rivera P, Cruces F (2018) Achnanthidium exiguum (Bacillariophyta): nuevas citas para localidades andinas del norte de Chile. Gayana Bot 75(2):646–649. https://doi.org/10.4067/S0717-66432018000200646

Riveros J, Vila I, Mendez M (2012) Nicho trófico de Orestias agassii (Cuvier & Valenciennes, 1846) del sistema de arroyos del salar de Huasco (20°05′S; 68°15′W). Gayana 76(2):79–91. https://doi.org/10.4067/S0717-65382012000300001

Ruiz-Gonzalez C, Simó R, Sommaruga R, Gasol JM (2013) Away from darkness: a review on the effects of solar radiation on heterotrophic bacterioplankton activity. Front Microbiol 4:131. https://doi.org/10.3389/fmicb.2013.00131

Sathiyanarayanan G, Filippidou S, Junier T, Muñoz Rufatt P, Jeanneret N, Wunderlin T, Sieber N, Dorador C, Junier P (2016) Manganese-II oxidation and Copper-II resistance in endospore forming Firmicutes isolated from uncontaminated environmental sites. AIMS Environ Sci 3(2):220–238. https://doi.org/10.3934/environsci.2016.2.220

Scheihing K, Tröger U (2018) Local climate change induced by groundwater overexploitation in a high Andean arid watershed, Laguna Lagunillas basin, northern Chile. Hydrogeol J 26(3):705–719. https://doi.org/10.1007/s10040-017-1647-4

Schiwitza S, Arndt H, Nitsche F (2018) Four new choanoflagellate species from extreme saline environments: indication for isolation-driven speciation exemplified by highly adapted Craspedida from salt flats in the Atacama Desert (northern Chile). Eur J Protistol 66:86–96. https://doi.org/10.1016/j.ejop.2018.08.001

Slade D, Radman M (2011) Oxidative stress resistance in Deinococcus radiodurans. Microbiol Mol Biol Rev 75(1):133–191. https://doi.org/10.1128/MMBR.00015-10

Sogin ML, Morrison HG, Huber JA, Welch DM, Huse SM, Neal PR, Arrieta JM, Herndl GJ (2006) Microbial diversity in the deep sea and the underexplored “rare biosphere”. Proc Natl Acad Sci U S A 103(32):12115–12120. https://doi.org/10.1073/pnas.0605127103

Strahsburger E, Zapata F, Pedroso I, Fuentes D, Tapia P, Poncea R, Valdes J (2018) Draft genome sequence of Exiguobacterium aurantiacum strain PN47 isolate from saline ponds, known as “Salar del Huasco”, located in the Altiplano in the north of Chile. Braz J Microbiol 49:7–9. https://doi.org/10.1016/j.bjm.2017.03.011

Takai K, Moser DP, DeFlaun M, Onstott TC, Fredrickson JK (2001) Archaeal diversity in waters from deep South African gold mines. Appl Environ Microbiol 67:5750–5760. https://doi.org/10.1128/AEM.67.21.5750-5760.2001

Teske A (2006) Microbial communities of deep marine subsurface sediments: molecular and cultivation surveys. Geomicrobiol J 23:357–368. https://doi.org/10.1080/01490450600875613

Troncoso VA, Daneri G, Cuevas LA, Jacob B, Montero P (2003) Bacterial carbon flow in the Humboldt Current System off Chile. Mar Ecol Prog Ser 250:1–12. https://doi.org/10.3354/meps250001

Vejar D (2019) Diversidad de bacterias fotótrofas anoxigénicas de un tapete microbiano del Salar de Huasco, utilizando una aproximación molecular funcional. Master thesis, Facultad de Ciencias de la Salud, Universidad de Antofagasta, Antofagasta, Chile, 93pp

Vetriani C, Jannasch HW, MacGregor BJ, Stahl DA & Reysenbach AL (1999) Population structure and phylogenetic characterization of marine benthic Archaea in deep-sea sediments. Appl Environ Microbiol 65: 4375–4384

Zhou Z, Liu Y, Lloyd K, Pan J, Yang Y, Gu J-D, Li M (2019) Genomic and transcriptomic insights into the ecology and metabolism of benthic archaeal cosmopolitan, Thermoprofundales (MBG-D archaea). ISME J 13:885–901. https://doi.org/10.1038/s41396-018-0321-8

Chapter 10
Prokaryotic Diversity at the Hypersaline Laguna Tebenquiche in the Salar de Atacama, Chile

Ana Beatriz Fernández, Pieter T. Visscher, María Cecilia Rasuk, Manuel Contreras Leiva, and María Eugenia Farías

10.1 Introduction

Hypersaline aquatic systems are mainly characterized by a high salt concentration. In addition, extreme values of physicochemical properties such as temperature, pH, pressure, oxygen, nutrient availability, solar radiation, and the presence of heavy metals and other toxic compounds are often associated with hypersaline lakes, limiting their biodiversity (Rodríguez-Valera 1988). Athalassohaline lakes are among hypersaline aquatic environments, which have an ionic composition different from seawater. Athalassohaline waters have disparate anion compositions (e.g., predominantly sulfate instead of chlorine) and frequently different cation composition (e.g., calcium and/or magnesium instead of sodium) (Hutchinson 1957). Prokaryotes inhabiting athalassohaline environments are poorly known and may represent a large, unexplored reservoir of microorganisms.

The athalassohaline Laguna Tebenquiche is located in the northern part of the Salar de Atacama, in the Antofagasta region of Chile, at 2300 m above sea level (Fig. 10.1) (Demergasso et al. 2008). In 2018 it was declared a Santuario de la

A. B. Fernández · M. C. Rasuk
Laboratorio de Investigaciones Microbiológicas de Lagunas Andinas (LIMLA), Planta Piloto de Procesos Industriales Microbiológicos (PROIMI), CCT, CONICET, Tucumán, Argentina

P. T. Visscher
Department of Marine Sciences, University of Connecticut, Groton, CT, USA

Department of Geosciences, University of Connecticut, Storrs, CT, USA

Australian Centre for Astrobiology, University of New South Wales, Sydney, NSW, Australia

M. C. Leiva
Centro de Ecología Aplicada (CEA), Ñuñoa, Santiago, Chile

M. E. Farías (✉)
LIMLA-PROIMI-CONICET, Tucuman, Argentina

M. E. Farías (ed.), *Microbial Ecosystems in Central Andes Extreme Environments*, https://doi.org/10.1007/978-3-030-36192-1_10

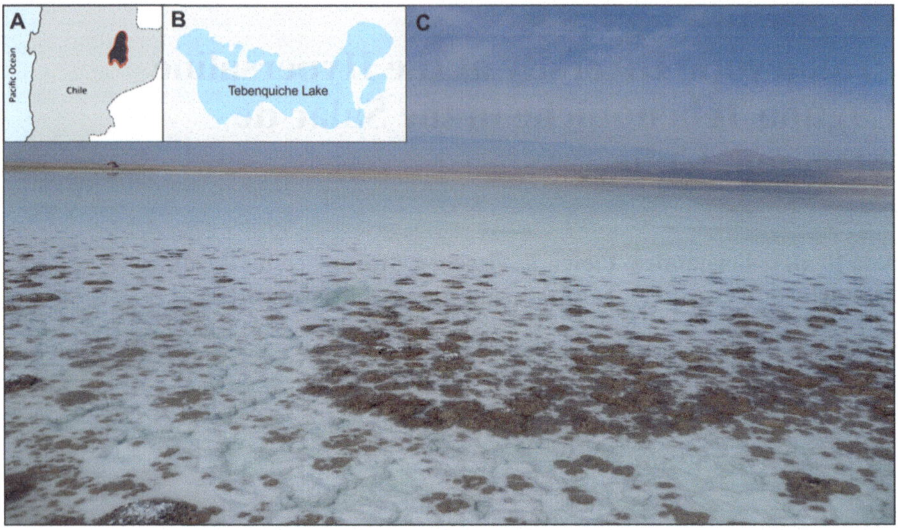

Fig. 10.1 Tebenquiche Lake. (**a**) Location in the north of Salar de Atacama in Chile. (**b**) Shape of Laguna Tebequiche Lake. (**c**) Picture of Laguna Tebenquiche

Naturaleza by the Chilean government for its diversity of microbial extremophiles. This protection was requested by the native Coyo's Aiyu community, supported by scientific findings (Fernandez et al. 2016; Farías et al. 2014). This marked a milestone where a coordinated effort of indigenous communities and scientists requested the protection of the extreme environment.

Laguna Tebenquiche is predominantly fed by groundwater which contains leached Tertiary and Quaternary volcanic material (Alonso and Risacher 1996). Laguna Tebenquiche is hypersaline, and the ionic composition of the water is dominated by sodium and chloride ions but with a high sulfate concentration in which ionic dominance is Na > K > Mg > Ca:Cl > SO_4^{2-} > HCO_3^- + CO_3^{2-} (Zúñiga et al. 1991). Prokaryotic microorganisms inhabiting this lake are subjected to extreme conditions, such as high solar radiation (53.4 UV-BA W/m^2 reached at noon), extreme changes in salinity caused by net evaporation (62–117 g/L), and high lithium (~145–445 mg/L), boron (~158–543 mg/L), and arsenic (~1–5 As mg/L) concentrations in the water due to volcanic events (Lara et al. 2012; Farías et al. 2014; Fernandez et al. 2016). Despite "extreme" environmental conditions in Laguna Tebenquiche, the lake supports the development of microbial mats (Van Gemerden 1993; Rothschild and Mancinelli 2001; Dupraz and Visscher 2005), including both non-lithifying and lithifying microbial mats as well as other organo-sedimentary ecosystems such as endoevaporites and rhizome-associated concretions (Farías et al. 2014; Fernandez et al. 2016).

10.2 Culture-Dependent Studies in Laguna Tebenquiche

Early ecological studies of Laguna Tebenquiche were based on the isolation and characterization of microorganisms in pure cultures. The first investigations of Laguna Tebenquiche's microbial communities focused on moderately halophilic bacteria, which grow optimally in media containing 3–15% NaCl (Kushner and Kamekura 1988) and are widely distributed in various natural habitats (Rodríguez-Valera 1988). The results of these studies suggested that Gram-negative rod-shaped bacteria comprised the most abundant group of heterotrophic moderately halophilic bacteria (Prado et al. 1991) and heterotrophic moderately halophilic Gram-positive cocci isolated accounted for a minor fraction (Valderrama et al. 1991). The moderately halophilic bacterial isolates were distributed among the genera *Vibrio*, *Halomonas* (including *Deleya* and *Volcaniella*), *Marinococcus*, *Micrococcus,* and *Bacillus* (Valderrama et al. 1991; Prado et al. 1993).

Consecutive investigations increased the knowledge of the biodiversity in Laguna Tebenquiche to include extremely halophilic archaea following the isolation of strains in media containing 25% (w/v) total salt. These isolates were assigned to the family *Halobacteriaceae* (Prado et al. 1993). The first taxonomical study focusing on extreme halophilic archaea isolated from the Laguna Tebenquiche revealed a high biodiversity of members of the genera *Halorubrum*, *Haloarcula*, *Halobacterium,* and *Haloferax* (Lizama et al. 2001). Surprisingly, these extreme halophilic archaea did not require Mg^{2+} for optimal growth at a saturated NaCl concentration. The lack of magnesium requirement for these strains is conceivably a consequence of lower Mg^{2+} concentrations in Salar de Atacama than other hypersaline environments in relation to Na^{+} or Cl^{-} concentration (Lizama et al. 2001). Acetate, pyruvate, and glycerol were used as the sole source of carbon and energy by these organisms. Of particular interest are the studies describing novel taxa as the moderately halophilic bacterial species *Chromohalobacter nigrandesensis* (Prado et al. 2006) and the two extremely halophilic archaeal species *Halorubrum tebenquichense* and *Halomicrobium katesii* (Lizama et al. 2002; Kharroub et al. 2008) isolated from Laguna Tebenquiche. For these archaeal species, the authors demonstrated that magnesium was not required for their growth (Lizama et al. 2002; Kharroub et al. 2008).

10.3 Culture-Independent Studies in Laguna Tebenquiche

Despite the elaborate studies using a range of specific culture media and incubation conditions for the isolation of microorganisms from extreme environments in pure cultures, most cases do not yield the dominant microorganisms (from these habitats). Some authors estimate that only from 0.1% to 1% of the total microbial population can be cultured in standard laboratory conditions (Amann et al. 1995; Rappé and Giovannoni 2003), while others demonstrate that well-designed experiments can support cultivation of 40% of the total population (Stewart 2012). However, this wide range demonstrates the need of alternative approaches for the study of the microbial diversity in extreme environments.

10.3.1 Dominant Prokaryotic Groups

Culture-dependent studies did not yield *Bacteroidetes* although *Bacteroidetes* have been frequently observed in aquatic environments such as freshwater (Kirchman 2002), as well as hypersaline environments (Bowman et al. 2000; Humayoun et al. 2003), marine waters (Kirchman 2002), and high-altitude aquatic environments (Demergasso et al. 2004, 2008; Dong et al. 2006; Jiang et al. 2006; Wu et al. 2006; Liu et al. 2006; Dorador 2007; Zhang et al. 2007). Thus, this phylum was thought to be absent from the extreme environments of the Atacama Desert. However, the onset of the wide use of molecular approaches revealed the opposite.

Early studies performed in Laguna Tebenquiche employing 16S rRNA gene amplification, PCR-denaturing gradient gel electrophoresis (DGGE), and sequencing of the most interesting 16S rDNA DGGE bands focused on the biodiversity of the planktonic microorganisms during winter and summer (Demergasso et al. 2004). Bacterial and archaeal DGGE bands recovered from Laguna Tebenquiche water revealed that the bacterial proportion of the total community was much higher than of archaea (Demergasso et al. 2004). The dominant DGGE bands were assigned to *Bacteroidetes*, specifically to the cultured strain *Psychroflexus torquis*, which had been previously isolated from the aquatic environment of Antarctic sea ice (Bowman et al. 1998). Less abundant yet significant bands were associated with relatives of the *Euryarchaeota* phyla (i.e., *Halobacteriaceae*), with the best match for the cultured strain *Haloarcula marismortui*, originally isolated from the Dead Sea. Different DGGE banding patterns were observed between winter and summer samples (Demergasso et al. 2004), the contribution to the total community of DGGE bands designated to Cytophaga–Flavobacterium–Bacteroides (CFB) taxa and cultured haloarchaea taxa decreased in summer. In contrast, the DGGE bands attributed to *Proteobacteria* and uncultured *Euryarchaeota* increased in the summer when compared to the winter.

Demergasso et al. carried out a more detailed study of the bacterioplankton composition, covering spatial (six different points in the lake were sampled) and temporal (samples were taken in summer, winter, and spring seasons) heterogeneity by genetic fingerprinting of the bacterial 16S rRNA gene (Demergasso et al. 2008). The most abundant groups detected included *Bacteroidetes* phylum showing a remarkable degree of novelty (16S rDNA genes very distantly related to genera as *Psychroflexus* or *Salinibacter* or closely related to clones obtained from hypersaline endoevaporitic microbial mat in Eilat, Israel (Sørensen et al. 2005)), followed by the class of *Gammaproteobacteria*. In addition, this study revealed a heterogeneous composition of the bacterial community because of changes in space and salinity (ranged between 1% and 30% w/v). Within the *Bacteroidetes*, *Salinibacter* relatives were recovered in an area of the lake with the highest salinity (29.60%) and clusters distantly related to *Psychroflexus* spp. were found in areas with lower salinity (3–15%). The most abundant cluster within the *Gammaproteobacteria* was related to uncultured bacteria from Mono Lake (CA, USA), which is also a high altitude, hypersaline lake. The remaining phylogenetic groups were represented only by a few clones, which were related to clones from the candidate division OD1 (recently reclassified within the phylum *Parcubacteria* (Nelson and Stegen 2015)) that were isolated from a uranium-contaminated aquifer and Guaymas Basin hydrothermal vent (Gulf of California, Mexico); sequences affiliated with uncultivated clones CS_B020 and BD1-5 from hydrothermal sediments in the Guaymas Basin (Gulf of California, Mexico) and in deep-sea sediments, respectively; and related to sequences of the KB1 group, present in sediments from Yellowstone hot spring (USA) and brine-seawater in the Red Sea.

The diversity of *Bacteroidetes* was further investigated using both 16S rRNA gene clone libraries created with targeted *Bacteroidetes*-specific primers and separation of specifically amplified gene fragments by denaturing gradient gel electrophoresis (DGGE) (Dorador et al. 2009). A greater richness for these microorganisms and the presence of different taxa was observed when compared to Salar de Huasco and Salar de Ascotán. The different *Bacteroidetes* communities between Laguna Tebenquiche and the salares of Huasco and Ascotán were probably a result of different conditions and selective pressures for microorganisms such as higher evaporation rate and lower water input in Laguna Tebenquiche (Risacher et al. 2003), among other factors. Most of the sequences analyzed in Laguna Tebenquiche were related to the family *Flavobacteriaceae*, which are described as psychrophilic and saline microorganisms, and that clustered with genera from marine habitats such as *Psychroflexus*, *Muricauda*, *Tenacibaculum*, and *Brumimicrobium*, and to the family *Rhodothermaceae* that clustered with *Salinibacter*, found in hypersaline habitats. In addition, several groups of unclassified sequences were identified. The possible Role of *Bacteroidetes* in high-altitude lakes, including hypersaline sites, could be associated with degradation of high molecular weight compounds (Dorador et al. 2009).

10.3.1.1 Phototrophic Bacteria

Phototrophic bacteria are considered the most important primary producers in high-altitude salt lakes, but in Laguna Tebenquiche a low Chl*a* concentration was observed (Demergasso et al. 2008). In addition, a study based on culturable diversity or ribosomal gene library analyses did not recover colonies or sequences of anoxygenic phototrophic bacteria, with the exception of a single clone related to the aerobic phototrophic purple bacteria of the *Roseobacter* clade (Demergasso et al. 2008). This unexpected result led Thiel et al. (2010) to investigate the phototrophic bacterial diversity in a microbial mat from Laguna Tebenquiche using group-specific primers for functional genes involved in the biosynthesis of the photosynthetic apparatus in phototrophic purple bacteria (*pufL* and *pufM*) and in green sulfur bacteria (*fmoA*) (Thiel et al. 2010). This study found a large number of *pufLM* phylotypes affiliated with a novel monophyletic lineage of anoxygenic phototrophic *Gammaproteobacteria* that was dominant in Laguna Tebenquiche. Other phylotypes were related to known halophilic *Chromatiaceae* and *Ectothiorhodospiraceae*. Green sulfur bacteria were not recovered from the mats in Laguna Tebenquiche, indicating that their natural abundance was below the detection limit of molecular methods.

10.4 Metagenomic Studies in Laguna Tebenquiche

Molecular genetic approaches offer an assessment of biodiversity prearranged for a given environment since most of these genetic typing methods are based on PCR and cloning, introducing different biases, and therefore estimate the natural abundance of the sequences incorrectly (von Wintzingerode et al. 1997; Casamayor et al. 2000, 2002). Additional problems include that some bands obtained by genetic typing may correspond to artifacts, e.g., increasing their number artificially due to the concurrent presence of several closely-related 16S rRNA fragments, which can easily result in heteroduplex formation (Espejo et al. 1998). Furthermore, different sequences can be stopped at the same position in the gradient of a gel, having the same restriction length or different bands could belong to different operons in the same organism (Nübel et al. 1996; Casamayor et al. 2002). An improved, alternative metagenomic approach is based on the analysis of the collective genome of a microbial community, extracted directly from an environmental sample without the need to isolate and culture microorganisms (Handelsman et al. 1998; Riesenfeld et al. 2004). As such, metagenomics has revolutionized the analyses of microbial communities from aquatic environments, soils, sediments, and other habitats (Handelsman et al. 2002). Several metagenomic studies have been carried out in Laguna Tebenquiche (Farías et al. 2014; Fernandez et al. 2016; Ordoñez et al. 2018), and differences from previous studies based on culture-dependent or other culture-independent molecular techniques have been confirmed.

10.4.1 Bacterial Diversity

"Extreme" environmental conditions characteristic of Laguna Tebenquiche are favorable for the development of microbial mats (Fig. 10.2) (Rothschild and Mancinelli 2001; Dupraz and Visscher 2005). These laminated organosedimentary structures in which the microbial community is embedded in an organic matrix (Dupraz et al. 2004, 2009) are, when associated with halite and gypsum crusts, commonly referred to as endoevaporites (Rothschild et al. 1994). Farías and colleagues reported for the first time widespread distribution of microbial mats and gypsum evaporites in Laguna Tebenquiche (Farías et al. 2014). A detailed analysis of the

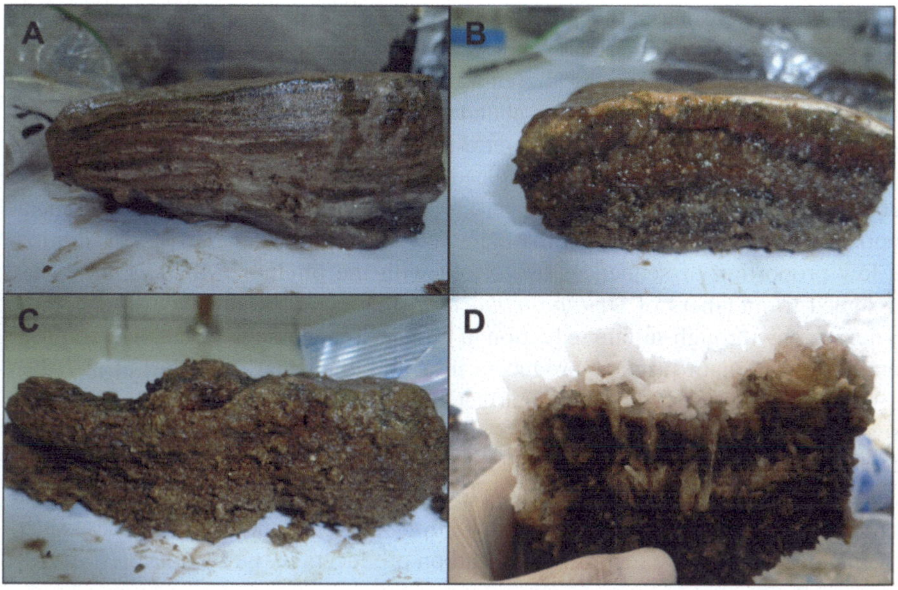

Fig. 10.2 Organosedimentary structures samples from Tebenquiche Lake. (**a**, **b**) Microbial mats, (**c**) rhizome-associated lithified concretions, and (**d**) endoevaporite

bacterial diversity of various sedimentary microbial ecosystems in this habitat was related to the mineral composition and abundance (Farías et al. 2014). In this study, halite was the most abundant mineral in non-lithifying microbial mats, with minor contributions of calcite, gypsum, and aragonite. The non-lithifying mats thrived at a location with a lower salinity (62 g/L) along the shoreline of the lake. In contrast, lithified gypsum domes, which were formed solely of $CaSO_4 \cdot 2H_2O$, developed in areas of the lake with a salinity of 116 g/L. In these domes, the bacterial diversity was analyzed using pyrosequencing of the V4 region of the bacterial 16S rDNA gene. The most abundant amplicons in these endoevaporites were *Bacteroidetes* (*Rhodothermaceae*), followed by *Proteobacteria* (*Alphaproteobacteria* and *Deltaproteobacteria*). As previously observed in a hypersaline lake in Eleuthera, Bahamas (Baumgartner et al. 2009), *Bacteroidetes* increased their presence with increasing salinity in Laguna Tebenquiche, and this could be associated with lithifying systems. The carbonate precipitation observed in some of Laguna Tebenquiche's microbial ecosystems could be explained by active sulfate reduction by *Deltaproteobacteria,* leading to the formation of $CaCO_3$ crystals observed within some microbial mats (Visscher et al. 1998). In addition, the low concentration of chlorophyll a (*Chla*) detected was consistent with the low cyanobacterial diversity detected in these samples, yet oxygen production could be detected with microelectrodes (Farías et al. 2014).

A recent investigation analyzed prokaryotic diversity from evaporite domes, microbial mats, and rhizome-associated lithified concretions in Laguna Tebenquiche by amplifying and pyrosequencing the V4 region of the 16S rDNA (Fernandez et al.

2016). Previous bacterial diversity studies reported the prevalence of *Bacteroidetes* and *Proteobacteria* (Demergasso et al. 2008; Farías et al. 2014); however, the pyrosequencing approach covering the entire microbial community (i.e., bacteria and archaea), revealed that the most abundant bacterial phyla were *Plantomycetes*, *Firmicutes*, *Acetotermia*, and *Chloroflexi* (Fernandez et al. 2016). Most OTUs classified as *Planctomycetes* were associated with the class *Phycisphaerae*, which is composed of very diverse organoheterotrophic aerobes that may play a key metabolic role in microbial mat and rhizome-associated lithified concretions. In addition, a low proportion of sequences associated with the phylum *Proteobacteria* were assigned to members of *Deltaproteobacteria* that could be involved in anaerobic heterotrophy through sulfate reduction in Laguna Tebenquiche. Cyanobacteria generally are the main primary producers in a microbial mat (Visscher et al. 1991, 1992; Van Gemerden 1993; Baumgartner et al. 2009); however, their presence was scarce in Tebenquiche samples, and other organisms likely contribute to carbon fixation. The presence of OTUs related to the family *Chlorothixaceae* and the class *Anaerolinaeae*, both belonging to the phylum *Chloroflexi*, and OTUs related to the order *Chromatiales*, belonging to *Gammaproteobacteria*, indicates that these anoxygenic phototrophs are major organic carbon producers in the Laguna.

10.4.2 Archaeal Diversity

The majority of studies performed in mineral-associated ecosystems describe communities almost exclusively formed by bacteria with scarcity or even complete absence of archaea. In spite of the documented importance of methanogenesis in microbial mats and hypersaline sediments (Visscher and van Gemerden 1991; Buckley et al. 2008), little attention has been given to archaeal diversity. Analysis of bacterial and archaeal diversity in organosedimentary ecosystems from Laguna Tebenquiche (Fernandez et al. 2016), surprisingly, showed *Euryarchaeota* as the main phylum of the total diversity. Most euryarchaeotal OTUs were classified as *Halobacteria* and methanogenic anaerobic archaea, suggesting an important role of (an)aerobic heterotrophy and the potential for methanogenesis in these ecosystems, respectively. The prokaryotic communities were investigated in discrete depth horizons (layers) of a microbial mat and an endoevaporite. *Euryarchaeota* were found in all layers and represented the most abundant phylum in both ecosystems. Noteworthy was that the community in the endoevaporite sample was almost exclusively composed of the class *Halobacteria*. *Halobacteria* were also most abundant in the upper layer of the microbial mat, where euryarcheotal classes *Methanobacteria* and *Thermoplasmata* were the dominant groups in the two bottom layers. Furthermore, in this study, the presence of *Crenarchaeota* was detected in microbial mat and rhizome-associated lithified concretions, which marked a first-time observation in high-altitude Andean lake ecosystems. Curiously, *Crenarchaeota* were not detected in the surface layer of the microbial mat sample, possibly due to high levels of UV radiation or reactive oxygen or hydroxyl species at this altitude.

10.5 Arsenic Metabolism in Tebenquiche Lake

Arsenic concentrations in microbial mats and evaporites from Laguna Tebenquiche varied between 3 and 5 mg/L (Fernandez et al. 2016). Based on this observation and an earlier study in Laguna Diamante, another high altitude Andean lake (Rascovan et al. 2016), Ordoñez et al. (2018) postulated that *Halobacteria* inhabiting these lakes may use arsenic for bioenergetic purposes. Total and genomic DNA and RNA from microbial mats and evaporites from Laguna Tebenquiche and from a biofilm from Laguna Diamante were screened for the presence of *aioA* and *arrA* genes. Novel haloarchaeal strains were isolated from the same samples previously mentioned from Laguna Tebenquiche and Laguna Diamante Lake and were evaluated for the As-cycling genes as well as amplification of putative *aioA* and *arrA* genes by PCR and their expression by RT-PCR using degenerate primers. Haloarchaeal strains isolated from microbial mat and evaporite in Laguna Tebenquiche in an enrichment WJK medium (Maldonado 2015) containing arsenite (As[III]) and arsenate (As[V]) at 1 and 20 mM, respectively, belonged to the genus *Halorubrum*. These strains were able to grow in culture media containing up to the tested maximum of 250 mM arsenate; however, not all the strains grew to the tested maximum of 5 mM arsenite, but they grew at least until 3 mM arsenite.

The expression of *aioA* genes, responsible for arsenite oxidation, was found in several haloarchaeal strains in a microbial mat from Laguna Tebenquiche and a biofilm from Laguna Diamante (Rascovan et al. 2016; Ordoñez et al. 2018). In contrast, the (putative) *arrA* gene (coding for the arsenate reductase gene) could neither be detected in isolates nor in environmental samples from Laguna Tebenquiche. In summary, arsenic, considered as a potentially toxic element, in its form of arsenite can be reduced as a bioenergetic substrate by the microorganisms inhabiting Laguna Tebenquiche.

10.6 Future Perspectives

Several studies evaluated the microbial diversity from Laguna Tebenquiche; initially, these were based on culture-dependent methods, and consecutively on culture-independent molecular approaches that recently included metagenomics. Particularly, the metagenomic approach increased our knowledge of the microbial diversity of this lake. However, it is important to study the metabolic activity and role in element cycling of these microorganisms in order to understand the microbial ecology. To help facilitate this, a metagenomic approach can identify genes involved in different metabolic pathways and thereby contribute to unraveling the functional distribution and understanding survival under the extreme conditions of this environment. Potentially, metagenomic information could aid in development of culture media for the isolation of uncultured microorganisms.

References

Alonso H, Risacher F (1996) Geoquímica del Salar de Atacama, parte 1: origen de los componentes y balance salino. Andean Geol 23:123–134

Amann RI, Ludwig W, Schleifer KH (1995) Phylogenetic identification and in situ detection of individual microbial cells without cultivation. Microbiol Rev 59:143–169

Baumgartner LK, Spear JR, Buckley DH et al (2009) Microbial diversity in modern marine stromatolites, Highborne Cay, Bahamas. Environ Microbiol 11:2710–2719. https://doi.org/10.1111/j.1462-2920.2009.01998.x

Bowman JP, McCammon SA, Lewis T et al (1998) Psychroflexus torquis gen. nov., sp. nov. a psychrophilic species from Antarctic sea ice, and reclassification of Flavobacterium gondwanense (Dobson et al. 1993) as Psychroflexus gondwanense gen. nov., comb. nov. Microbiology 144:1601–1609. https://doi.org/10.1099/00221287-144-6-1601

Bowman JP, McCammon SA, Rea SM, McMeekin TA (2000) The microbial composition of three limnologically disparate hypersaline Antarctic lakes. FEMS Microbiol Lett 183:81–88

Buckley DH, Baumgartner LK, Visscher PT (2008) Vertical distribution of methane metabolism in microbial mats of the Great Sippewissett Salt Marsh. Environ Microbiol 10:967–977. https://doi.org/10.1111/j.1462-2920.2007.01517.x

Casamayor EO, Calderón-Paz JI, Pedrós-Alió C (2000) 5S rRNA fingerprints of marine bacteria, halophilic archaea and natural prokaryotic assemblages along a salinity gradient. FEMS Microbiol Ecol 34:113–119

Casamayor EO, Massana R, Benlloch S et al (2002) Changes in archaeal, bacterial and eukaryal assemblages along a salinity gradient by comparison of genetic fingerprinting methods in a multipond solar saltern. Environ Microbiol 4:338–348

Demergasso C, Casamayor EO, Chong G et al (2004) Distribution of prokaryotic genetic diversity in athalassohaline lakes of the Atacama Desert, Northern Chile. FEMS Microbiol Ecol 48:57–69. https://doi.org/10.1016/j.femsec.2003.12.013

Demergasso C, Escudero L, Casamayor EO et al (2008) Novelty and spatio-temporal heterogeneity in the bacterial diversity of hypersaline Lake Tebenquiche (Salar de Atacama). Extremophiles 12:491–504. https://doi.org/10.1007/s00792-008-0153-y

Dong H, Zhang G, Jiang H et al (2006) Microbial diversity in sediments of saline Qinghai Lake, China: linking geochemical controls to microbial ecology. Microb Ecol 51:65–82. https://doi.org/10.1007/s00248-005-0228-6

Dorador C (2007) Microbial communities in high altitude altiplanic wetlands in northern Chile: phytogeny, diversity and function. Christian-Albrechts-Universitat, Kiel, Germany

Dorador C, Meneses D, Urtuvia V et al (2009) Diversity of Bacteroidetes in high-altitude saline evaporitic basins in northern Chile. J Geophys Res Biogeosci 114:35–44. https://doi.org/10.1029/2008JG000837

Dupraz C, Reid RP, Braissant O et al (2009) Processes of carbonate precipitation in modern microbial mats. Earth Sci Rev 96:141–162. https://doi.org/10.1016/j.earscirev.2008.10.005

Dupraz C, Visscher PT (2005) Microbial lithification in marine stromatolites and hypersaline mats. Trends Microbiol 13:429–438. https://doi.org/10.1016/j.tim.2005.07.008

Dupraz C, Visscher PT, Baumgartner LK, Reid RP (2004) Microbe-mineral interactions: early carbonate precipitation in a hypersaline lake (Eleuthera Island, Bahamas). Sedimentology 51:745–765. https://doi.org/10.1111/j.1365-3091.2004.00649.x

Espejo RT, Feijoo CJ, Romero J, Vasquez M (1998) PAGE separation of the heteroduplex formed between PCR amplified 16S ribosomal RNA genes allows estimation of phylogenetic relatedness between isolates and assessment of bacterial diversity. Microbiology 144:1611–1617

Farías ME, Contreras M, Rasuk MC et al (2014) Characterization of bacterial diversity associated with microbial mats, gypsum evaporites and carbonate microbialites in thalassic wetlands: Tebenquiche and La Brava, Salar de Atacama, Chile. Extremophiles 18:311–329. https://doi.org/10.1007/s00792-013-0617-6

Fernandez AB, Rasuk MC, Visscher PT et al (2016) Microbial diversity in sediment ecosystems (evaporites domes, microbial Mats, and crusts) of Hypersaline Laguna Tebenquiche, Salar de Atacama, Chile. Front Microbiol 7:1284. https://doi.org/10.3389/fmicb.2016.01284

Handelsman J, Liles M, Mann D et al (2002) Cloning the metagenome: culture-independent access to the diversity and functions of the uncultivated microbial world. Methods Microbiol 33:241–255. https://doi.org/10.1016/S0580-9517(02)33014-9

Handelsman J, Rondon MR, Brady SF et al (1998) Molecular biological access to the chemistry of unknown soil microbes: a new frontier for natural products. Chem Biol 5:R245–R249

Humayoun SB, Bano N, Hollibaugh JT (2003) Depth distribution of microbial diversity in Mono Lake, a meromictic soda lake in California. Appl Environ Microbiol 69:1030–1042

Hutchinson GE (1957) A treatise on limnology. Wiley & Sons, Inc., New York

Jiang H, Dong H, Zhang G et al (2006) Microbial diversity in water and sediment of Lake Chaka, an athalassohaline lake in northwestern China. Appl Environ Microbiol 72:3832–3845. https://doi.org/10.1128/AEM.02869-05

Kharroub K, Lizama C, Aguilera M et al (2008) Halomicrobium katesii sp. nov., an extremely halophilic archaeon. Int J Syst Evol Microbiol 58:2354–2358. https://doi.org/10.1099/ijs.0.65662-0

Kirchman DL (2002) The ecology of Cytophaga-Flavobacteria in aquatic environments. FEMS Microbiol Ecol 39:91–100. https://doi.org/10.1111/j.1574-6941.2002.tb00910.x

Kushner DJ, Kamekura M (1988) Physiology of halophilic eubacteria. In: Rodríguez-Valera F (ed) Halophilic bacteria. CRC Press, Boca Raton, pp 109–138

Lara J, Escudero González L, Ferrero M et al (2012) Enrichment of arsenic transforming and resistant heterotrophic bacteria from sediments of two salt lakes in Northern Chile. Extremophiles 16:523–538. https://doi.org/10.1007/s00792-012-0452-1

Liu Y, Yao T, Jiao N et al (2006) Microbial community structure in moraine lakes and glacial meltwaters, Mount Everest. FEMS Microbiol Lett 265:98–105. https://doi.org/10.1111/j.1574-6968.2006.00477.x

Lizama C, Monteoliva-Sánchez M, Prado B et al (2001) Taxonomic study of extreme halophilic archaea isolated from the "Salar de Atacama", Chile. Syst Appl Microbiol 24:464–474. https://doi.org/10.1078/0723-2020-00053

Lizama C, Monteoliva-Sánchez M, Suárez-García A et al (2002) Halorubrum tebenquichense sp. nov., a novel halophilic archaeon isolated from the Atacama Saltern, Chile. Int J Syst Evol Microbiol 52:149–155

Maldonado J (2015) Biodiversidad de arqueas en ambientes extremos de los andes. In: Sistemas de resistencia al estrés. Universidad Nacional de Tucumán, Tucumán

Nelson WC, Stegen JC (2015) The reduced genomes of Parcubacteria (OD1) contain signatures of a symbiotic lifestyle. Front Microbiol 6:713. https://doi.org/10.3389/fmicb.2015.00713

Nübel U, Engelen B, Felske A et al (1996) Sequence heterogeneities of genes encoding 16S rRNAs in Paenibacillus polymyxa detected by temperature gradient gel electrophoresis. J Bacteriol 178:5636–5643

Ordoñez OF, Rasuk MC, Soria MN et al (2018) Haloarchaea from the Andean Puna: biological role in the energy metabolism of arsenic. Microb Ecol 76:695–705. https://doi.org/10.1007/s00248-018-1159-3

Prado B, del Moral A, Campos V (1993) Distribution and types of Heterotrophyc Halophilic Flora from Salar de Atacama, Chile. Toxicol Environ Chem 38:163–166. https://doi.org/10.1080/02772249309357887

Prado B, Del Moral A, Quesada E et al (1991) Numerical taxonomy of moderately halophilic Gram-negative rods isolated from the Salar de Atacama, Chile. Syst Appl Microbiol 14:275–281. https://doi.org/10.1016/S0723-2020(11)80381-4

Prado B, Lizama C, Aguilera M et al (2006) Chromohalobacter nigrandesensis sp. nov., a moderately halophilic, Gram-negative bacterium isolated from Lake Tebenquiche on the Atacama Saltern, Chile. Int J Syst Evol Microbiol 56:647–651. https://doi.org/10.1099/ijs.0.63983-0

Rappé MS, Giovannoni SJ (2003) The uncultured microbial majority. Annu Rev Microbiol 57:369–394. https://doi.org/10.1146/annurev.micro.57.030502.090759

Rascovan N, Maldonado J, Vazquez MP, Eugenia Farías M (2016) Metagenomic study of red biofilms from Diamante Lake reveals ancient arsenic bioenergetics in haloarchaea. ISME J 10:299–309. https://doi.org/10.1038/ismej.2015.109
Riesenfeld CS, Schloss PD, Handelsman J (2004) Metagenomics: genomic analysis of microbial communities. Annu Rev Genet 38:525–552. https://doi.org/10.1146/annurev.genet.38.072902.091216
Risacher F, Alonso H, Salazar C (2003) The origin of brines and salts in Chilean salars: a hydrochemical review. Earth Sci Rev 63:249–293. https://doi.org/10.1016/S0012-8252(03)00037-0
Rodríguez-Valera F (1988) Characteristics and microbial ecology of hypersaline environments. In: Rodríguez-Valera F (ed) Halophilic bacteria. CRC Press, Boca Raton, pp 3–30
Rothschild LJ, Giver LJ, White MR, Mancinelli RL (1994) Metabolic activity of microorganisms in evaporites. J Phycol 30:431–438
Rothschild LJ, Mancinelli RL (2001) Life in extreme environments. Nature 409:1092–1101. https://doi.org/10.1038/35059215
Sørensen KB, Canfield DE, Teske AP, Oren A (2005) Community composition of a hypersaline endoevaporitic microbial mat. Appl Environ Microbiol 71:7352–7365. https://doi.org/10.1128/AEM.71.11.7352-7365.2005
Stewart EJ (2012) Growing unculturable bacteria. J Bacteriol 194:4151–4160. https://doi.org/10.1128/JB.00345-12
Thiel V, Tank M, Neulinger SC et al (2010) Unique communities of anoxygenic phototrophic bacteria in saline lakes of Salar de Atacama (Chile): evidence for a new phylogenetic lineage of phototrophic Gammaproteobacteria from pufLM gene analyses. FEMS Microbiol Ecol 74:510–522. https://doi.org/10.1111/j.1574-6941.2010.00966.x
Valderrama MJ, Prado B, del Moral A et al (1991) Numerical taxonomy of moderately halophilic gram-positive cocci isolated from the Salar de Atacama (Chile). Microbiologia 7:35–41
Van Gemerden H (1993) Microbial mats: a joint venture. Mar Geol 113:3–25
Visscher PT, Beukema J, Van Gemerden H (1991) In situ characterization of sediments: measurements of oxygen and sulfide profiles with a novel combined needle electrode. Limnol Oceanogr 36:1476–1480. https://doi.org/10.4319/lo.1991.36.7.1476
Visscher PT, Prins RA, Gemerden H (1992) Rates of sulfate reduction and thiosulfate consumption in a marine microbial mat. FEMS Microbiol Lett 86:283–294. https://doi.org/10.1111/j.1574-6968.1992.tb04820.x
Visscher PT, Reid RP, Bebout BM et al (1998) Formation of lithified micritic laminae in modern marine stromatolites (Bahamas); the role of sulfur cycling. Am Mineral 83:1482–1493
Visscher PT, van Gemerden H (1991) Production and consumption of dimethylsulfoniopropionate in marine microbial mats. Appl Environ Microbiol 57:3237–3242
von Wintzingerode F, Göbel UB, Stackebrandt E (1997) Determination of microbial diversity in environmental samples: pitfalls of PCR-based rRNA analysis. FEMS Microbiol Rev 21:213–229
Wu QL, Zwart G, Schauer M et al (2006) Bacterioplankton community composition along a salinity gradient of sixteen high-mountain lakes located on the Tibetan Plateau, China. Appl Environ Microbiol 72:5478–5485. https://doi.org/10.1128/AEM.00767-06
Zhang G, Niu F, Ma X et al (2007) Phylogenetic diversity of bacteria isolates from the Qinghai-Tibet Plateau permafrost region. Can J Microbiol 53:1000–1010. https://doi.org/10.1139/W07-031
Zúñiga LR, Campos V, Pinochet H, Prado B (1991) A limnological reconnaissance of Lake Tebenquiche, Salar de Atacama, Chile. Hydrobiologia 210:19–24. https://doi.org/10.1007/BF00014320

Chapter 11
Complete Characterization of Stratified Ecosystems of the Salar de Llamara (Atacama Desert)

María Cecilia Rasuk, Manuel Contreras Leiva, Daniel Kurth, and María Eugenia Farías

11.1 Background

Based on the aridity index of 0.05, the Atacama Desert is considered a hyperarid area with extremely dry conditions (Wierzchos et al. 2012). Because of its aridity and UV incidence, it is largely compared to Mars, and it is also at the dry limit of microbial life (Navarro-Gonzalez et al. 2003). In addition, it is the Earth's largest modern evaporitic regions, which comprises a large number of salt flats, with a combination of evaporitic crusts and saline lakes or playa-lakes (locally referred to as "lagunas") (Stoertz and Ericksen 1974; Risacher et al. 2003). The Salar de Llamara is located in the north of the Atacama Desert (Fig. 11.1a). This is one of the aridest parts of the desert span toward northern Chile, and it is situated between the rain shadows of the Andes Mountains and the Coastal Cordillera (Michalski et al. 2004). Llamara 1 (L1) and Llamara 2 (L2) are two shallow wetlands at an approximately 500 m distance, containing two kinds of complex and diverse stratified microbial communities, microbial mats (Demergasso et al. 2003; Rasuk et al. 2016), and evaporitic domes (Rasuk et al. 2014, ongoing publication) (Fig. 11.1).

The occurrence of microbial life associated with these desert environments opens up new perspectives regarding how communities adapt to and tolerate extreme environmental conditions and increases our understanding of microbial ecology and evolution. Thus, the present book chapter comprises all the available information about the stratified microbial ecosystems located in the Salar de Llamara.

M. C. Rasuk · D. Kurth
Laboratorio de Investigaciones Microbiológicas de Lagunas Andinas (LIMLA), Planta Piloto de Procesos Industriales Microbiológicos (PROIMI), CCT, CONICET, Tucumán, Argentina

M. C. Leiva
Centro de Ecología Aplicada (CEA), Ñuñoa, Santiago, Chile

M. E. Farías (✉)
LIMLA-PROIMI-CONICET, Tucuman, Argentina

M. E. Farías (ed.), *Microbial Ecosystems in Central Andes Extreme Environments*, https://doi.org/10.1007/978-3-030-36192-1_11

Fig. 11.1 (**a**) Red square shows the location of the Salar de Llamara. (**b**) Location of the two water bodies L1 and L2 (marked with red), and the holes with mats (marked with yellow). (**c**) Photograph showing L1 lagoon. (**d**) Photograph showing the domes inside L2 lagoon. (**e**) Photograph of domes inside L1. (**f**) Photograph of the holes with microbial mats

11.2 Physicochemical Parameters in the Water from the Salar the Llamara

Physicochemical parameters in the Salar the Llamara were determined in different field campaigns (Table 11.1) (Rasuk et al. 2014, 2016, ongoing publication). Overall, similar values were obtained, and these measurements showed that the water is characterized by being moderately alkaline, having pH around 8, and a temperature between 25–30 °C. The conductivity, which is given by the charged ions dissolved in the water, is high, making these lakes hypersaline. In this sense, concentrations of the major ions were elevated in the three samples, presenting in the following order, chloride > sodium > sulfate > potassium > magnesium > calcium, except in the holes where calcium > magnesium. It is also important to highlight that nutrients including total organic nitrogen, organic matter, and phosphorus concentrations have low values in all cases; although, in L2 total phosphorous and orthophosphate had the highest values, 13.8 and 12.6 mg/L, respectively. Additionally, dissolved oxygen was markedly lower in L2 (0.7 mg/L) and in the hole waters (LL1 0.7 and LL2 1.8 mg/L) compared with L1 where the values reached 7.9 mg/L.

High arsenic content was detected in these water samples, reaching 24.6 mg/L in L2, as seems to be a feature of the Andean lakes. In fact, the major quantities of arsenic around the world were registered in those lakes (Farías et al. 2013; Rasuk et al. 2014, 2016; Rascovan et al. 2016; Ordoñez et al. 2018).

Table 11.1 Physicochemical parameters from water in L1 (summer and winter), L2, and the two holes sampled

Parameter	Unit	Llamara 1		Llamara 2	Hole LL1	Hole LL2
		Summer	Winter	Summer		
Conductivity	mS/cm	177	267	191	152	143
Dissolved oxygen	mg/L	3.45	7.9	0.7	0.7	1.8
pH	–	7.88	8	8	7.6	7.6
Salinity	g/L	–	76	70	–	–
Temperature	°C	18.6	20.5	28.6	36.5	27.4
Chlorophyll	μg/L	1.5	0.4	1.5	–	–
Total alkalinity	mgCaCO$_3$/L	263	200	9010	159	191
Hardness	mg/L	6701	5143	15,967	4213	3855
Total phosphorus	mg/L	4.4	1.9	13.8	4.2	0.8
Orthophosphate (P-PO$_4$)	mg/L	2.6	1.8	12.6	0.4	0.8
Dissolved silica (SiO$_2$)	mg/L	103	49	85	98	94
Nitrate (N-NO$_3$)	mg/L	<0.05	0.7	0.04	0.1	0.1
Nitrite (N-NO$_2$)	mg/L	<0.1	<0.1	<0.1	<0.1	<0.1
Total organic nitrogen	mg/L	0.4	0.3	0.6	0.3	0.9
Organic matter	%	4.2	3.9	3.2	3.3	2.1
Calcium	mg/L	796	1149	657	1088	1054
Sodium	mg/L	58,005	33,540	53,690	20,990	17,330
Magnesium	mg/L	1145	552	3480	364	297
Potassium	mg/L	2021	926	5138	668	552
Chloride	mg/L	86,693	41,738	85,888	31,586	25,467
Dissolved sulfur	mg/L	6917	–	7144	5119	4033
Sulfate	mg/L	17,605	17,368	18,395	13,540	9780
Total sulfide	mg/L	<0.2	<0.2	<0.2	–	–
Dissolved arsenic	mg/L	5.0	4.7	–	–	–
Total arsenic	mg/L	9.1	6	24.6	2	1.2
Total iron	mg/L	0.468	0.19	0.18	–	–
Total boron	mg/L	85	52	163	–	–

11.3 Unravelling the Microbial World Harboring Stratified Ecosystems in the Salar de Llamara

Two kinds of stratified microbial ecosystems were reported in the Salar de Llamara, microbial mats and evaporites.

11.3.1 Microbial Mats

Microbial mats are laminated structures, which are controlled by environmental factors such as light, temperature, salinity, dissolved oxygen, and the presence of sulfides. These laminations have different colorations as a result of the development of photosynthetic bacteria that contain photosynthetic pigments and, therefore, different patterns of available light. The presence of these mats is restricted to extreme environments that include coastal marine environments and hypersaline environments (Bauld 1984; Van Gemerden 1993), thermal springs (Castenholz 1984; Jorgensen and Nelson 1988), and alkaline lakes (Brock 1978). Microbial mats are considered the modern counterparts of the ancient laminated microbialites, and studying them gives interesting insights about ancient microbial life.

Up to now, four studies about the mats of the laguna Llamara were conducted:

Demergasso et al. (2003) analyzed the bacterial community of microbial mats at different sampling sites of the Salar de Llamara through microscopic and spectrophotometric techniques. They reported four heterogeneous samples along the sampling site, where the lamination of mats was different during winter and summer. Three kinds of stratification were detected. The first one was characterized by only one green layer, the second had a green and an orange layer, and the last had an additional layer with purple color. The orange layer was characterized by diatoms, the green by cyanobacteria, and the purple by anoxygenic phototrophic bacteria similar to cells of the genera *Chromatium* and *Thiocapsa*. Additionally, abundant non-photosynthetic microorganisms were detected in the mats, including unidentified cocci and bacilli. It is also important to highlight that sulfate-reducing bacteria were also present in all the sampled mats and the black layer above them was attributed to the oxidation of the sulfide generated by their metabolism.

This lamination was more developed in winter than summer, as evidenced by increased pigment and protein content, and attributed to the higher water level in winter. In addition, the pigments from the mat samples were spectrophotometrically analyzed, revealing that chlorophyll *a* and bacteriochlorophyll *a*, were the most abundant pigments. Chlorophyll *a* was predominant in the green layer, whereas bacteriochlorophyll *a* in the purple layer.

High-throughput sequencing methods enable detailed, semiquantitative analysis of entire communities in large sample sets. In addition, they provide ecological information that extends far beyond that provided by previous methods in terms of detail and magnitude. In this sense, Rasuk et al. (2016) analyzed the microbial diversity of the microbial mats using high-throughput sequencing technology and correlated it with the physicochemical parameters from water columns. They compared the samples from Llamara with sediment from other lagoons from the Atacama Desert. In this work, two kinds of mat were recognized, which were situated within holes next to L1 (Fig. 11.1b, f). One of the holes had a white suspension named LL1. In a second hole, a mat with a different appearance is noticed, with purple spots on an ocher background, this was called LL2.

In the two samples analyzed, *Proteobacteria*, *Bacteroidetes*, *Spirochaetes*, and sequences that could not be assigned to any taxa were the most prevalent phyla. *Proteobacteria* and *Spirochaetes* were the most abundant in LL1, while *Bacteroidetes* and *Caldithrix*, unique to LL2, were predominant in this sample. The *Rhodothermaceae* family was the major member of the *Bacteroidetes* found in both samples. It was widely detected in microbial ecosystems of Atacama Desert and other regions of the Andean Puna (Demergasso et al. 2004; Dorador et al. 2009,

2013; Farías et al. 2013; Rasuk et al. 2014; Fernandez et al. 2016; Toneatti et al. 2017; Kurth et al. 2017). Members of this family such as the genus *Salinibacter* sp. are characterized as requiring high salt concentrations for their growth (Anton et al. 2002; Makhdoumi-Kakhki et al. 2012). Finally, the small proportion of sequences affiliated with *Cyanobacteria* not only in these mat samples but also in similar geographically different samples indicate that they might not play an essential role related to carbon fixation in these extreme environments.

When they compared the diversity coupled with the physicochemical parameters between the mats and sediment samples, a higher diversity in sediment than in mat was found. They attributed this feature to the water conditions such as higher DO and more organic matter and phosphorous in the water where sediments were taken. Also, a lower conductivity was found in water with sediment samples. Finally, in the mat samples, they found to a more specific level, a community compounded by extremophiles, especially halophilic organisms, showing how environmental conditions influence the microbial composition of an ecosystem.

A more recent article (Saghaï et al. 2017) studied microbial mat fragments collected along 30 cm of a pond from the Salar de Llamara. Those fragments were obtained along strong physicochemical gradients (depth, salinity, oxygen, and temperature), and the structure of archaea, bacteria, and protist communities applying 16S/18S rRNA metabarcoding approaches was also characterized. They found that the mats were highly diverse, with a number of OTUs comparable to the most diverse environment types (soil or sediments), they included known eukaryotic and prokaryotic taxa as well as many novel lineages. Bacterial candidate divisions were more abundant in deeper layers (almost 50% of sequences), and Archaea represented up to 40% of sequences in some mat layers. Mats situated in the oxic zone were mostly composed of relatively well characterized bacterial phyla, including *Bacteroidetes*, *Cyanobacteria*, *Proteobacteria*, and *Verrucomicrobia*. The fact that these researchers found, in general, the same major phyla as the works mentioned above should be highlighted. Below the chemocline defining an oxic/anoxic and salinity transition, bacterial candidate divisions and archaea accounted for up to 75% of the sequences. Molecular phylogenetic analyses revealed six novel deeply divergent archaeal groups.

Cyanobacteria and *Alphaproteobacteria*, which are potential phototrophic groups were represented by 5–10% of the sequences and located in surface layers. On the other hand, potential anoxygenic phototrophs, mainly *Rhodobacterales* and *Rhodospirillales*, were detected in significant proportions in the middle zone.

Candidate divisions were particularly abundant in the two deeper mats, where the majority was *Aminicenantes* (former OP8), *Parcubacteria* (former OD1), and TA06. Also, a few OTUs from Candidate divisions were abundant and affiliated with *Latescibacteria* (former WS3), *Saccharibacyeria* (TM7), and *Gracilibacteria* (BD1-5). They also detected halophilic and halotolerant microbial eukaryotes, found among a variety of taxa, including especially *Stramenopiles* (e.g., *Bacillariophyta*, *Bicosoecida*, and *Chrysophyceae*), *Alveolata* (e.g., *Ciliophora*), and Fungi, being more abundant and diverse in the oxic zone mats.

Although located a few centimeters away, each studied mat fragment developed under a set of specific abiotic factors, with oxygen values dropping to zero and salinity and temperature values increased from the surface to the bottom of the pond. All these differences in environmental parameters correlated with the different structure of the prokaryotic communities.

The same mats were analyzed by WGS sequencing in the latest work (Gutiérrez-Preciado et al. 2018). This allowed further insight into the biogeochemical processes potentially taking place in the mats. The authors proposed a space-for-time substitution modeling for these mats, thus simulating the transition from early Earth conditions in the deeper mats to current oxygenic conditions in the upper mats.

Inferences from the taxonomic composition were confirmed by gene abundances. Anoxygenic photosynthesis, represented by *Chloroflexi*, *Chlorobi*, and *Alpha/Gammaproteobacteria* lineages, was more abundant than oxygenic photosynthesis, which could be related to *Cyanobacteria* and eukaryotic algae. Surprisingly, this occurred both in oxic and anoxic layers. These organisms are likely in charge of primary production in these mats. Photosynthesis genes decreased with depth, and in the lower mats, which bear increased diversity and rare phyla, primary production might be supplemented by other autotrophic metabolisms.

Carbon fixation pathways showed an inversion from the deepest, anoxic mats up to the upper oxygen-exposed mats. Ancient anoxygenic pathways, such as dicarboxylate/hydrobutyrate, 3-hydroxiproprionate/4-hydrobutyrate and, most importantly, Wood–Ljungdahl increased their abundance with depth, while the Calvin–Benson cycle markers had reduced abundance in the deeper layers.

11.3.2 Evaporitic Domes

Evaporites are widespread in arid locations in the world. Due to the inhospitable conditions in the desert (aridity, UV radiation, brine chemistry, and oligotrophic conditions), microbial life has been induced to search out the microhabitats most suitable for life (Albarracín et al. 2015) such as gypsum or halite pores.

Microorganisms form aggregates occupying the pore spaces inside halite, where microbial interactions occur. In this exceptional, salty, porous habitat, microbial consortia with a community structure probably acclimated to the environmental conditions occupy special microhabitats with physical and chemical properties that promote their survival (de Los Ríos et al. 2010). On the other hand, inside gypsum pores UV radiation is quenched by the selenite crystal, and it is highly hygroscopic, creating a wet UV-protected microenvironment with high access to O_2 and light (Oren et al. 1995; Stivaletta et al. 2010; Farías et al. 2014). These features create a stratified system very similar to that found in microbialites and microbial mats. Thus, in hyperarid deserts, microbial life is essentially present in the form of microorganisms that take refuge in such endolithic habitats (Wierzchos et al. 2012). This kind of lifestyle is represented in the Salar de Llamara with microbial communities stratified into evaporitic structures having domal shapes.

de Los Ríos et al. (2010) characterized microbial communities inside halite evaporites from different parts of the Atacama Desert (Yungay, Salar de Llamara, and

Salar Grande) using denaturing gradient gel electrophoresis (DGGE) and microscopy. Their analysis revealed that the endolithic communities harboring evaporitic halite rocks are made up predominantly of cyanobacteria, along with heterotrophic bacteria (uncultured Bacteroidetes bacterium clone from sediments of hypersaline lakes) and archaea (uncultured unidentified archaea).

11.4 Diversity of the Evaporitic Gypsum Domes Along of the Stations in Llamara 1

In Salar de Llamara, there are two shallow wetlands dominated by microbial ecosystems formed by dome-shaped bioherms, which presented a stratified distribution of microbial communities in color sections. Some of these domal structures are partially submerged defining an air-exposed surface and the embedded section, creating (at least) two microbial niches.

Rasuk et al. (2014) performed, for the first time, a detailed description of the microbial diversity of a dome from L1 along the stations using enhanced techniques of DNA sequencing and supported this information with the analysis of the pigments using high-performance liquid chromatography (HPLC). They also utilized SEM to view the dome sample and determined its mineralogical nature with X-ray diffraction.

The main mineral found in these domes was gypsum. Regarding biodiversity, the communities associated with these structures were analyzed by amplicon sequencing and compared between winter and summer seasons. In general, sequences related to *Bacteroidetes* and *Proteobacteria* (mainly *Alfa* and *Gammaproteobacteria*) remained as the main phylogenetic groups, and the diversity duplicated in winter (determined using Chao index, 502 in winter vs 275 in summer). The comparison of the upper (air-exposed) and bottom section (water-submerged) between the seasons showed slight variation. The upper region was dominated by *Chromatiales* (*Gammaproteobacteria*), *Rhodospirillales* (*Alphaproteobacteria*), both of them characterized by having anoxygenic photosynthesizers. Also, *Sphingobacteriales* (*Bacteroidetes*), which have halophile members such as *Salinibacter* sp., were significant. They have been found in microbial communities harboring salt waters and were also found in ecosystems from the Atacama Desert (Baati et al. 2008; Dorador et al. 2009, 2013; Schneider et al. 2013; Simachew et al. 2016; Toneatti et al. 2017). However, the submerged part showed marked differences between seasons, being dominated in summer by *Alpha* and *Gammaproteobacteria* and a good representation of *Verrucomicrobia*, but winter showed a remarkable difference being more diverse (evidenced by Chao index 435 vs 282 in summer). Even though scarce by sequence, *Cyanobacteria* were visually identified by SEM, which also revealed the presence of diatoms. Photosynthetic pigments related to anoxygenic bacteria, bacteriochlorophyll *e*, and *c*, together with a good representation of sequences related to *Alfa* and *Gammaproteobacteria* and a very low proportion of *Cyanobacteria* were found. These findings suggest that a phylognetical group other than *Cyanobacteria* seems to be significantly involved in carbon uptake in the Llamara domes. This was also found in the other Andean microbialite ecosystems such as Brava and Tebenquiche lakes (Fernandez et al. 2016; Farias et al. 2017). Moreover, widespread pigments such as β-carotene, lycopene, and fucoxanthin, as well as pigments related to specific groups of microorganism such as diatoxanthin (associated with diatoms) and astaxanthin (associated with *Cyanobacteria*) were also detected. All of them were consistent with the diversity obtained by sequencing. They proposed that the higher abundance of pigments and diversity metrics in the top layer compared to the bottom is because the upper section of the dome presents advantages over the bottom. UV would be quenched by the selenite crystals in gypsum crust and exposure in microbial communities would be minimized. In addition, gypsum is highly hygroscopic, creating a wet, UV protected microenvironment with high access to O_2 and light (Oren et al. 1995). In contrast, in the bottom layer, since conditions during summer and winter are less stable, conductivity and O_2 availability changes are reflected in the differences in diversity.

Although conductivity, salinity, and phosphorous were higher in summer, whereas dissolved oxygen was higher in winter (Table 11.1), it was found that there were not considerable changes in the temperature, depth, and pH of the water.

11.5 Characterization of the Evaporitic Gypsum Domes in Llamara 2

A detailed characterization of an evaporitic dome and the physicochemical parameters of the water from L2 were performed by Rasuk et al. (ongoing publication). Domes were similar to those found in Llamara 1 and also composed of gypsum. The study of the microbial diversity aimed to analyze the different colored horizons, and this was addressed using a combination of techniques, including amplicon sequencing of the 16S rRNA metagenomics analysis, pigment determination, and electron microscopy.

The major proportion of sequences belonged to the Bacteria domain, and just a few sequences were assigned to Archaea and Eukarya domains. Regarding phylum level, the results achieved by both approaches were consistent. In this sense, *Proteobacteria*, *Bacteroidetes*, and *Firmicutes* were the prevalent phyla, and rare phyla represented by unclassified bacteria and candidate phyla were also significant. On the other hand, a small amount of cyanobacterial sequences were found. The distribution of the community along the layers showed a pronounced diversity in the upper section (Fig. 11.2a), *Proteobacteria* (*Alpha* and *Deltaproteobacteria*) being the major component in it; in agreement with the results obtained by Rasuk et al. (2014). This also correlated with pigment analysis, where diversity was maximal in the upper photosynthetic layers. As the major member of *Alphaproteobacteria*, the family *Rhodobacteraceae* was detected (18%) in the middle section, followed by *Rhodospirillaceae* (11%), which was distributed in the three sections of the dome and more abundant in the middle. Another interesting finding was that the functional analysis demonstrated that anoxygenic photosynthesis was more represented

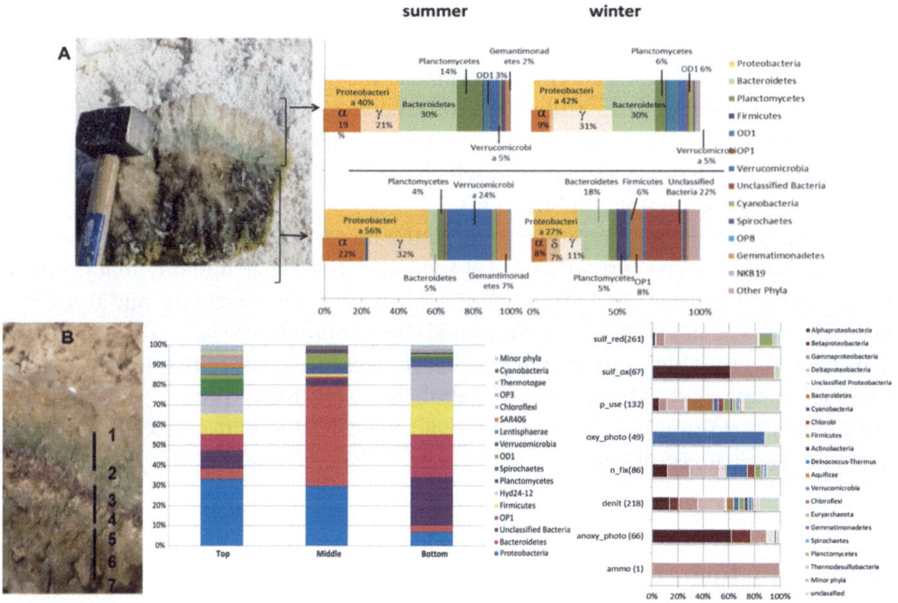

Fig. 11.2 **(a)** (*Left*) Bacterial diversity represented by percentages of sequences of phylum on the top, middle, and bottom layers in the gypsum dome. (*Right*) Bars showing the number of reads by the analyzed functional genes and their affiliation. (Adapted from Rasuk et al. ongoing publication)

than the oxygenic and mainly by *Alphaproteobacteria* (*Rhodobacteraceae* and *Rhodospirillaceae*). Sulfur oxidation genes were affiliated with the same taxa (Fig. 11.2b), suggesting that sulfur oxidation could be coupled with anoxygenic photosynthesis. *Rhodobacteraceae* and *Rhodospirillaceae* are purple non-sulfur bacteria (PNS) that generally use electrons from molecular hydrogen for photosynthesis, and they were thought not to use hydrogen sulfide as an electron donor while growing photoautotrophically. However, some PNS have been found using sulfide or organic compounds at lower concentrations than purple sulfur bacteria (Madigan et al. 2003).

An unexpected result found in this work was that sulfate-reduction appears to be mainly done by *Deltaproteobacteria*, even in the upper layers of the dome, since a high abundance not only of 16S rRNA genes was found but also the majority of genes related to this metabolism were affiliated with them. This result has also been found in the oxic zones of hypersaline microbial mats in Guerrero Negro, Solar Lake, Kiritimati Atoll, the Bahamas, and Shark Bay and in microbial ecosystems in the Salar de Atacama (Canfield and Des Marais 1991; Teske et al. 1998; Dupraz et al. 2004; Glunk et al. 2011; Arp et al. 2012; Pages et al. 2014; Farias et al. 2017).

On the other hand, the bottom layers were characterized by rare phyla, represented by sequences that could not be affiliated with any phyla (unclassified bacteria) and candidate phyla such as Hyd24-12, OP1, OP3, OD1, and SAR406. However, *Firmicutes*, *Proteobacteria*, and *Lentisphaerae* were also present.

Moreover, a large amount of genes related to alternative pathways of carbon fixation, such as the ancient reductive acetyl CoA, were revealed. Since the carbon monoxide dehydrogenase/acetyl-CoA synthase (key enzyme of the pathway) was even more represented than RubisCO from the classic Calvin–Benson cycle, the identified reductive acetyl-CoA pathway sequences were affiliated mainly with the sulfate-reducing *Deltaproteobacteria*, which are obligate anaerobic hydrogen oxidizers (Rasuk et al. ongoing publication).

This study gave a good representation of the taxonomic diversity and metabolic potentials thought to be currently used at Llamara. It might be unlikely that primary production is supported solely by *Cyanobacteria*, given their scarcity, and from this analysis anoxygenic photosynthesizers and *Deltaproteobacteria* could be significant contributors. This supports the hypothesis presented before that in these environments many of the dominating metabolisms were likely present during early earth, when conditions of the primitive land were hostile and scarce in nutrients (especially organic compounds).

11.6 Conclusion

This chapter presented the available studies about biodiversity in one of the driest and most irradiated environments on Earth, the Salar de Llamara. This oligotrophic system has an unexpected diversity, where oxygenic photosynthesis is not the main primary metabolism. Here, other dynamics are at work. An elevated number of genes associated with anoxygenic photosynthesis coupled with a large number of genes related to alternative carbon fixation pathways were observed. These results support the hypothesis that these poly-extreme environments would be a good model for early Earth conditions and open a universe of possibilities to elucidate these ancient metabolic pathways that provide critical information about early life in extreme environments.

References

Albarracín VH, Kurth D, Ordoñez OF et al (2015) High-up: a remote reservoir of microbial extremophiles in central Andean wetlands. Front Microbiol 6:1404. https://doi.org/10.3389/fmicb.2015.01404

Anton J, Oren A, Benlloch S et al (2002) Salinibacter ruber gen. nov., sp. nov., a novel, extremely halophilic member of the Bacteria from saltern crystallizer ponds. Int J Syst Evol Microbiol 52:485–491

Arp G, Helms G, Karlinska K et al (2012) Photosynthesis versus exopolymer degradation in the formation of microbialites on the atoll of Kiritimati, Republic of Kiribati, Central Pacific. Geomicrobiol J 29:29–65. https://doi.org/10.1080/01490451.2010.521436

Baati H, Guermazi S, Amdouni R et al (2008) Prokaryotic diversity of a Tunisian multipond solar saltern. Extremophiles 12:505–518. https://doi.org/10.1007/s00792-008-0154-x

Bauld J (1984) Microbial mats in marginal marine environments: Shark Bay, Western Australia, and Spencer Gulf, South Australia. In: Cohen Y, Castenholz R, Halvorson H (eds) Microbial mats: stromatolites. Alan R. Liss, New York, p 498

Brock TD (1978) Thermophilic microorganisms and life at high temperatures. Springer Berlin Heidelberg, New York

Canfield D, Des Marais D (1991) Aerobic sulfate reduction in microbial mats. Science 251:1471–1473. https://doi.org/10.1126/science.11538266

Castenholz R (1984) Composition of hot springmicrobial mats: a summary. In: Cohen Y, Castenholz R, Halvorson H (eds) Microbial mats: stromatolites. Alan R. Liss, New York, pp 107–119

de Los Ríos A, Valea S, Ascaso C et al (2010) Comparative analysis of the microbial communities inhabiting halite evaporites of the Atacama Desert. Int Microbiol 13:79–89

Demergasso C, Casamayor EO, Chong G et al (2004) Distribution of prokaryotic genetic diversity in athalassohaline lakes of the Atacama Desert, Northern Chile. FEMS Microbiol Ecol 48:57–69. https://doi.org/10.1016/j.femsec.2003.12.013

Demergasso C, Chong G, Galleguillos P et al (2003) Tapetes microbianos del Salar de Llamará, norte de Chile. Rev Chil Hist Nat 76:485–499. https://doi.org/10.4067/S0716-078X2003000300012

Dorador C, Meneses D, Urtuvia V et al (2009) Diversity of Bacteroidetes in high-altitude saline evaporitic basins in northern Chile. J Geophys Res 114:G00D05. https://doi.org/10.1029/2008JG000837

Dorador C, Vila I, Witzel K-P, Imhoff JF (2013) Bacterial and archaeal diversity in high altitude wetlands of the Chilean Altiplano. Fundam Appl Limnol/Arch für Hydrobiol 182:135–159. https://doi.org/10.1127/1863-9135/2013/0393

Dupraz C, Visscher PT, Baumgartner LK, Reid RP (2004) Microbe-mineral interactions: early carbonate precipitation in a hypersaline lake (Eleuthera Island, Bahamas). Sedimentology 51:745–765. https://doi.org/10.1111/j.1365-3091.2004.00649.x

Farías ME, Contreras M, Rasuk MC et al (2014) Characterization of bacterial diversity associated with microbial mats, gypsum evaporites and carbonate microbialites in thalassic wetlands: Tebenquiche and La Brava, Salar de Atacama, Chile. Extremophiles 18:311–329. https://doi.org/10.1007/s00792-013-0617-6

Farías ME, Rascovan N, Toneatti DM et al (2013) The discovery of stromatolites developing at 3570 m above sea level in a high-altitude volcanic lake Socompa, Argentinean Andes. PLoS One 8:e53497. https://doi.org/10.1371/journal.pone.0053497

Farias ME, Rasuk MC, Gallagher KL et al (2017) Prokaryotic diversity and biogeochemical characteristics of benthic microbial ecosystems at La Brava, a hypersaline lake at Salar de Atacama, Chile. PLoS One 12:e0186867. https://doi.org/10.1371/journal.pone.0186867

Fernandez AB, Rasuk MC, Visscher PT et al (2016) Microbial diversity in sediment ecosystems (evaporites domes, microbial mats, and crusts) of Hypersaline Laguna Tebenquiche, Salar de Atacama, Chile. Front Microbiol 7:1284. https://doi.org/10.3389/fmicb.2016.01284

Glunk C, Dupraz C, Braissant O et al (2011) Microbially mediated carbonate precipitation in a hypersaline lake, Big Pond (Eleuthera, Bahamas). Sedimentology 58:720–736. https://doi.org/10.1111/j.1365-3091.2010.01180.x

Gutiérrez-Preciado A, Saghaï A, Moreira D et al (2018) Functional shifts in microbial mats recapitulate early Earth metabolic transitions. Nat Ecol Evol. 2(11):1700–1708. https://doi.org/10.1038/s41559-018-0683-3

Jorgensen BB, Nelson DC (1988) Bacterial zonation, photosynthesis, and spectral light distribution in hot spring microbial mats of Iceland. Microb Ecol 16:133–147. https://doi.org/10.1007/BF02018909

Kurth D, Amadio A, Ordoñez OF et al (2017) Arsenic metabolism in high altitude modern stromatolites revealed by metagenomic analysis. Sci Rep 7:1024. https://doi.org/10.1038/s41598-017-00896-0

Madigan MT, Martinko JM, Parker J (2003) Brock biology of microorganisms. Prentice Hall/Pearson Education, Upper Saddle River, NJ

Makhdoumi-Kakhki A, Amoozegar MA, Ventosa A (2012) Salinibacter iranicus sp. nov. and Salinibacter luteus sp. nov., isolated from a salt lake, and emended descriptions of the genus Salinibacter and of Salinibacter ruber. Int J Syst Evol Microbiol 62:1521–1527. https://doi.org/10.1099/ijs.0.031971-0

Michalski G, Böhlke JK, Thiemens M (2004) Long term atmospheric deposition as the source of nitrate and other salts in the Atacama Desert, Chile: new evidence from mass-independent oxygen isotopic compositions. Geochim Cosmochim Acta 68:4023–4038. https://doi.org/10.1016/J.GCA.2004.04.009

Navarro-Gonzalez R, Rainey FA, Molina P et al (2003) Mars-like soils in the Atacama Desert, Chile, and the dry limit of microbial life. Science 302:1018–1021. https://doi.org/10.1126/science.1089143

Ordoñez OF, Rasuk MC, Soria MN et al (2018) Haloarchaea from the Andean Puna: biological role in the energy metabolism of arsenic. Microb Ecol 76:1–11. https://doi.org/10.1007/s00248-018-1159-3

Oren a, Kuhl M, Karsten U (1995) An endoevaporitic microbial mat within a gypsum crust: zonation of phototrophs, photopigments, and light penetration. Mar Ecol Prog Ser 128:151–159. https://doi.org/10.3354/meps128151

Pages A, Welsh DT, Teasdale PR et al (2014) Diel fluctuations in solute distributions and biogeochemical cycling in a hypersaline microbial mat from Shark Bay, WA. Mar Chem 167:102–112. https://doi.org/10.1016/j.marchem.2014.05.003

Rascovan N, Maldonado J, Vazquez MP, Eugenia Farías M (2016) Metagenomic study of red biofilms from Diamante Lake reveals ancient arsenic bioenergetics in haloarchaea. ISME J 10(2):299–309. https://doi.org/10.1038/ismej.2015.109

Rasuk MC, Fernández AB, Kurth D et al (2016) Bacterial diversity in microbial mats and sediments from the Atacama Desert. Microb Ecol 71:44–56. https://doi.org/10.1007/s00248-015-0649-9

Rasuk MC, Kurth D, Flores MR et al (2014) Microbial characterization of microbial ecosystems associated to evaporites domes of gypsum in Salar de Llamara in Atacama desert. Microb Ecol 68:483–494. https://doi.org/10.1007/s00248-014-0431-4

Risacher F, Alonso H, Salazar C (2003) The origin of brines and salts in Chilean salars: a hydrochemical review. Earth Sci Rev 63:249–293. https://doi.org/10.1016/S0012-8252(03)00037-0

Saghaï A, Gutiérrez-Preciado A, Deschamps P et al (2017) Unveiling microbial interactions in stratified mat communities from a warm saline shallow pond. Environ Microbiol 19:2405–2421. https://doi.org/10.1111/1462-2920.13754

Schneider D, Arp G, Reimer A et al (2013) Phylogenetic analysis of a microbialite-forming microbial mat from a hypersaline lake of the Kiritimati atoll, Central Pacific. PLoS One 8:e66662. https://doi.org/10.1371/journal.pone.0066662

Simachew A, Lanzén A, Gessesse A, Øvreås L (2016) Prokaryotic community diversity along an increasing salt gradient in a soda ash concentration pond. Microb Ecol 71:326. https://doi.org/10.1007/s00248-015-0675-7

Stivaletta N, López-García P, Boihem L et al (2010) Biomarkers of endolithic communities within gypsum crusts (southern Tunisia). Geomicrobiol J 27:101–110. https://doi.org/10.1080/01490450903410431

Stoertz GE, Ericksen GE (1974) Geology of salars in northern Chile. US Geological Survey professional paper, Washington, DC

Teske A, Ramsing NB, Habicht K et al (1998) Sulfate-reducing bacteria and their activities in cyanobacterial mats of solar lake (Sinai, Egypt). Appl Environ Microbiol 64:2943–2951

Toneatti DM, Albarracín VH, Flores MR et al (2017) Stratified bacterial diversity along physicochemical gradients in high-altitude modern stromatolites. Front Microbiol 8:646. https://doi.org/10.3389/fmicb.2017.00646

Van Gemerden H (1993) Microbial mats: a joint venture. Mar Geol 113:3–25

Wierzchos J, de los Ríos A, Ascaso C (2012) Microorganisms in desert rocks: the edge of life on Earth. Int Microbiol 15:173–183. https://doi.org/10.2436/20.1501.01.170

Chapter 12
Microbial Diversity in Athalassohaline Argentinean Salterns

Merit del Rocío Mora-Ruiz and Carlos Díaz-Gil

12.1 Athalassohaline Environments and Their Principal Representatives

Microorganisms comprise the wealth of life on the planet, and research based on them has dramatically expanded our understanding of biodiversity. Particularly, the microbiology of extreme environments, such as acidic or hypersaline habitats, has been a candent scientific topic of research because it resembles the hypothesized conditions of the origin of life or even as models of hypothetical life outside the Earth (Rampelotto 2013). Hypersaline habitats are considered extreme environments; however, life finds its way and organisms from the three domains of life—*Bacteria*, *Archaea*, and *Eukarya*—are present in high numbers in these habitats.

In general, hypersaline environments can be divided into thalassohaline and athalassohaline. Thalassohaline systems are those having a similar ionic composition as the ocean. They originated from the evaporation of seawater or by dissolution of evaporites (Ventosa 2006; Ventosa et al. 2011). Typical examples of thalassohaline systems are natural coastal lagoons and human-made solar salterns, which are located along tropical and subtropical coasts (e.g., Exportadora de Sal, México; The Petchaburi salterns, Thailand; Salinas de Levante and Santa Pola salterns, Spain). During evaporation there is a serial precipitation of salts due to the different coefficient of saturation (Ω) of each salt. By contrast, athalassohaline systems have different ionic composition than seawater, as they are a consequence of the dissolution of different geochemical properties of rocks and substrates of each specific location which, through watershed, tend to concentrate in endorheic lagoons (Rodríguez-Valera et al. 1998). Several examples of athalassohaline systems are the Dead Sea in Jordan, Israel, and Palestine; the Great Salt Lake in the USA; cold

M. d. R. Mora-Ruiz (✉) · C. Díaz-Gil
Department of Ecology and Marine Resources, Mediterranean Institute for Advanced Studies (IMEDEA UIB-CSIC), Esporles, Illes Balears, Spain

M. E. Farías (ed.), *Microbial Ecosystems in Central Andes Extreme Environments*, https://doi.org/10.1007/978-3-030-36192-1_12

Fig. 12.1 Landscape of Salar de Antofalla. (Photo courtesy of Farías, M.)

hypersaline lakes in Antarctica, Tirez lagoon in Spain, Lake Magadi or the lakes of Wadi Natrun and the Atacama's lakes in South America (Rodríguez-Valera 1988; Javor 1989; Grant et al. 1998; Demergasso et al. 2004; Moreira et al. 2006; Ventosa 2006; Montoya et al. 2013).

Although saline water masses are widely distributed in arid regions around the world, the high-altitude hypersaline water masses are characterized not only by their elevated salinity but also by their location at high altitude. Typically, those water masses have been classified as lakes, lagoons, and salares, but hereafter all of them will be grouped under the term of high-altitude hypersaline Andean Lakes (HAAL) (Albarracín et al. 2015). This chapter presents a broad perspective regarding the knowledge of the microbial diversity in HAAL focusing on those from the Argentinean Altiplano (Fig. 12.1).

12.2 Athalassohaline Environments in the Argentinean Altiplano and Their Physicochemical Composition

The Argentinean Altiplano is an area located in the central Andes in South America in the Puna region. This mountainous plateau presents a rigorous climate: marked daily temperature fluctuations, high solar radiation (165% higher than sea level), frequent strong winds, low rate of precipitation (<200 mm per annum), and low xerophylous vegetation (De Mitrovich et al. 2005; Ordoñez et al. 2009). The geography of the Argentinean Altiplano has generated a series of high-mountain saline lakes found in close proximity and most of them are far away from any significant urban population (Farías et al. 2013; Fig. 12.1). The lakes formed during the Quaternary geological period have been considered as athalassohalines due to their ionic composition, which is provided by the geology, topography, and climatic conditions of each specific location (Rodríguez-Valera et al. 1998; Alonso 2006; Albarracín et al. 2010).

HAAL included in this chapter are distributed in an altitude range from 3000 to 6000 m above sea level (m a.s.l.). They usually present salinities higher than 28% with a dominance of Cl^- as anion and Na^+ as cation. Although, the values of Li^+ in most cases is lower than 1 ppm, its presence is relevant in HAAL compared to other saline systems (e.g., salterns) where this ion is usually under detection level (Di Meglio et al. 2016; Mora-Ruiz et al. 2017). This is not unexpected since Li^+ is relatively abundant in the region. In fact, the intensive extraction of Li^+ in the past few years has led Argentina to be considered as the most dynamic producer of lithium derivatives. Particularly, Salar del Hombre Muerto in Catamarca and Salar de Olaroz in Jujuy are the most important sources in the country (Delbuono et al. 2017).

Besides the geological studies performed in HAAL and their relevant location (e.g., its closeness to mines or volcanoes), there is an important lack of information about the microbial composition of these sites. We will focus on 16 lakes several hundred kilometers apart from each other but with a very homogeneous climate and vegetation. Despite this, the different HAAL included in this chapter present important differences mostly in terms of salinity, temperature, pH, and geological characteristics, which are summarized in Table 12.1 (De Mitrovich et al. 2005).

12.3 Microbial Diversity in Argentinean Lakes

The harsh environmental characteristics of the Argentinean Altiplano have led to this region being considered as an area with a low biodiversity. However, the microbial diversity in the Argentinean Altiplano has demonstrated to be far more complex than previously expected (Farías et al. 2013; Filker et al. 2017). The group of saline lagoons generated in this region is an excellent system for the analysis of extreme microbial communities due to the multitude of extreme conditions of the area (as mentioned above). HAAL are a rich source where poly-extremophiles flourish and those environments are excellent models for understanding the response of the microbial community confronted with perturbations since they are partially isolated from the stochastic effects present in open waters (Celussi et al. 2010; Main et al. 2015; Suh et al. 2015; Bowman et al. 2017).

Several studies have been performed with the aim of understanding the microbial communities in different saline lakes in Argentina and in the altiplanic water masses of other countries using culture-dependent and culture-independent approaches (De Mitrovich et al. 2005; Pagaling et al. 2009; Di Meglio et al. 2016; Fernández et al. 2016). Nonetheless, research keeps finding unexpected microbial diversity; for example, the study of Farías et al. (2013) reported active stromatolites in the Socompa Lake at an altitude of 3570 m a.s.l., which is the highest altitude where

Table 12.1 Environmental characteristics of the high–altitude hypersaline Andean Lakes (HAAL) in Argentina

HAAL	Altitude	Coordinates		Salinity	Cl^-	Na^+	Mg^{2+}	Ca^{2+}	K^+	Li+
Salar de COIPA	3650	−24.52	−68.21	39	82,696	73,306	6151	11,795	9511	0.062
Laguna Santa María	3508	−24.09	−67.36	43	24,782	18,198	586.3	4208.8	3060	0.071
Ojo Naranja	3338	−25.57	−67.60	29	177,070	120,201	210	196.3	12,162	0.082
Ojo Blanco	3338	−25.56	−67.59	35	915,851	600,277	1406	26,301.3	1927	0.007
Ojo Seco	3338	−25.55	−67.57	34	49,442.5	29,250	486	48,786.3	317	0.009
Laguna Diamante	4560	−26.03	−67.04	32	170,348	102,081	3630	7451.3	10,178	0.082
Laguna Cabe	4255	−26.25	−67.06	35	193,632	111,347	6913	1230	14,267	0.082
Salar de Pocitos	3673	−24.37	−66.98	43	146,721	136,601	293	845	36,996	0.008
Tolar Grande	3508	−24.55	−67.49	40	142,837	103,825	418	116.3	41,258	0.859
Salar de Llullialiaco	3677	−24.8	−68.29	36	190,227	113,901	1250	1213.8	7721	0.062
Laguna Negra	1615	−27.63	−68.55	35	106,406	62,748	1266	2236	6536	0.021
La Lagunita	3800	−26.71	−67.00	>17	—	—	—	—	—	—
Pasto Ventura	3746	−26.83	−66.01	6.52	—	—	—	—	—	—
Salinas Grande	3300	−27.80	−66.01	>17	100,000	—	2066	—	9289	775
Salar Guayatayoc	3660	−23.60	−65.87	—	194,992	112,651	1482	—	6764	66.7
Laguna Verde	4100	−27.58	−68.65	21.1	—	—	—	—	—	—

those microbialites have been found. The microbiota inhabiting the stromatolites was atypical in comparison with other modern stromatolites and dominated by representatives of *Deinococcus–Thermus*, *Rhodobacteraceae*, *Desulfobacterales*, and *Spirochaetes* reflecting the peculiarity of those structures.

12.4 Microbial Diversity in HAAL

Although several microbial studies have been conducted in other high-altitude lakes of South America, principally in the Salar de Atacama (Chile) (Lizama et al. 2002; Demergasso et al. 2008; Thiel et al. 2010), the microbiological studies in the Puna region have been developed principally in the past decade (Ordoñez et al. 2009; Pagaling et al. 2009; Farías et al. 2013; Filker et al. 2017; Mora-Ruiz et al. 2017). Initial surveys attempted to obtain cultures of microorganisms from water of wetlands with salinities up to 11.7% (Zenoff et al. 2006; Ordoñez et al. 2009).

The abundances of microorganisms reported in HAAL are in the range previously reported for other hypersaline systems, such as solar salterns and other saline lakes (Demergasso et al. 2004; Di Meglio et al. 2016; Boujelben et al. 2012; Gomariz et al. 2014). The dominance of *Archaea* members is marked and followed by *Bacteria*, while the *Eukarya* domain remains in low relative abundances (Boujelben et al. 2012; Fernández et al. 2014; Podell et al. 2014; Di Meglio et al. 2016).

In general, salterns have been represented by a high diversity of lineages, where the most common halophiles are affiliated with the phyla *Rhodothermaeota*, *Bacteroidetes*, *Proteobacteria*, *Firmicutes*, and *Actinobacteria*. These lineages include relevant genera such as *Alkalilimnicola*, *Owenweeksia*, *Desulfuromonas*, *Halovibrio*, and *Salinibacter*, the latter considered as one of the most relevant members of the known halophilic microbial communities (Antón et al. 2002; Benlloch et al. 2002; Rosselló-Mora et al. 2008; Oren 2008; Peña et al. 2010; Gomariz et al. 2014; Munoz et al. 2016). In the specific case of HAAL, a recent study by Mora-Ruiz et al. (2017) analyzed the bacterial and archaeal diversity from 11 different HAAL using the operational phylogenetic unit (OPU) as a biological entity (Mora-Ruiz et al. 2016). They analyzed sediments and brines of each one of these HAAL and found dominance of *Proteobacteria*, *Firmicutes*, and *Bacteroidetes* phyla in both environments (sediments and brines); similar to athalassohaline lakes in the Tibetan Plateau (China) at an altitude of 3200 m a.s.l. (Lake Chaka and Lake Quinghai) (Dong et al. 2006; Jiang et al. 2006). In general, *Gammaproteobacteria* and *Deltaproteobacteria* are the most abundant class of *Proteobacteria*, while *Betaproteobacteria* is usually less represented with abundances no higher than 6% (Dong et al. 2006; Ordoñez et al. 2009; Mora-Ruiz et al. 2017) and principally represented by the *Comamonadaceae* family similar to other altiplanic lakes in Chile (Ortiz-Dorador 2007).

Furthermore, in the above-mentioned study (Mora-Ruiz et al. 2017), the bacterial richness in sediments was composed of a total of 683 OPUs included in 151 families with a notable dominance of the *Halanaerobiaceae* family as well as its representative genus *Halanaerobium*. Additionally, the richness in brines was included in 557 OPUs with the same pattern of dominance at family and genus level as in sediments (*Halanaerobiaceae* and *Halanaerobium*, respectively). Those results differ from those observed by Pagaling et al. (2009), who detected principally bacteria affiliated with *Salinibacter*, *Thiohalosphira*, *Desulbobulbus* and in lesser proportion *Halanaerobium* genera in brines of Salar Guayatayoc (Jujuy, Argentina).

As we have previously mentioned, the environmental conditions in HAAL produce a constant pressure on the microbial community, specially related to the extreme salinity and UV radiation. Some bacterial strains found in HAAL present a high resistance to UV-B (280–310 nm) exposure, the most resistant have been affiliated with *Pseudomonas* sp., *Acinetobacte*r sp., *Stenotrophomonas maltophilia*, *Halomonas* sp., *Halomonas* sp. (Ordoñez et al. 2009). However, the UV-B resistance of the microbiota in the Altiplano is not only restricted to hypersaline environments. This capability is well-distributed among bacterial species present in high-altitude wetlands, including those with low salinity (e.g., Laguna Aparejos in Catamarca (Argentina), salinity: 0.004%) (Zenoff et al. 2006; Ordoñez et al. 2009). However, not only are salinity, solar radiation, and high altitude extreme conditions in HAAL but also the concentration of some metalloids such as arsenic, which is toxic for most life forms, is elevated. Resistance to arsenic As(V) and As(III) was observed in strains of *Exiguobacterium* inhabiting stromatolites in Laguna Socompa. A set of 25 proteins up or down-regulated have been related to this arsenic resistance as well as the presence of *acr3 and arsB* genes (Belfiore et al. 2013; Ordoñez et al. 2015).

As in other hypersaline environments, *Archaea* represents the dominant group in the microbiota of the HAAL, where the richness of *Archaea* is lower, but the abundances are higher than *Bacteria* (Boujelben et al. 2012; Fernández et al. 2014; Podell et al. 2014; Di Meglio et al. 2016). Mora-Ruiz et al. (2017) reported a richness for *Archaea* in HAAL comprising 259 and 157 OPUs in sediments and brines, respectively. Most archaeal richness in HAAL has been affiliated with *Euryarchaeota* phylum and with *Halobacteriaceae* and *Haloferacaceae* families as in other hypersaline brines and sediments (Hollister et al. 2010; Dillon et al. 2013; López-López et al. 2013; Podell et al. 2013; Fernández et al. 2014; Vavourakis et al. 2016). A remarkable aspect of Argentinean HAAL is the dominance of halophilic *Euryarchaeota*, while in other high-altitude lakes a high presence of non-halophilic *Euryarchaeota* has been reported (Demergasso et al. 2004; Albarracín et al. 2015).

The phylum *Nanohaloarcheota* has been reported as the second phyla in abundance both in sediments and brines in HAAL (Gomariz et al. 2014; Andrei et al. 2015; Mora-Ruiz et al. 2017). The presence of *Nanohaloarchaeota* in several hypersaline environments has been reported in the past decade (Narasingarao et al. 2012; Andrade et al. 2015; Andrei et al. 2015). However, these were reported as aerobes exclusively occurring in brines (Narasingarao et al. 2012; Podell et al. 2013) and the presence and abundance of *Nanohaloarchaeota* in sediments of HAAL could suggest a versatile metabolism in this group.

With more taxonomic detail, the most important archaeal families in HAAL seem to be *Halobacteriaceae* and *Haloferacaceae* with a notable dominance of *Halorubrum*, *Natronomonas*, and *Halorhabdus* genera. As in *Bacteria*, brines have shown a lower richness than sediments (102 OPUs less as reported by Mora-Ruiz et al. (2017)) with the same dominance pattern of phyla, families, and genera. Those dominant families and genus are typically represented in other hypersaline environments (Grant et al. 1998; Maturrano et al. 2006; Ventosa 2006; Burns et al. 2007; Oh et al. 2010; Oren and Hallsworth 2014). However, the relative abundances of other taxa are different in HAAL, for example, the genus *Haloferax* has been detected with higher abundances in HAAL than in other hypersaline environments (Mora-Ruiz et al. 2017). Furthermore, in HAAL, *Halorubrum* has been found as the dominant genus over *Haloquadratum* (Mora-Ruiz et al. 2017), the last one often reported as the major component in brines of salterns around the world (Maturrano et al. 2006; Oh et al. 2010; Boujelben et al. 2012; Gomariz et al. 2014; Podell et al. 2014).

The presence of archaeas has also been demonstrated in other benthic systems, calcareous microbialites. Microbialites in Laguna Diamante present red biofilms principally conformed by members of the *Halorubrum* genus. An anaerobic respiration based on the use of substrates such as arsenate has been suggested as a principal metabolism of the archaeas inhabiting those microbialites due to the oxygen limitation (Albarracín et al. 2015; Rascovan et al. 2016).

Typically, the abundances of eukaryotes in hypersaline environments has been considered a minority in comparison with the other two domains (*Archaea* and *Bacteria*), and in general the knowledge on the whole protistan community diversity and ecology in hypersaline waters is still very scarce (Filker et al. 2017). The eukaryotic diversity is usually represented by algae such as *Dunaliella salina* (Ventosa and Arahal 2009) and *Chlamydomonas* spp.; flagellates such as *Pleurostomum flabellatum* (De Jonckheere et al. 2009); fungi such as *Alternaria* spp. or *Trimmatostroma salinum* (Butinar et al. 2005; Liu et al. 2014); and the brine shrimp *Artemia* spp. It has been suggested that microbial eukaryotes have greater difficulties reproducing due to the selective effect of high salinity, resulting in a large decrease in the number of species as salinity increases (Benlloch et al. 2002). However, as most information about microbial eukaryote diversity in such environments derives from microscopy and fingerprinting approaches, the true extent of their diversity in these extreme habitats still remains unknown. Recent studies including the description of eukaryotic diversity in HAAL evidenced that diversity of microbial eukaryotes in such environments is much higher than previously described and also that different salt regimes harbor distinct ecosystems (Casamayor et al. 2013; Stoeck et al. 2014; Filker et al. 2015, 2017).

A study of zooplankton in Laguna Blanca, La Lagunita, and Pasto Ventura found that the diversity is principally dominated by rotifers (genus *Brachionus*, *Lecane*, and *Notholca*), cladocerans (e.g., *Daphnia pulex*) and copepods (especially *Diacyclops andinus* and *Boeckella* spp.) (De Mitrovich et al. 2005). The number of species for each one of these groups ranged from 3 to 15 species per lagoon and with weights in the range of 422–1.805 μg. Moreover, representative members of *Simocephalus* have been detected in Laguna Blanca (De Mitrovich et al. 2005).

Although, over the past century the presence of other planktonic organisms such as ostracods was considered null (Bayly 1993), their presence has been reported with important abundances in some HAAL including Laguna Blanca (De Mitrovich et al. 2005).

Recently, Filker et al. (2017) analyzed the eurkaryotic diversity in other HAAL (Salar Coipa, Ojo Naranja, Ojo Blanco, and Laguna Diamante). The diversity in those environments was dominated by members of phyla *Ciliophora*, *Chlorophyta*, and *Colpedellida* with a high degree of novel genetic diversity at lower taxonomic levels (OTUs). Although the most abundant phyla were the three previously mentioned, the novel diversity was occurred principally in the unicellular flagellates *Bicosoecida* and the parasitic *Apicomplexa.* Additionally, Filker et al. (2017) detected an effect of the geographical distance in prostistan communities separated by more than 500 km. In the same way, diatoms have been detected in HAAL both in hypersaline water as in sediments conforming the main algal group (Farías et al. 2014; Albarracín et al. 2015; Filker et al. 2017). The presence of diatoms has been reported in important abundances in microbialites of Socompa Lake, even in parts of the stromatolites where light in the visible spectra range is minimum (up to 15 mm depth), the genus *Navicula* being the most abundant (Farías et al. 2013).

On other hand, other organisms can also be found in HAAL, such as nematodes and larvae of chironomid (De Mitrovich et al. 2005) and flamingos (including *Phoenicoparrus andinus*, *Phoenicoparrus jamesi*, and *Phoenicopterus chilensis*) that use the species of copepod *Boeckella* as their principal food resource (De Mitrovich et al. 2005; Caziani et al. 2007; Belluscio 2010).

12.5 Benthic Communities in HAAL

HAAL are environments with shallow water where benthic communities have a close interaction with the planktonic communities (Albarracín et al. 2015). Some recent studies, which included both brines and sediments from HAAL, have exhibited the high connectivity among the sediment and brines in the Altiplano Lakes (Filker et al. 2017; Mora-Ruiz et al. 2017). As we mentioned in the previous section, sediments and brines have exhibited a similar pattern of abundances at phyla, family, and genera level for *Bacteria*, *Archaea*, and also for protists (Filker et al. 2017; Mora-Ruiz et al. 2017). Even at the species level (OTUs or OPUs) a large part of taxa has been observed in both habitats, for example, 84.4% of the total archaeal sequences in the study of Mora-Ruiz et al. (2017) were present in sediments and brines, suggesting a strong connection between both habitats. The presence of anaerobic groups has been observed in brines, probably due to the oxygen deficit or hypoxia when the respiration exceeds primary productivity (Javor 1989; Williams et al. 1995; De Mitrovich et al. 2005) and also because the high summer temperatures reduce the oxygen solubility (Tromans 1998), making the presence of facultative anaerobes feasible as part of their community. Moreover, the important presence of aerobic microorganism have been detected in sediments of HAAL (suggesting

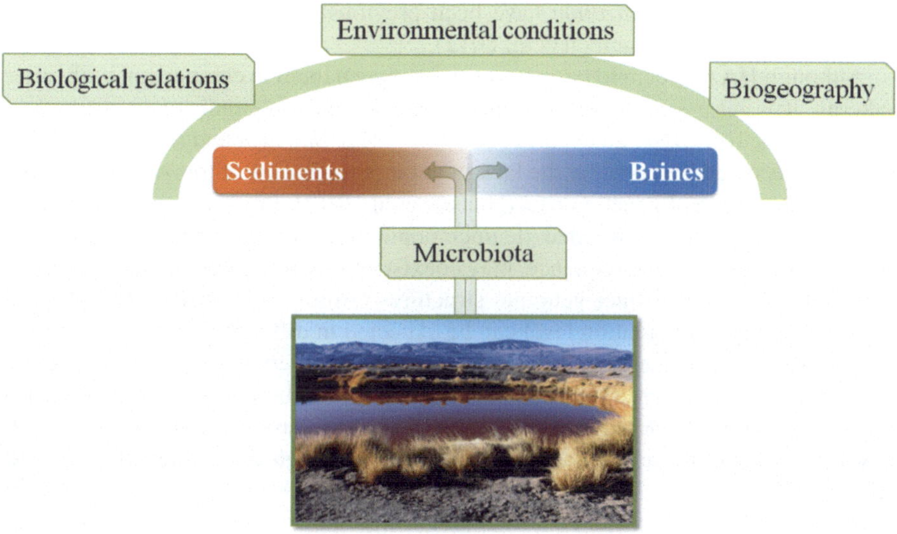

Fig. 12.2 Principal processes determining the structure of the microbial community in brines and sediments of HAAL: environmental conditions, biological relation (e.g., intra- and inter-species interactions), and biogeography

them as potentially relevant in the communities of sediments (Mora-Ruiz et al. 2017)).

On the other hand, although some studies have reported a high similarity of HAAL with other high altitude lakes (Dong et al. 2006; Jiang et al. 2006), this similarity is based on a comparison at high taxonomic levels (phylum and class). This diversity overlap is probably a reflection of some common environmental conditions, especially high salinity, which is considered the most relevant driving factor in the microbial communities (Lozupone and Knight 2007). With more detail to lower taxonomic levels, other studies have showed a high level of endemism in HAAL (Mora-Ruiz et al. 2017). The comparison of the microbial richness with other salterns exhibited that ~24.8% of the total species mainly affiliated with putative methanogens (*Methanobacteriaceae* and *Methanothermaceae*) were found exclusively in HAAL (Mora-Ruiz et al. 2017).

The differences detected in the diversity of HAAL with other hypersaline environments has not always been explained principally by the geographic distance. While the geographic distance has been reported as correlated with changes in the bacterial and archaeal communities (Mora-Ruiz et al. 2017; Fig. 12.2.), another study found this pattern in *Bacteria* but not in *Archaea* (Pagaling et al. 2009). However, in both studies, the concentration of specific ions (especially NO_3^-, Mg^{2+}, CO_3^{2-}, and HCO_3^-) were significant factors influencing the microbial communities. Some of these ions have also been correlated with abundances of specific groups, for example, the positive correlation between Mg^{2+} and the *Halorubellus* abundance (Mora-Ruiz et al. 2017). This patter is not unexpected since some important archaeal groups such as *Haloquadratum* have been previously described as related to the Mg^{2+} concentration (Boujelben et al. 2012).

Although HAAL seem to be very similar between them in comparison with other salterns, it is necessary to mention that most biodiversity studies have been performed using 16S rRNA gene partial sequences (for *Bacteria* and *Archaea)*. In this way, it is necessary to consider the lack of enough resolution of this marker to guarantee the detection of genetic drifts (Hanson et al. 2012). Previous studies in hypersaline environments shows that distinct populations of the same species, with identical 16S rRNA gene sequence, may coexist expressing different metabolic profiles and exhibiting distinct genomic structures (Antón et al. 2013). Therefore, a higher diversity at population level can be expected in different HAAL.

HAAL play a critical role in sustaining a unique diversity of rare and endemic biota in the Argentinean Altiplano. This microbiota has broadened our understanding of the microbial diversity, and the studies in HAAL have also evidenced the high novel diversity in those high-altitude wetlands. Although the studies of microbial communities of *Bacteria*, *Archaea*, and even of *Eukarya* have increased in recent years, research on viruses in HAAL is still scarce. Haloviruses (viruses that infect halophiles) are also an important component in the microbial community with abundances up to 10^9 virus-like particles per milliliter (VLPs)/mL (Santos et al. 2012). Evidence of infection in some archaeal groups such as *Haloquadratum*, *Halorubrum*, and *Halobacterium* by haloviruses has been shown in previous studies (Santos et al. 2012; Atanasova et al. 2015) as well as their relevant participation in biogeochemical cycles (Pina et al. 2011). New characterization of haloviruses showed that hypersaline environments contain a higher richness than previously assumed (Ventosa et al. 2014). HAAL offer excellent opportunities to deeply explore the virus diversity in the same way as it has been explored in other lakes in Argentina (Di Meglio et al. 2016).

As we mentioned previously, the multifactorial-extreme conditions of HAAL provide a great chance to explore the resistant factors presented in the microbiota that led them to survive in those high-mountain lakes. Some studies have been focused on the detection of UV exposure resistance, resistance to toxic compounds, and hypersalinity in some strains as well as the mechanism explaining that resistance (Dib et al. 2008; Ordoñez et al. 2009; Farias et al. 2011). Those investigations have produced a notable biotechnological interest related to the potential of those microorganisms for different applications, considering them as a novel source for bioactive compounds (Dib et al. 2008; Albarracín et al. 2010). An example is the evidence of antibiotic resistance in some strains isolated from HAAL (especially in the genus *Acinetobacter*). Researchers have suggested that this resistance is a consequence of a high frequency of mutational events, which may also be enhanced by the extreme conditions (e.g., UV radiation) (Albarracín et al. 2010). In the same way, the presence of hydrolases (proteases, amylases, lipases, cellulases, and nucleases) and bioactive molecules (e.g., surfactants) have been evidenced in some strains both in *Bacteria* and *Archaea* from salterns in Argentina. Therefore, HAAL are a wide reservoir for the biotechnological exploration that shelters multiple applications (Nercessian et al. 2015).

Finally, as in other hypersaline environments, most studies in HAAL have been focused on the description of the microbial diversity but in a punctual instant (Benlloch et al. 2002; Oh et al. 2010; Dillon et al. 2013), and just a few studies involve temporal analysis (Podell et al. 2014; Atanasova et al. 2015). Other microbiology topics have clear examples of temporal variation in microbiota such as humans (Sharon et al. 2013) and wastewater (Vaquer-Sunyer et al. 2016). Nevertheless, the exploration of adaptive processes of microbial communities, including changes in the total microbial structure and on specific groups remains challenging, because it requires a pertinacious effort to obtain detailed, accurate measurements of both physical and biological factors (Podell et al. 2014). The analysis of communities sampled at sequential moments can lend insight toward understanding the underlying environmental drivers causing the shifts in the communities' structure over time (Habel et al. 2014). Frequent and long-term observations of the biotic and abiotic components in HAAL are required, especially to know the seasonal variations of the microbiota in those extreme environments (Farías et al. 2013). However, we have to agree with Pagaling et al. (2009) who mentioned that "*in an ideal world, sequential sampling over an extended time period would be appropriate, but logistical and financial considerations preclude repeated visits to these remote sites*".

References

Albarracín V, Dib JR, Ordoñez OF, Farías ME (2010) A harsh life to indigenous proteobacteria at the Andean mountains: microbial diversity and resistance mechanisms towards extreme conditions. In: Sezenna M (ed) Phylogeny, metabolic diversity and ecological effects. Nova Science Publishers, Inc., New York, pp 1–24

Albarracín VH, Kurth D, Ordoñez OF et al (2015) High-up: a remote reservoir of microbial extremophiles in central Andean wetlands. Front Microbiol 6:1404

Alonso RN (2006) Ambientes evaporíticos continentales de Argentina. Correl Geol 21:155–170

Andrade K, Logemann J, Heidelberg KB et al (2015) Metagenomic and lipid analyses reveal a diel cycle in a hypersaline microbial ecosystem. ISME J 9:2697–2711. https://doi.org/10.1038/ismej.2015.66

Andrei A-Ş, Robeson MS, Baricz A et al (2015) Contrasting taxonomic stratification of microbial communities in two hypersaline meromictic lakes. ISME J 9:2642–2656. https://doi.org/10.1038/ismej.2015.60

Antón J, Lucio M, Peña A et al (2013) High metabolomic microdiversity within co-occurring isolates of the extremely halophilic bacterium Salinibacter ruber. PLoS One 8:e64701. https://doi.org/10.1371/journal.pone.0064701

Antón J, Oren A, Benlloch S et al (2002) Salinibacter ruber gen. nov., sp. nov., a novel, extremely halophilic member of the Bacteria from saltern crystallizer ponds. Int J Syst Evol Microbiol 52:485–491. https://doi.org/10.1099/00207713-52-2-485

Atanasova NS, Demina TA, Buivydas A et al (2015) Archaeal viruses multiply: temporal screening in a solar saltern. Viruses 7:1902–1926. https://doi.org/10.3390/v7041902

Bayly IAE (1993) The fauna of athalassic saline waters in Australia and the Altiplano of South America: comparisons and historical perspectives. Hydrobiologia 267:225–231. https://doi.org/10.1007/BF00018804

Belfiore C, Ordoñez OF, Farías ME (2013) Proteomic approach of adaptive response to arsenic stress in Exiguobacterium sp. S17, an extremophile strain isolated from a high-altitude Andean lake stromatolite. Extremophiles 17:421–431. https://doi.org/10.1007/s00792-013-0523-y

Belluscio A (2010) Hostile volcanic lake teems with life. Nature:10–12. https://doi.org/10.1038/news.2010.161

Benlloch S, Lopez-Lopez A, Casamayor EO et al (2002) Prokaryotic genetic diversity throughout the salinity gradient of a coastal solar saltern. Environ Microbiol 4:349–360. https://doi.org/10.1046/j.1462-2920.2002.00306.x

Boujelben I, Gomariz M, Martínez-García M et al (2012) Spatial and seasonal prokaryotic community dynamics in ponds of increasing salinity of Sfax solar saltern in Tunisia. Antonie Van Leeuwenhoek 101:845–857. https://doi.org/10.1007/s10482-012-9701-7

Bowman JS, Amaral-Zettler LA, J Rich J et al (2017) Bacterial community segmentation facilitates the prediction of ecosystem function along the coast of the western Antarctic Peninsula. ISME J 11:1460–1471. https://doi.org/10.1038/ismej.2016.204

Burns DG, Janssen PH, Itoh T et al (2007) Haloquadratum walsbyi gen. nov., sp. nov., the square haloarchaeon of Walsby, isolated from saltern crystallizers in Australia and Spain. Int J Syst Evol Microbiol 57:387–392. https://doi.org/10.1099/ijs.0.64690-0

Butinar L, Sonjak S, Zalar P et al (2005) Melanized halophilic fungi are eukaryotic members of microbial communities in hypersaline waters of solar salterns. Bot Mar 48:73–79. https://doi.org/10.1515/BOT.2005.007

Casamayor EO, Triadó-Margarit X, Castañeda C (2013) Microbial biodiversity in saline shallow lakes of the Monegros desert, Spain. FEMS Microbiol Ecol 85:503–518. https://doi.org/10.1111/1574-6941.12139

Caziani SM, Rocha Olivio O, Rodríguez Ramírez E et al (2007) Seasonal distribution, abundance, and nesting of Puna, Andean, and Chilean flamingos. Condor 109:276. https://doi.org/10.1650/0010-5422(2007)109[276:SDAANO]2.0.CO;2

Celussi M, Bergamasco A, Cataletto B et al (2010) Water masses' bacterial community structure and microbial activities in the Ross Sea, Antarctica. Antarct Sci 22:361–370. https://doi.org/10.1017/S0954102010000192

De Jonckheere JF, Baumgartner M, Opperdoes FR, Stetter KO (2009) Marinamoeba thermophila, a new marine heterolobosean amoeba growing at 50°C. Eur J Protistol 45:231–236. https://doi.org/10.1016/j.ejop.2009.01.001

De Mitrovich CL, De Gamundi AV, Juárez J, Ceraolo M (2005) Características limnológicas y zooplancton de cinco lagunas de la Puna – Argentina. Ecol Boliv 40:10–24

Delbuono V, Such T, Toleto E, Jerez D (2017) Informe especial – mercado de litio. Argentina

Demergasso C, Casamayor EO, Chong G et al (2004) Distribution of prokaryotic genetic diversity in athalassohaline lakes of the Atacama Desert, Northern Chile. FEMS Microbiol Ecol 48:57–69. https://doi.org/10.1016/j.femsec.2003.12.013

Demergasso C, Escudero L, Casamayor EO et al (2008) Novelty and spatio-temporal heterogeneity in the bacterial diversity of hypersaline Lake Tebenquiche (Salar de Atacama). Extremophiles 12:491–504. https://doi.org/10.1007/s00792-008-0153-y

Di Meglio L, Santos F, Gomariz M et al (2016) Seasonal dynamics of extremely halophilic microbial communities in three Argentinian salterns. FEMS Microbiol Ecol 92:1–15. https://doi.org/10.1093/femsec-fiw184

Dib J, Motok J, Zenoff VF et al (2008) Occurrence of resistance to antibiotics, UV-B, and arsenic in bacteria isolated from extreme environments in high-altitude (above 4400 m) Andean wetlands. Curr Microbiol 56:510–517. https://doi.org/10.1007/s00284-008-9103-2

Dillon JG, Carlin M, Gutierrez A et al (2013) Patterns of microbial diversity along a salinity gradient in the Guerrero Negro solar saltern, Baja CA Sur, Mexico. Front Microbiol 4:1–13. https://doi.org/10.3389/fmicb.2013.00399

Dong H, Zhang G, Jiang H et al (2006) Microbial diversity in sediments of saline Qinghai Lake, China: linking geochemical controls to microbial ecology. Microb Ecol 51:65–82. https://doi.org/10.1007/s00248-005-0228-6

Farías ME, Contreras M, Rasuk MC et al (2014) Characterization of bacterial diversity associated with microbial mats, gypsum evaporites and carbonate microbialites in thalassic wetlands: Tebenquiche and La Brava, Salar de Atacama, Chile. Extremophiles 18:311–329. https://doi.org/10.1007/s00792-013-0617-6

Farías ME, Rascovan N, Toneatti DM et al (2013) The discovery of stromatolites developing at 3570 m above sea level in a high-altitude volcanic Lake Socompa, Argentinean Andes. PLoS One 8:e53497. https://doi.org/10.1371/journal.pone.0053497

Farias ME, Revale S, Mancini E et al (2011) Genome sequence of Sphingomonas sp. S17, isolated from an alkaline, hyperarsenic, and hypersaline volcano-associated lake at high altitude in the Argentinean Puna. J Bacteriol 193:3686–3687. https://doi.org/10.1128/JB.05225-11

Fernández AB, Rasuk MC, Visscher PT et al (2016) Microbial diversity in sediment ecosystems (evaporites domes, microbial mats, and crusts) of hypersaline Laguna Tebenquiche, Salar de Atacama, Chile. Front Microbiol 7:1284. https://doi.org/10.3389/fmicb.2016.01284

Fernández AB, Vera-Gargallo B, Sánchez-Porro C et al (2014) Comparison of prokaryotic community structure from Mediterranean and Atlantic saltern concentrator ponds by a metagenomic approach. Front Microbiol 5:1–12. https://doi.org/10.3389/fmicb.2013.00196

Filker S, Forster D, Weinisch L et al (2017) Transition boundaries for protistan species turnover in hypersaline waters of different biogeographic regions. Environ Microbiol 19:3186–3200. https://doi.org/10.1111/1462-2920.13805

Filker S, Gimmler A, Dunthorn M et al (2015) Deep sequencing uncovers protistan plankton diversity in the Portuguese Ria Formosa solar saltern ponds. Extremophiles 19:283–295. https://doi.org/10.1007/s00792-014-0713-2

Gomariz M, Martínez-García M, Santos F et al (2014) From community approaches to single-cell genomics: the discovery of ubiquitous hyperhalophilic Bacteroidetes generalists. ISME J 9:16–31. https://doi.org/10.1038/ismej.2013.95

Grant WD, Gemmel RT, McGenity TJ (1998) In: Horikoshi K, Grant WD (eds) Extremophiles: microbial life in extreme environments. Wiley-Liss, New York, pp 93–132

Habel JC, Husemann M, Finger A et al (2014) The relevance of time series in molecular ecology and conservation biology. Biol Rev 89:484–492. https://doi.org/10.1111/brv.12068

Hanson CA, Fuhrman JA, Horner-Devine MC, Martiny JBH (2012) Beyond biogeographic patterns: processes shaping the microbial landscape. Nat Rev Microbiol 10:497–506. https://doi.org/10.1038/nrmicro2795

Hollister EB, Engledow AS, Hammett AJM et al (2010) Shifts in microbial community structure along an ecological gradient of hypersaline soils and sediments. ISME J 4:829–838. https://doi.org/10.1038/ismej.2010.3

Javor B (1989) Hypersaline environments. Springer Berlin Heidelberg, Berlin, Heidelberg

Jiang H, Dong H, Zhang G et al (2006) Microbial diversity in water and sediment of Lake Chaka, an athalassohaline lake in northwestern China. Appl Environ Microbiol 72:3832–3845. https://doi.org/10.1128/AEM.02869-05

Liu K, Ding X, Wang HF et al (2014) Eukaryotic microbial communities in hypersaline soils and sediments from the alkaline hypersaline Huama Lake as revealed by 454 pyrosequencing. Antonie van Leeuwenhoek. Int J Gen Mol Microbiol 105:871–880. https://doi.org/10.1007/s10482-014-0141-4

Lizama C, Monteoliva-Sánchez M, Suárez-García A et al (2002) Halorubrum tebenquichense sp. nov., a novel halophilic archaeon isolated from the Atacama Saltern, Chile. Int J Syst Evol Microbiol 52:149–155. https://doi.org/10.1099/00207713-52-1-149

López-López A, Richter M, Peña A et al (2013) New insights into the archaeal diversity of a hypersaline microbial mat obtained by a metagenomic approach. Syst Appl Microbiol 36:205–214. https://doi.org/10.1016/j.syapm.2012.11.008

Lozupone CA, Knight R (2007) Global patterns in bacterial diversity. Proc Natl Acad Sci U S A 104:11436–11440. https://doi.org/10.1073/pnas.0611525104

Main CE, Ruhl HA, Jones DOB et al (2015) Hydrocarbon contamination affects deep-sea benthic oxygen uptake and microbial community composition. Deep-Sea Res I Oceanogr Res Pap 100:79–87. https://doi.org/10.1016/j.dsr.2013.12.008

Maturrano L, Santos F, Rosselló-Mora R, Antón J (2006) Microbial diversity in Maras salterns, a hypersaline environment in the Peruvian Andes. Appl Environ Microbiol 72:3887–3895. https://doi.org/10.1128/AEM.02214-05

Montoya L, Vizioli C, Rodríguez N et al (2013) Microbial community composition of Tirez lagoon (Spain), a highly sulfated athalassohaline environment. Aquat Biosyst 9:19. https://doi.org/10.1186/2046-9063-9-19

Mora-Ruiz MR, Cifuentes A, Font-Verdera F et al (2017) Biogeographical patterns of bacterial and archaeal communities from distant hypersaline environments. Syst Appl Microbiol 41:139. https://doi.org/10.1016/j.syapm.2017.10.006

Mora-Ruiz MDR, Font-Verdera F, Orfila A et al (2016) Endophytic microbial diversity of the halophyte Arthrocnemum macrostachyum across plant compartments. FEMS Microbiol Ecol 92:1–10. https://doi.org/10.1093/femsec/fiw145

Moreira D, Rodríguez-Valera F, López-García P (2006) Metagenomic analysis of mesopelagic Antarctic plankton reveals a novel deltaproteobacterial group. Microbiology 152:505–517. https://doi.org/10.1099/mic.0.28254-0

Munoz R, Rosselló-Móra R, Amann R (2016) Revised phylogeny of Bacteroidetes and proposal of sixteen new taxa and two new combinations including Rhodothermaeota phyl. nov. Syst Appl Microbiol 39:491–492. https://doi.org/10.1016/j.syapm.2016.08.006

Narasingarao P, Podell S, Ugalde JA et al (2012) De novo metagenomic assembly reveals abundant novel major lineage of Archaea in hypersaline microbial communities. ISME J 6:81–93. https://doi.org/10.1038/ismej.2011.78

Nercessian D, Di Meglio L, De Castro R, Paggi R (2015) Exploring the multiple biotechnological potential of halophilic microorganisms isolated from two Argentinean salterns. Extremophiles 19:1133–1143. https://doi.org/10.1007/s00792-015-0785-7

Oh D, Porter K, Russ B et al (2010) Diversity of Haloquadratum and other haloarchaea in three, geographically distant, Australian saltern crystallizer ponds. Extremophiles 14:161–169. https://doi.org/10.1007/s00792-009-0295-6

Ordoñez OF, Flores MR, Dib JR et al (2009) Extremophile culture collection from Andean lakes: extreme pristine environments that host a wide diversity of microorganisms with tolerance to UV radiation. Microb Ecol 58:461–473. https://doi.org/10.1007/s00248-009-9527-7

Ordoñez OF, Lanzarotti E, Kurth D et al (2015) Genome comparison of two Exiguobacterium strains from high altitude andean lakes with different arsenic resistance: identification and 3D modeling of the Acr3 efflux pump. Front Environ Sci 3:1–12. https://doi.org/10.3389/fenvs.2015.00050

Oren A (2008) Microbial life at high salt concentrations: phylogenetic and metabolic diversity. Saline Systems 4:2. https://doi.org/10.1186/1746-1448-4-2

Oren A, Hallsworth JE (2014) Microbial weeds in hypersaline habitats: the enigma of the weed-like Haloferax mediterranei. FEMS Microbiol Lett 359:134–142. https://doi.org/10.1111/1574-6968.12571

Ortiz-Dorador CI (2007) Microbial communities in high altitude altiplanic wetlands in northern Chile: phylogeny, diversity and function. Christian-Albrechts-Universität, Kiel

Pagaling E, Wang H, Venables M et al (2009) Microbial biogeography of six salt lakes in Inner Mongolia, China, and a salt lake in Argentina. Appl Environ Microbiol 75:5750–5760. https://doi.org/10.1128/AEM.00040-09

Peña A, Teeling H, Huerta-Cepas J et al (2010) Fine-scale evolution: genomic, phenotypic and ecological differentiation in two coexisting Salinibacter ruber strains. ISME J 4:882–895. https://doi.org/10.1038/ismej.2010.6

Pina M, Bize A, Forterre P, Prangishvili D (2011) The archeoviruses. FEMS Microbiol Rev 35:1035–1054. https://doi.org/10.1111/j.1574-6976.2011.00280.x

Podell S, Emerson JB, Jones CM et al (2014) Seasonal fluctuations in ionic concentrations drive microbial succession in a hypersaline lake community. ISME J 8:979–990. https://doi.org/10.1038/ismej.2013.221

Podell S, Ugalde JA, Narasingarao P et al (2013) Assembly-driven community genomics of a hypersaline microbial ecosystem. PLoS One 8:e61692. https://doi.org/10.1371/journal.pone.0061692

Rampelotto PH (2013) Extremophiles and extreme environments. Life (Basel) 3:482–485. https://doi.org/10.3390/life3030482

Rascovan N, Maldonado J, Vazquez MP, Eugenia Farías M (2016) Metagenomic study of red biofilms from Diamante Lake reveals ancient arsenic bioenergetics in haloarchaea. ISME J 10:299–309. https://doi.org/10.1038/ismej.2015.109

Rodríguez-Valera F (1988) In: Rodríguez-Valera F (ed) Halophilic bacteria. CRC Press, Boca Raton, FL, pp 3–30

Rodríguez-Valera F, Acinas SG, Antón J (1998) In: Oren A (ed) Microbiology and biochemistry of hypersaline environments. CRC Press, Boca Raton, FL, pp 27–38

Rosselló-Mora R, Lucio M, Peña A et al (2008) Metabolic evidence for biogeographic isolation of the extremophilic bacterium Salinibacter ruber. ISME J 2:242–253. https://doi.org/10.1038/ismej.2007.93

Santos F, Yarza P, Parro V et al (2012) Culture-independent approaches for studying viruses from hypersaline environments. Appl Environ Microbiol 78:1635–1643. https://doi.org/10.1128/AEM.07175-11

Sharon I, Morowitz MJ, Thomas BC et al (2013) Time series community genomics analysis reveals rapid shifts in bacterial species, strains, and phage during infant gut colonization. Genome Res 23:111–120. https://doi.org/10.1101/gr.142315.112

Stoeck T, Filker S, Edgcomb V et al (2014) Living at the limits: evidence for microbial eukaryotes thriving under pressure in deep anoxic, hypersaline habitats. Adv Ecol 2014:1–9. https://doi.org/10.1155/2014/532687

Suh S-S, Park M, Hwang J et al (2015) Seasonal dynamics of marine microbial community in the South sea of Korea. PLoS One 10:e0131633. https://doi.org/10.1371/journal.pone.0131633

Thiel V, Tank M, Neulinger SC et al (2010) Unique communities of anoxygenic phototrophic bacteria in saline lakes of Salar de Atacama (Chile): evidence for a new phylogenetic lineage of phototrophic Gammaproteobacteria from pufLM gene analyses. FEMS Microbiol Ecol 74:510–522. https://doi.org/10.1111/j.1574-6941.2010.00966.x

Tromans D (1998) Temperature and pressure dependent solubility of oxygen in water: a thermodynamic analysis. Hydrometallurgy 48:327–342. https://doi.org/10.1016/S0304-386X(98)00007-3

Vaquer-Sunyer R, Reader HE, Muthusamy S et al (2016) Effects of wastewater treatment plant effluent inputs on planktonic metabolic rates and microbial community composition in the Baltic Sea. Biogeosciences 13:4751–4765. https://doi.org/10.5194/bg-13-4751-2016

Vavourakis CD, Ghai R, Rodriguez-Valera F et al (2016) Metagenomic insights into the uncultured diversity and physiology of microbes in four hypersaline Soda Lake brines. Front Microbiol 7:211

Ventosa A (2006) Unusual micro-organisms from unusual habitats: hypersaline environments. In: SGM Symp 66 prokaryotic divers – mech significance, pp 223–255. https://doi.org/10.1017/CBO9780511754913.015

Ventosa A, Arahal DR (2009) Extremophiles. In: Gerday C, Glansdorff, N (eds) Extremophiles vol. Encyclopedia of life support systems. EOLSS Publishers Company Limited, Oxford. pp 233–246

Ventosa A, Oren A, Ma Y (2011) Halophiles and hypersaline environments. Springer Heildelberg Dordrecht London New York, pp 480

Ventosa A, Fernández AB, León MJ et al (2014) The Santa Pola saltern as a model for studying the microbiota of hypersaline environments. Extremophiles 18:811–824. https://doi.org/10.1007/s00792-014-0681-6

Williams WD, Carrick TR, Bayly IAE et al (1995) Invertebrates in salt lakes of the Bolivian Altiplano. Int J Salt Lake Res 4:65–77. https://doi.org/10.1007/BF01992415

Zenoff VF, Heredia J, Ferrero M et al (2006) Diverse UV-B resistance of culturable bacterial community from high-altitude wetland water. Curr Microbiol 52:359–362. https://doi.org/10.1007/s00284-005-0241-5

Part V
Carbonatic Microbialites at Central Andean Lakes

Chapter 13
Characterization of Microbialites and Microbial Mats of the Laguna Negra Hypersaline Lake (Puna of Catamarca, Argentina)

Flavia Jaquelina Boidi, Estela Cecilia Mlewski, Fernando Javier Gomez, and Emmanuelle Gérard

13.1 Introduction

The Puna-Altiplano region (including northwest Argentina, northern Chile, and southern Bolivia) has become a region of interest to study microbial mineralization processes. The convergence of factors such as extreme environmental conditions, particularly strong negative hydrological balance, favors the development of these mineralizing systems and of extremophile microbial communities. This, together with a favorable local geology and geomorphology, facilitate the formation of closed lakes and groundwater springs that are saturated in minerals, including carbonates. Thus, a highly mineralizing setting together with physicochemical and microbiological processes trigger carbonate precipitation and produce a set of diverse organo-sedimentary structures usually referred to as microbialites (cf. Burne and Moore 1987). Some of the Puna-Altiplano microbialites resemble those observed in the ancient geological record, for example, stromatolites, which are the oldest evidence of life on Earth (Allwood et al. 2006). Thus, these microbialite systems provide a unique opportunity to apply an integrated geobiological approach to gain insight to better understand microbial mineralization processes and biosignatures preservation.

The Laguna Negra (LN) is an interesting system that caught our attention 10 years ago due to the presence of an extensive plain with abundant carbonate precipitation (Fig. 13.1). The evaporative sedimentary environment, the diversity of

F. J. Boidi (✉) · E. C. Mlewski · F. J. Gomez
CICTERRA-Centro de Investigaciones en Ciencias de la Tierra, CONICET-UNC, Córdoba, Argentina

E. Gérard
Institut de Physique du Globe de Paris, Université de Paris, CNRS, UMR 7154, Paris, France

M. E. Farías (ed.), *Microbial Ecosystems in Central Andes Extreme Environments*, https://doi.org/10.1007/978-3-030-36192-1_13

Fig. 13.1 Left: Location map of the study area (red square). Center: Panoramic view of the LN, the Stromatolite Belt (SB), and the Saline Plain. Right: Spatial zoning of the LN Stromatolite Belt

microbial mat types, and the combination of physicochemical and microbial processes influencing carbonate precipitation and carbonates geochemistry makes this place an excellent natural laboratory that shows strong similarities with some ancient microbial mineralizing systems, including the Precambrian microbialites of the Tumbiana Formation (Australia, Buick 1992; Awramik and Buchheim 2009) and the Strelley Pool Chert Formation (Australia, Allwood et al. 2006). Here, some recent findings are summarized to discuss the combination of microbial and physicochemical processes involved in carbonate precipitation and microbialite formation.

13.2 Geological and Environmental Setting

The LN is a shallow (≤2 m) hypersaline lake located at 4200 m above sea level with an area of ~8.63 km^2 and, as most lakes in the region, a strongly negative water balance. It is part of the Laguna Verde Complex, which is in northwest Argentine, at the southern end of the Puna Plateau (Catamarca province), a high-altitude plateau in the Andean region (Fig. 13.1). The high-altitude, hot and dry climate combine to favor the precipitation of evaporites (halite, gypsum, calcite, etc.) and set extreme conditions where microbial mats adapted to hypersalinity as well as high UV radiation influx develop.

The microbial mats distribution and carbonate precipitation are mostly spatially restricted to the region where groundwater springs mix with the main lake, a region of ~0.3 km^2 that it is called the Stromatolite Belt (Gomez et al. 2014). Geochemical modeling suggests that this water mixing increases the saturation state of carbonate minerals favoring abundant carbonate precipitation (Gomez et al. 2014). The presence of microbial mats also influences carbonate precipitation processes, the resulting textures, and geochemical signatures (Gomez et al. 2018; Buongiorno et al. 2018).

13.3 Microbialite Spatial Zoning in the Laguna Negra

The Stromatolite Belt presents three main distinct zones mostly defined by a combination of sedimentary environment, microbialite morphology, water depth, and salinity:

Zone 1: a proximal belt that is colonized by salt-marsh grass in the region of freshwater input,

Zone 2: an intermediate zone consisting of microbial ponds that lack both oncolitic structures and mineralization,

Zone 3: the main belt of carbonate oncoidal microbialites and crusts. This zone can also be separated into four zones: *Zone 3A* where *laminar crusts* are common, *Zone 3B* with centimeter-scale carbonate gravel aggregates, *Zone 3C* where oncoids are primarily concentrated, and *Zone 3D* is represented mainly by peloidal to micritic carbonate sediment, locally interlayered with gypsum or organic-rich laminae. The transition between Zones 3C and 3D is where stromatolites are typically represented, and these were not previously described by Gomez et al. (2014) or Gomez et al. (2018).

13.4 Microbialites Macro-Morphologies

Previous work enabled us to recognize and to characterize the different carbonate microbialites morphologies within the LN (Gomez et al. 2014, 2018; Mlewski et al. 2018; Buongiorno et al. 2018). Three main types of macro-morphologies (at the decimeter-size scale) have been documented in the LN, and here their main macroscopic features are summarized (Figs. 13.2–13.6, and Table 13.1). For details about the different microbialites micro-textures see Table 13.1 (for a deeper analysis see Gomez et al. 2014, 2018; Mlewski et al. 2018). These structures include Laminar Crusts, Oncoids, and Stromatolites.

Laminar crusts are present in the region that is better connected with the main lake, on the northwest side of the Stromatolite Belt. These are represented by mil-

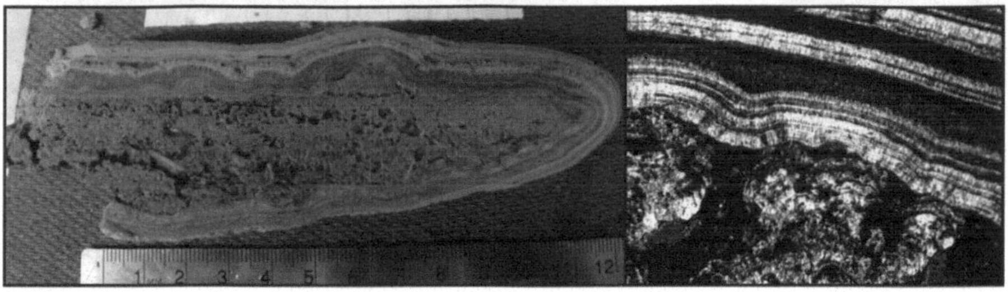

Fig. 13.2 Polished slab (left) and thin-section (right) of the LN laminar crusts

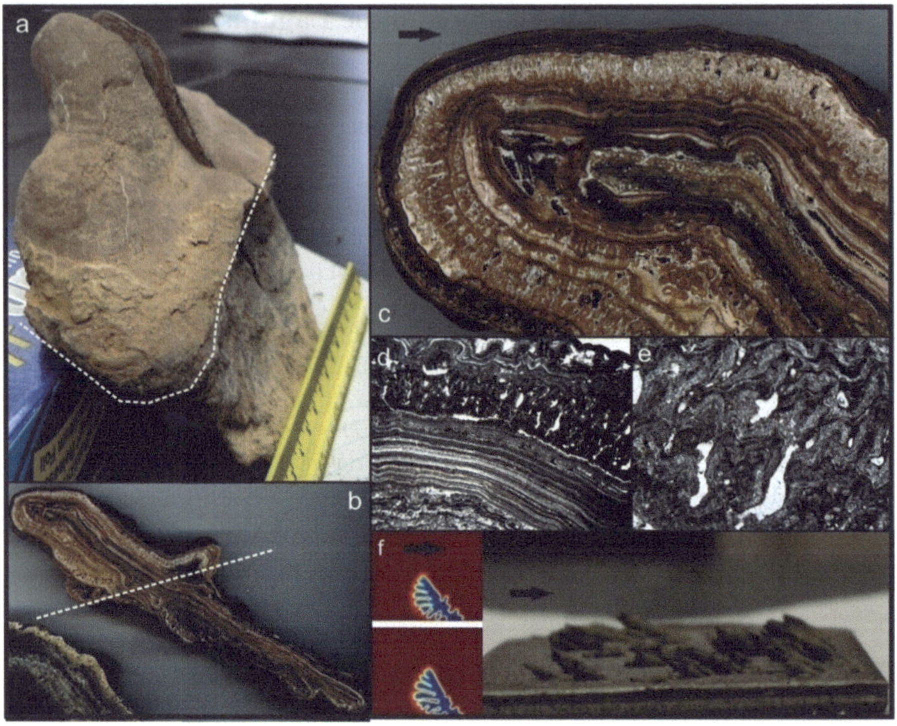

Fig. 13.3 Current-oriented structures observed in Zone 3A. (**a**) Hand-sample, (**b**) Polished slab in cross section. The white dotted line marks the location of the sediment–water interface during sampling. (**c**) Closer view of the polished slab showing the complex micro-columnar to dendritic internal structure with preferential growth on the side affected by currents. The black arrows indicate the current transport direction. (**d, e**) Thin-sections showing the micro-columnar to dendritic micritic to micro-spar-rich textures that alternate with more regular laminar, smooth isopachous laminae. (**f**) The left side panel shows the resulting modeled columnar to dendritic structures developed due to mineral precipitation under unidirectional flow, and the right side shows an example of silica micro-columns and dendrites (both pictures taken from Hawkins et al. 2013, 2014)

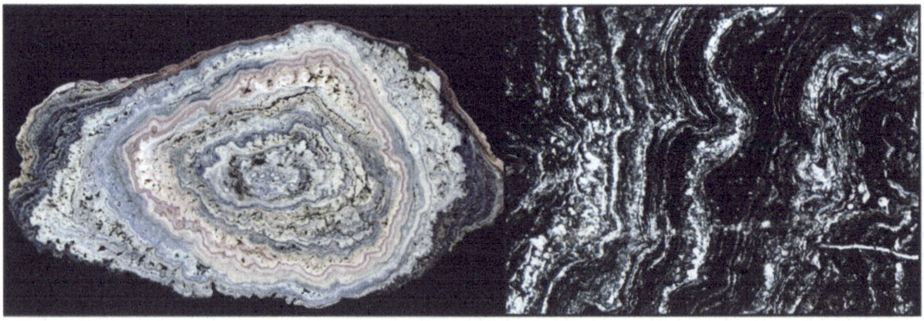

Fig. 13.4 Polished slab (left) and thin-section (right) of the LN oncoids

Fig. 13.5 (**a**) Carbonate oncoids of the Laguna Negra Stromatolite Belt where hydrogenetic and diagenetic growth zones are observed and marked by the sediment–water interface (white dotted lines). (**b**) Manganese nodules with the same growth patterns shown for comparative purposes and suggesting comparable growth mechanisms (from Baker and Beaudoin 2013)

Fig. 13.6 Well-laminated planar to mini-columnar stromatolite nucleated over an oncoidal structure showing concentric growth

Table 13.1 Summary of the main micro-textures preserved in the LN microbialites (for details see Gomez et al. 2014, 2018; Mlewski et al. 2018; Buongiorno et al. 2018)

Microbialite type	Microfabric elements	Observations
Laminar crusts—mm to cm thick laminated crusts forming patchy to laterally extensive pavements	Isopachous laminae	Microbial mats are typically absent. Remarkable regular isopachous lamina with high degree of inheritance during lamina accretion. This results in translation and gradual smoothing of surface morphology. Lamina composed of closely spaced acicular calcite crystals, individual laminae (50–100 μm thick) are separated by micrite isopahcous laminae (10–50 μm thick) composed of irregularly shaped anhedral to subhedral calcite crystals. Smaller scale lamination (≤5 μm) given by luminescence changes related to variable trace elements content
Oncoids—cm to dm-scale concentrically laminated discs, spheres, and flattened domes	Alternating micritic and botryoidal laminae (cf. Gomez et al. 2014)	Mostly developed in ponds with pinkish to orange stratified mats. Alternating micritic and botroidal lamina are usually irregular with variable thickness (50–500 μm thick). These are the most common lamina types in these structures. Micritic: represented by nanometer-scale spherical, globular, or spherulitic calcite (up to 300 nm) or more irregular globular to anhedral calcite. Botryoidal: individual or stacked micro-laminated botryoids (300 μm wide and 100 μm tall) or radial fibrous crystal bundles (50–100 μm wide 400–600 μm tall) associated with bacteria remains and diatom frustules
	Locally microspar lamina with preserved *Rivularia*-like filaments	Laminae bearing *Rivularia*-like filaments: irregular laminae with tufted dark brown to yellowish vertically oriented filaments (diameter 15–20 μm) in micro-spar translucent carbonates. Filaments form a paintbrush-like (cf. Reitner et al. 1996) palisade fabric. Alternates with micrite or botryoidal laminae
	Whitish irregular granular laminae	White granular precipitates with the presence of diatom and forming irregular laminae alternating with other lamina types. It shows characteristics remarkably similar to the granular texture and diatom–bacteria aggregates observed in the stratified pinkish-orange microbial mats thus being a fossilized equivalent
	Spar to microspar laminae with oriented pennate diatoms	Irregular and translucent microspar laminae, occasionally micritic. Parallel-oriented pennate diatom frustules. This has also been recorded during diatom blooming events in the areas where the pinkish-orange microbial mats are common
Stromatolites—cm to dm-scale flat to columnar stromatolites	Micritic laminae	Within ponds where stromatolites are covered by dark green to gray colored biofilms. These mini-stromatolites are columnar at the cm-scale and usually encrusting previous oncoids. Micritic laminae are irregular (100–300 μm thick) and shows variably preserved organic remains (degraded coccoid clusters) which are evident by fluorescence under UV light microscopy

limeter to decimeter carbonate crusts encrusting volcanic rocks, carbonate sediments, and other microbial structures as well as organic remains (e.g., flamingoes feathers) (Fig. 13.2). These can have a patchy distribution or form laterally continuous crusts (at a meter scale), covering loose or cemented carbonate sediments (peloidal and nodular carbonates). Carbonate plates formed by laminar crusts can coalesce to form more complex structures and can have overgrowths and show different growth stages due to movement and rotation by currents or cryoturbation processes. Laminar crusts can also develop dome-shaped morphologies, occasionally showing concentric growth patterns (Fig. 13.2). In addition, oriented and elongated structures are common, where these develop according to the main currents related to the prevailing winds (from northwest to southeast) (Fig. 13.3). Although regular isopachous laminae (Fig. 13.2) are the main building blocks of the laminar crusts (see details in Table 13.1), it is worth mentioning that the wind-oriented structures, in cross-section, develop more complex micro-textures. These include columnar, shrub-like to dendritic and micro-stromatolite microfabrics that resemble microbially influenced structures but are related to abiotic carbonate precipitation. These are formed by preferential growth where advective–diffusive flux provides calcium and carbonate ions (Fig. 13.3), analogous to silica and carbonate dendrites developed under unidirectional flow (see for example numerical models by Hawkins et al. 2013, 2014). These dendrite-shrubs-rich layers alternate with the previously described smooth laminar crusts.

Interpretation: Given the absence of microbial mats, and the macro-morphologies and micro-textures described (e.g., lamina regularity and degree of inheritance, lack of organic remains within the lamina), these structures have been interpreted as predominantly chemically precipitated carbonates, triggered by oversaturation related to water mixing (Gomez et al. 2014), strong CO_2 degassing, and evaporation (see Gomez et al. 2014; Buongiorno et al. 2018; Beeler et al. 2020).

Oncoids are typically located in the central area of the Stromatolite Belt and are composed of concentrically laminated, centimeter to decimeter spheroidal structures (up to ~35 cm). Morphologically, these are represented by discs, spheres, and flattened domes that can coalesce to form more complex, composite structures (Fig. 13.4). Oncoids grow by the accretion of smooth to irregular and overlapping laminae, showing lateral protrusions, typically formed at the sediment–water and the air–water interface. The external surface can be smooth or can show pillar-like to shrub-shaped millimeter scale protrusions and ornamentations (Fig. 13.4), particularly on the side affected by wind and currents. Oncoid rotation, particularly by cryoturbation, is also common, producing more complex overlapping overgrowths. Oncoids are typically associated with well-stratified pinkish to orange microbial mats (Fig. 13.8) but can also be colonized by cyanobacteria-rich microbial mats (*Rivularia halophila*) (Mlewski et al. 2018; Shalygin et al. 2018), which produce a different set of carbonate micro-textures and lateral overgrowths (Table 13.1, see Gomez et al. 2018; Mlewski et al. 2018 for details). Although oncoids are subspherical in shape, they can show asymmetric growth, occasionally expanded below the sediment–water interface, thus showing both hydrogenetic and diagenetic growth (above and below the sediment–water interface, respectively) (Fig. 13.5) analogous to manganese nodules observed in the deep sea (Fig. 13.5, right), being morphologically similar and suggesting similar diffusive–reactive processes involved in oncoids growth (Baker and Beaudoin 2013).

Interpretation: Given that oncoids are closely associated with microbial mats, the recorded morphologies and the diverse set of micro-textures observed in previous works (Table 13.1), these structures are interpreted as microbially influenced structures. The microbial influence is particularly observed in the development of microfabrics as well as the geochemical signatures preserved in the carbonates (Gomez et al. 2014; Buongiorno et al. 2018). Despite this, physicochemical processes (mostly water mixing, CO_2 degassing, and evaporation) are also particularly important to trigger carbonate precipitation (Gomez et al. 2014, 2018; Buongiorno et al. 2018; Beeler et al. 2020). The observed differential growth patterns, where oncoids grow bigger in the diagenetic zone (below the sediment–water interface), suggest that at least in some cases growth rates are higher in the anoxic zones. This is probably related to the influence of anoxic microbial metabolisms (e.g., sulfate reduction) that are known to increase alkalinity and thus carbonate precipitation.

Stromatolites here represent centimeter to decimeter-scale laminated structures (up to 25 cm) that typically have a planar to columnar shape (Fig. 13.6) resembling classic stromatolites. They are observed on the east side of the Stromatolite Belt (transition between Zones 3C and D), associated with dark greenish to black microbial mats and biofilms and usually are encrusting previous structures (typically oncoids) (Fig. 13.6), as well as loose to cemented sediments and other carbonate crusts. The columnar structures are usually centimeter-sized and represent ministromatolites (Fig. 13.6). The main difference with oncoids and laminar structures is in the shape (planar to columnar) and in the micro-textures, predominantly irregular, crenulated micritic to micro-peloidal crusts that preserve abundant organic remains (Fig. 13.6 and Table 13.1).

Interpretation: Given the crenulate and overlapping, irregularly shaped micrite- to micro-spar-rich laminae, which also preserve abundant organic remains, these are also interpreted as microbially influenced structures, as typically observed in other microbialites (Riding 2008). Since these structures are usually nucleated on oncoids (Fig. 13.6), it suggests changes in growth patterns probably triggered by local environmental changes as well as in the microbial mat type.

13.5 Stable Isotopes of the Laguna Negra Microbialites (Carbon and Oxygen)

Previous studies of $\delta^{13}C$ and $\delta^{18}O$ in the LN carbonates showed strong isotopic enrichment, particularly of the $\delta^{13}C$ of the carbonates reaching values up to 18‰ (Gomez et al. 2014; Buongiorno et al. 2018). This enrichment is similar to what has been observed in other saline lakes in the Puna–Altiplano region (Valero-Garces et al. 1999, 2000). Figure 13.7 shows a cross-plot of the $\delta^{13}C$ and $\delta^{18}O$ showing covariation, as typically observed in closed lakes where evaporation and CO_2 degassing are important (Talbot 1990) usually following a Rayleigh distillation pattern (Valero-Garces et al. 2000; Buongiorno et al. 2018).

In addition to the covariation pattern, Buongiorno et al. (2018) showed that both $\delta^{13}C$ and $\delta^{18}O$ values decreased over time when observed from the core to the outer edge of the oncoids and laminar crusts. These can show a change of up to 8‰ in $\delta^{13}C$ and 4‰ in $\delta^{18}O$ toward lower values. This was interpreted as due to progressive freshening, the increase in the influx of freshwater carrying a lighter isotope signal (for details see Buongiorno et al. 2018). An increase in humidity between 2200 and 1800 years before present has been documented with different proxies, with an estimated increase of 15–20% (Boschetti et al. 2007). This pattern toward lighter isotopes is less clear in the laminar crusts (when compared with the oncoids), given that in Zone 3A the mixing rate with groundwater is lower so the evaporation signal

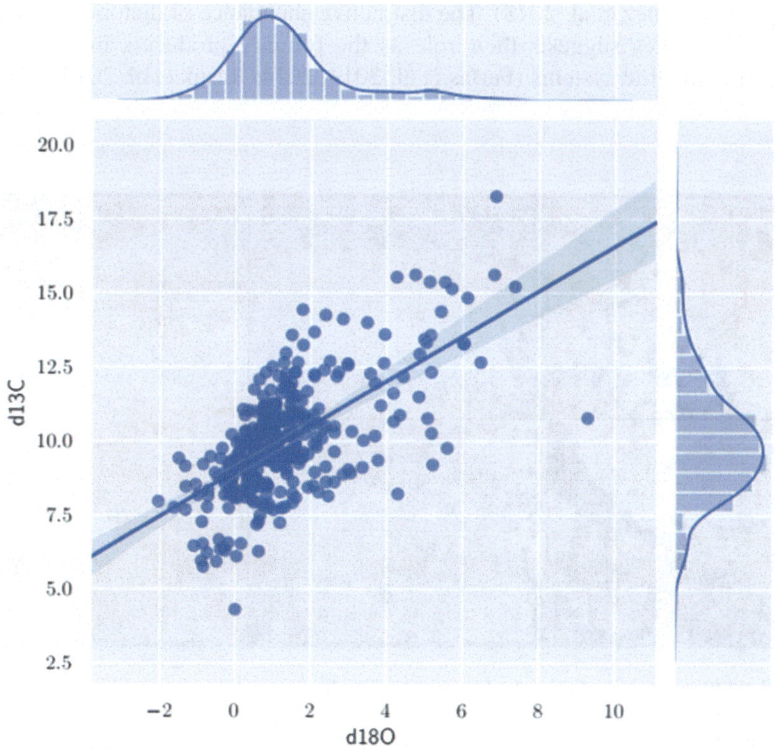

Fig. 13.7 Cross-plot of the carbon and oxygen stable isotopes of the LN microbialites and laminar crusts compiled from those published by Gomez et al. (2014) and Buongiorno et al. (2018). The adjusted trend line is $\delta^{13}C = 0.76 * \delta^{18}O + 8.9$ ($R = 0.65$)

is stronger, as suggested by heavier values in $\delta^{18}O$ of laminar crusts when compared with oncoids (Buongiorno et al. 2018).

13.6 Microbial Mats Diversity

The LN system allows studying the relationship between microorganisms and calcium carbonate precipitation. As stated, the environmental conditions, such as high UV exposure, high salinity, temperature fluctuations, and strong winds, restrict life mostly to bacteria, archaea, and unicellular eukaryotes. The LN mats consist of complex microbial communities that develop in the interface between sediment and water in the Stromatolite Belt. The presence of diatoms and bacteria–diatom–mineral aggregates represents the main component of the LN microbial mats. Pennate and centric diatoms are observed; although pennates appear to be more abundant. Some of the diatom groups recognized by morphological microscopy observation include *Achnanthes brevipes*, *Halamphora* sp., *Navicula* sp., *Surirella* sp., and *Striatula* sp. (Gomez et al. 2018). The distinctive abundance of diatoms reported in other Andean lakes suggests their role as the primary producers in these high-altitude microbialitic systems (Farías et al. 2013, 2014; Rasuk et al. 2014, 2015).

Fig. 13.8 Different types of microbial mats found in the Stromatolite Belt area, associated with carbonates, mostly oncoids. (**a**) Orange Stratified Mat (OSM), the most common microbial mat. Black Pustular Mat (BPM) is also observed surrounding partially exposed oncoids. (**b**) Certain zones exhibit a blackish microbial community, here called Black Patch (BP). (**c**) Pink Stratified Mat (PSM) near greenish patches (Green Patch-GP). (**d**) A distinct microbial community develops a greenish soft bubble floating mat (Green Mat-GM) associated with insipient oncoids. (**e**) Black Pustular Mat (BPM), a cyanobacterial dominated community that is located close to the air–water interface covering carbonate crusts

There is a variety of microbial mats inhabiting the Stromatolite Belt, with recognizable macroscopic differences in superficial coloration and textural aspects (Fig. 13.8). The following descriptions correspond to the most remarkable types of microbial mats found covering the sediments in the area and their related bacterial diversity according to 16S rDNA 454 pyrosequencing analysis.

13.6.1 Orange Stratified Mat (OSM)

The oncoids of the Stromatolite Belt are mainly associated with this type of microbial mat. These mats have an orange surface and a granular texture and show a stratified internal structure controlled by light and redox gradients (Teske and Stahl 2002), as commonly seen in microbial mats in hypersaline settings. There is a distinguishable orange top layer (1–2 cm thick), followed by a purple layer (2–5 mm thick), then a thin green layer and a black anoxic horizon at the very bottom (several cm to dm thick). Among the studied mats from the LN, this has the highest diversity based on observed species and Shannon index, and the lowest Dominance and highest Equitability, meaning that the most prominent groups are more evenly distributed in the community composition than in other mats (Gomez et al. 2018). The most abundant phylum is Proteobacteria (24%) (Fig. 13.9), mostly Alphaproteobacteria and

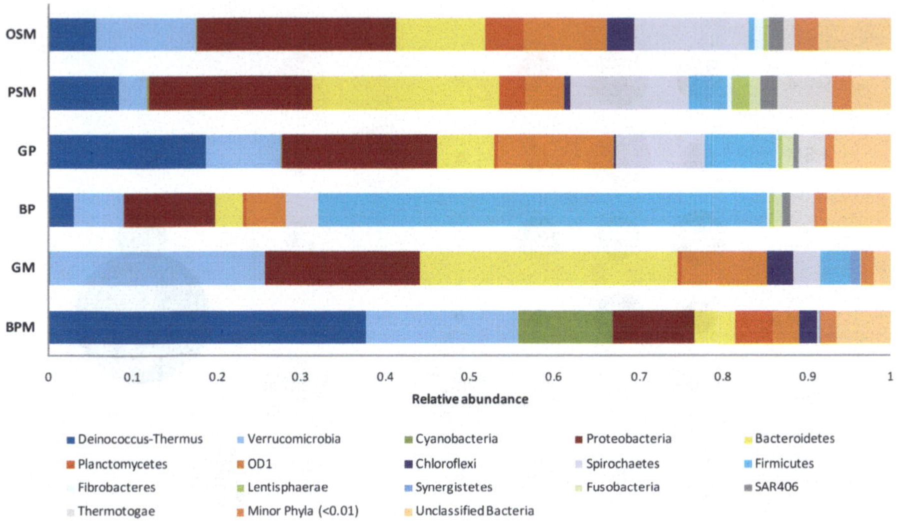

Fig. 13.9 Relative abundance of phylum-level bacterial composition derived from each type of microbial mat sampled. Note: Deinococcus–Thermus, Verrucomicrobia, Proteobacteria, Bacteroidetes, Spirochaetes, and Firmicutes are among the most representative phyla in almost all samples studied. Category "Minor Phyla (<0.01)" includes classified phyla present in less than 1% relative abundance: Acidobacteria, Actinobacteria, Armatinmonadetes, Chlamydiae, Chlorobi, Gemmatimonadetes, and Tenericutes

Deltaproteobacteria (Fig. 13.10). The most frequent Gammaproteobacteria operational taxonomic unit (OTU) matches with *Halochromatium* sp., a purple sulfur bacterium, and the most common OTU of Deltaproteobacteria matches with *Desulfobacula* sp., a marine sulfate-reducing bacterium with the capacity to degrade aromatic compounds. Other remarkable phyla found are Spirochaetes (14%), Verrucomicrobia (12%), Bacteroidetes (11%), and candidate phylum OD1 (currently named Parcubacteria, 10%) (Fig. 13.9), whose members were recently described as probably being ectosymbionts or parasites of other organisms (Nelson and Stegen 2015).

13.6.2 Pink Stratified Mat (PSM)

This mat presents a more pinkish coloration in the surface, a stratified internal structure, and a less granular texture. Bacteroidetes (22%) is the dominant phylum (Fig. 13.9). Proteobacteria is found second in abundance (19%) and most of it belongs to the Desulfobacteraceae family in the Deltaproteobacteria class (Fig. 13.10). The phylum Spirochaetes is also present (14%). Deinococcus–Thermus (8%) and Firmicutes (4%) are less abundant. These results show that heterotrophic saccharolytic bacteria are the main bacterial families detected and mostly include Rhodothermaceae (Park et al. 2014), Spirochaetaceae (Karami et al. 2014), and

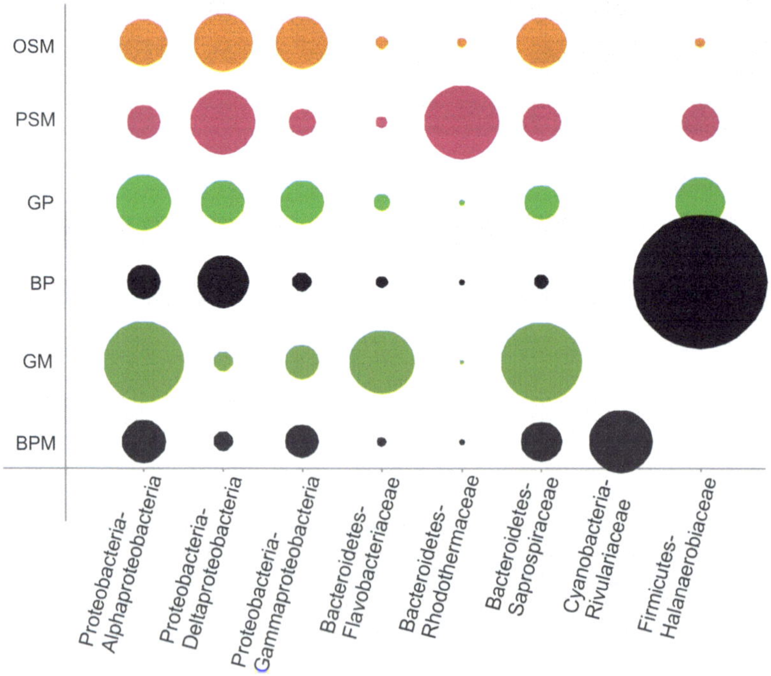

Fig. 13.10 Relative abundance of most remarkable subphylum level groups in each microbial mat sampled

Deinococcaceae (Makarova et al. 2007). The Chromatiales order (anoxygenic photosynthetic bacteria) that use hydrogen sulfide as electron donors and accumulate elemental sulfur in globules inside their cells (Imhoff 2014) are also detected in abundance within this mat (for more details see Gomez et al. 2018).

13.6.3 Green Patch (GP)

This mat represents an isolated greenish patch in between the PSM with a less stratified structure. The most abundant phylum present is Deinococcus–Thermus (19%), follow by Proteobacteria (18%), OD1 (Parcubacteria, 14%), Spirochaetes (11%), and Verrucomicrobia (9%) (Fig. 13.9). Among Proteobacteria, Rhodobacteraceae (3%) and Desulfobacteraceae (4%) are abundant. The green coloration of this mat could be explained by the abundance of diatoms, as the highest number of Stramenopiles chloroplast 16S rDNA sequences is recorded in this mat.

13.6.4 Black Patch (BP)

This is another patchy community, restricted to shallow ponds in between oncoids. The bacterial diversity record is clearly distinct. The BP had the lowest diversity with a lower Equitability and higher Dominance index in comparison with other spots and the stratified sites (Gomez et al. 2018) (Fig. 13.9). In this mat, half of the sequences belong to Firmicutes (53%), the dominant phylum. Within Firmicutes, nearly 44% belong to *Halanaerobium* sp., Halanaerobiaceae (98% identity Greengenes database) (Fig. 13.10). This genus may indicate the predominance of anaerobic fermentative halophilic communities. Other relatively abundant groups include Proteobacteria (11%) and Verrucomicrobia (6%).

13.6.5 Green Mat (GM)

This mat is observed closer to the groundwater springs, it is a green mat occasionally observed floating due to the presence of gas bubbles. The pyrosequencing analysis of the GM shows that most of the members belong to the Bacteroidetes phylum (30%) (Fig. 13.9). Roughly, 17% of them correspond to the Saprospiraceae family (Fig. 13.10), which is known to have an important role in the breakdown of complex organic compounds (McIlroy and Nielsen 2014). Besides, Flavobacteriaceae (11%) (Fig. 13.10), mostly *Winogradskyella* sp., is detected. Verrucomicrobia is abundant as well with a representation of 26%. Proteobacteria is well represented with 18%, where 83% of the relative abundance in this group belongs to Alphaproteobacteria (mostly Rhodobacteraceae). In comparison with all the others mats, the phylum Deinococcus–Thermus is not observed within the GM.

13.6.6 Black Pustular Mat (BPM)

There are clear macroscopic characteristics that make this mat unique. Cyanobacterial colonies develop a pustular to pinacular surface built by filaments, and it has a distinct black coloration that might be due to the scytonemin pigment from the cyanobacteria for protection against UV radiation. The BPM is commonly found in shallow ponds (a few cm) near the coast of the lake, close to the air–water interface, or colonizing the rims of partially exposed oncoids. This mat presents the highest Dominance index, and the lowest Equitability, meaning that the taxonomic groups present have uneven relative abundances and that only a few of them dominate the community (Gomez et al. 2018). Of the bacterial diversity, 37% belongs to Deinococcus–Thermus (all Deinococcaceae family), Verrucomicrobia (17%) with families like Spartobacteriaceae (10%) and Puniceicoccaceae (4%), and Proteobacteria (10%), mainly Alphaproteobacteria (4%, with Rhodobacteraceae and Rhodospirillaceae members) and Gammaproteobacteria (3%) (Fig. 13.9). Remarkably, in terms of diversity, the BPM is the only type of microbial mat in the LN that presents a significant abundance of the phylum Cyanobacteria (11%), mainly Rivulariaceae (10.5%) (Fig. 13.10). Most groups present in the mat are heterotrophic bacteria, except for Cyanobacteria, and the most abundant family (Deinococcaceae) includes saccharolytic bacteria (Makarova et al. 2007), which are probably able to degrade extracellular polymeric substances (EPS) produced by diatoms.

In summary, considering all the different mats, the diversity analysis shows that Proteobacteria, Verrucomicrobia, Bacteroidetes, and Deinococcus–Thermus are the most representative phyla in almost all samples studied. Other phyla like Spirochaetes and the gram positive Firmicutes are also abundant. All these groups are well known to be abundant in marine ecosystems and also in extreme environments, such as microbial mats from hypersaline systems such as Shark bay (Wong et al. 2015), Abu Dhabi (Abed et al. 2008), Guerrero Negro (Harris et al. 2013); with few differences such as the absence of Acidobacteria and Actinobacteria phyla. The LN microbial community seems to be similar at the phyla level with other Andean ecosystems, known as High Altitude Andean Lakes (HAALs). Deinococcus–Thermus is a group of bacteria known for their high UV radiation tolerance (Ivanova et al. 2011), and it is also recorded in other HAALs, such as Tebenquiche, La Brava (Farías et al. 2014; Fernandez et al. 2016), Socompa (Farías et al. 2013; Toneatti et al. 2017), Diamante (Rascovan et al. 2016), Cejar, Llamara, Jachucoposa, and Pujsa (Rasuk et al. 2016), where the UV radiation is one of the environmental challenges that organisms have to cope with. Cyanobacteria is one of the major groups in the BPM but is less represented or absent in the rest of the mats sampled in the LN and in other HAALs. However, confocal microscopy observations detected the presence of cyanobacteria in all analyzed samples from the LN, although not as a dominant group (Gomez et al. 2018). This distinction of the BPM is also reflected in PCoA analysis, where the BPM does not cluster with any other bacterial community studied. Something similar occurs with the non-stratified GM; this mat appears distant in the analysis, illustrating differences in the taxonomical composition of the community (Fig. 13.11).

To better characterize the BPM, the predominant filamentous cyanobacterium was collected, isolated, and cultured for a 16S–23S ITS phylogenetic and morphological characterization. The isolated strain presents unambiguous morphological characteristics, such as wider trichomes and filaments, uniquely branched trichomes, and mucilaginous pads at the bases of young trichomes. Additionally, based on the molecular phylogenetic analysis, the strain was found to be a unique and independent lineage on the evolutionary tree belonging to the genus *Rivularia* inside the Rivulariaceae family (Shalygin et al. 2018). As a result, considering the morphological and phylogenetic analyses, the cyanobacterial strain retrieved from the LN constitutes a new species (see Shalygin et al. 2018 for details) under requirements of the International Code of Nomenclature for Algae, Fungi and Plants, named *Rivularia halophila* sp. nov. (Fig. 13.12). It is worth mentioning that *Rivularia halophila* is the first species of the *Rivularia* genus reported from an inland, hypersaline aquatic environment.

Different types of carbonate laminae are identified within the oncoids and these seem to be partly associated with different microbial communities (for details see Gomez et al. 2014, 2018). Interestingly, laminae with calcified *Rivularia*-like cyanobacterial filaments showing tufted palisade fabrics are observed alternating with micritic and botryoidal laminae (see details in Table 13.1). Gomez et al. (2018) and Mlewski et al. (2018) showed that carbonate precipitation is not directly associated

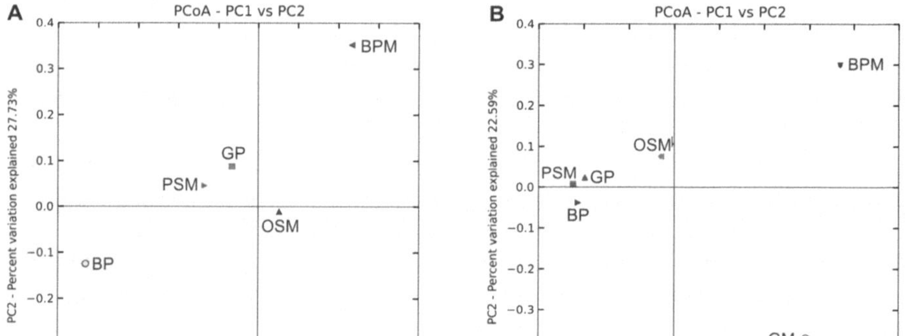

Fig. 13.11 Bacterial communities clustered using PcoA of the weighted (**a**) and unweighted (**b**) UniFrac distance matrices. Each point corresponds to a different microbial mat sampled. The percentages of variation explained by the plotted principal coordinates are indicated on the axes

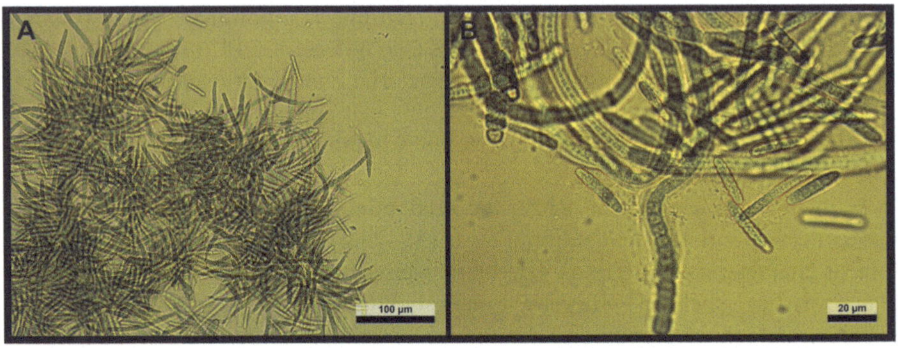

Fig. 13.12 Microscopy photographs of *Rivularia halophila*, the representative cyanobacterium of the BPM. (**a**) Almost spherical colonies, with radial arrangement of filaments. (**b**) Detail of filaments, where short juvenile stages are visible

with *Rivularia* itself (sheats are not usually calcified) but with the microbial consortia living around the *Rivularia* filaments, where diatoms and anoxygenic phototrophic bacteria are particularly abundant (for details see Mlewski et al. 2018). This is interesting given the fact that *Rivularia*-like filaments have been observed in the fossil stromatolite record, so this may have implications to better understand *Rivularia* mineralization and preservation. For this, a more detailed study focused on the microorganisms belonging to the consortia existing around the *Rivularia Halophila* filament was conducted (Mlewski et al. 2018).

Phylogenetic analyses (by Sanger sequencing) on the bulk BPM and whole genome amplification on laser microdissected filaments highlighted the presence of Bacteroidetes affiliated with *Marivirga* (Pagani et al. 2011), *Maribacter,* and *Winogradskyella*. All these genera constitute the most abundant epiphytic bacterial community associated with the *Rivularia* filaments (Mlewski et al. 2018). Proteobacteria and Verrucomicrobia are also among the most abundant phyla detected. Interestingly, members of Gammaproteobacteria found here have their closest relative in uncultivated bacteria (98% identity) retrieved from the Altiplano at Salar de Ascotan in Chile.

Interestingly, *Rivularia halophila* does not present any carbonate precipitation on its sheath itself. The microorganisms located on the cyanobacterial sheath may metabolically modify the local physicochemical conditions and induce or preclude carbonate precipitation. The analysis of the microorganisms specifically associated with *Rivularia halophila* sheaths revealed that these are affiliated with epiphytic members of Bacteroidetes phylum, more specifically to the *Maribacter* genus that includes heterotrophic bacteria (Nedashkovskaya et al. 2004). Thus, their activity possibly induces acidic conditions by producing CO_2 around the *Rivularia* sheaths, precluding carbonate precipitation (Dupraz and Visscher 2005; Dupraz et al. 2009) and explaining, in part, the absence of carbonation on the *Rivularia halophila* sheaths.

As previously stated, carbonate precipitation typically occurs on the EPS matrix around the *Rivularia* filaments and not on the sheaths (Mlewski et al. 2018), and this may have implications to better understand taphonomic aspects of cyanobacteria calcification in the rock record. EPS-related carbonate precipitation was also observed in all described LN mats, within the EPS matrix excreted by a diverse diatoms-bacterial consortium (Gomez et al. 2018; Mlewski et al. 2018). Diatoms and other microorganisms such as Myxococcales and methanogenic archaea are known to produce large amounts of EPS (Bapteste et al. 2005; Scholten et al. 2005). These EPS should serve as nucleation sites for carbonate precipitation following organo-mineralization. In addition to the photosynthetic activity of *Rivularia* (for the BPM case) and diatoms that promoted local alkalinization, some other bacteria identified in the whole microbial community potentially present metabolisms that favor carbonate precipitation. For example, some bacteria belonging to the Myxococcales order are known to favor mineral precipitation (Jimenez-López et al. 2007) by ammonification, enhancing the alkalinity of the medium (González-Muñoz et al. 2010). Besides, some of the aerobic anoxygenic phototrophic bacteria (AAnPB) affiliated with the marine Roseobacter clade are known to interact with marine phytoplankton, including diatoms, and this association allows microbes to use metabolic niches that would be inaccessible otherwise (Overmann and van Gemerden 2000; Schink 2002; Orphan et al. 2008). Most of the members of the Roseobacterclade are ureolysers and some are denitrifiers (Luo and Moran 2014). Both ureolysis (Zhu and Dittrich 2016) and denitrification metabolisms (Erşan et al. 2015) increase pH in the surrounding medium and favor carbonate precipitation. Hence, phototrophy, ureolysis, and denitrification associated with the activity of AAnPB may be important drivers of alkalinization and carbonate precipitation in the BPM.

13.7 Concluding Remarks

Carbonate precipitation is expected to occur in the LN, given that mixing between groundwater springs (alkalinity-rich) and the main lake waters (calcium-rich) increase the saturation state triggering carbonate precipitation (Gomez et al. 2014). In addition to mixing, other physicochemical processes such as evaporation and strong CO_2 degassing also contribute to increase saturation states, thus promoting carbonate precipitation (Beeler et al. 2020, see also Gomez et al. 2019, chapter 16). Despite this, the extensive microbial mat system also contributes, through its metabolic activity, to carbonate precipitation. This occurs by changing chemical equilibrium (producing alkalinity) and providing nucleation sites for mineral precipitation, particularly due to the presence of EPS. This is recorded in the carbonate micro-textures in the oncoids and stromatolites as well as in the microbial mats (Table 13.1). Some of the recognized microbes show metabolisms that potentially contribute to carbonate precipitation, including sulfate reduction, ammonification, oxygenic and anoxygenic phototrophic bacteria, uroelyzers, and denitrifiers (Gomez et al. 2018; Mlewski et al. 2018). This microbial influence has an impact on the texture of the microbial carbonates, which is evident when comparing laminar crusts (where no microbial mats are visible), oncoids, and stromatolites (where microbial mats and biofilms are present). By using carbon and oxygen isotope proxies, the LN carbonates have also proven to be useful for recording local environmental changes (Gomez et al. 2014; Buongiorno et al. 2018). These include variable mixing rates of groundwaters and lake waters, strong CO_2 degassing and evaporation, as well as regional environmental changes, for example, the progressive increase in moisture and freshening of lake waters is related to recent climate changes. Unraveling the relative influences of these different controls in carbonate precipitation, micro-textures, and geochemical signatures is challenging, even in these modern active systems, so care must be taken when working in the rock record. Thus, a combined multi-proxy approach in modern systems such as the LN provides some insights to better understand the biosphere–geosphere interactions and record in modern and ancient systems.

References

Abed RMM, Kohls K, Schoon R, Scherf AK, Schacht M, Palinska KA, Al-Hassani H, Hamza W, Rullkötter J, Golubic S (2008) Lipid biomarkers, pigments and cyanobacterial diversity of microbial mats across intertidal flats of the arid coast of the Arabian Gulf (Abu Dhabi, UAE). FEMS Microbiol Ecol 65(3):449–462. https://doi.org/10.1111/j.1574-6941.2008.00537.x

Allwood A, Walter MR, Kamber BS, Marshall CP, Burch IW (2006) Stromatolite reef from the Early Archaean era of Australia. Nature 441:714–718

Awramik SM, Buchheim HP (2009) A giant, late Archean lake system: the Meentheena Member (Tumbiana Formation; Fortescue Group), Western Australia. Precambrian Res 174:215–240

Baker E, Beaudoin Y (2013) In: Baker E, Beaudoin Y (eds) Deep sea minerals: manganese Nodules, a physical, biological, environmental, and technical review, vol 1B. Secretariat of the Pacific Community, Noumea

Bapteste E, Brochier C, Boucher Y (2005) Higher-level classification of the Archaea: evolution of methanogenesis and methanogens. Archaea 1:353–363. https://doi.org/10.1155/2005/859728

Beeler S, Gomez FJ, Bradley A (2020) Controls of extreme isotopic enrichment in modern microbialites and associated abiogenic carbonates. Geochim Cosmochim Acta 269:136–149

Boschetti T, Cortecci G, Barbieri M, Mussi M (2007) New and past geochemical data on fresh to brine waters of the Salar de Atacama and Andean Altiplano, northern Chile. Geofluids 7(1):33–50

Buick R (1992) The antiquity of oxygenic photosynthesis: evidences from stromatolites in sulphate-deficient Archean lakes. Science 255:74–77

Buongiorno J, Gomez FJ, Fike D, Kah LC (2018) Mineralized microbialites as archives of environmental evolution, Laguna Negra, Catamarca Province, Argentina. Geobiology 1:1–24

Burne R, Moore L (1987) Microbialites: organo-sedimentary deposits of benthic microbial communities. PALAIOS 2:241–254

Dupraz C, Reid RP, Braissant O, Decho AW, Norman RS, Visscher PT (2009) Processes of carbonate precipitation in modern microbial mats. Earth Sci Rev 96:141–152. https://doi.org/10.1016/j.earscirev.2008.10.005

Dupraz C, Visscher PT (2005) Microbial lithification in marine stromatolites and hypersaline mats. Trends Microbiol 13:429–438. https://doi.org/10.1016/j.tim.2005.07.008

Erşan YÇ, de Belie N, Boon N (2015) Microbially induced $CaCO_3$ precipitation through denitrification: an optimization study in minimal nutrient environment. Biochem Eng J 101:108–118. https://doi.org/10.1016/j.bej.2015.05.006

Farías ME, Contreras M, Rasuk MC, Kurth D, Flores MR, Poiré DG, Novoa F, Visscher PT (2014) Characterization of bacterial diversity associated with microbial mats, gypsum evaporites and carbonate microbialites in thalassic wetlands: Tebenquiche and La Brava, Salar de Atacama, Chile. Extremophiles 18:311–329. https://doi.org/10.1007/s00792-013-0617-6

Farías ME, Rascovan N, Toneatti DM, Albarracín VH, Flores MR et al (2013) The discovery of stromatolites developing at 3570 m above sea level in a high-altitude volcanic lake Socompa, Argentinean Andes. PLoS One 8(1):e53497. https://doi.org/10.1371/journal.pone.0053497

Fernandez AB, Rasuk MC, Visscher PT, Contreras M, Novoa F, Poire DG, Patterson MM, Ventosa A, Farías ME (2016) Microbial diversity in sediment ecosystems (evaporites domes, microbial Mats, and crusts) of hypersaline Laguna Tebenquiche, Salar de Atacama, Chile. Front Microbiol 7:1284. https://doi.org/10.3389/fmicb.2016.01284

Gomez FJ, Boidi FJ, Mlewski C, Gerard E (2019) Chapter 19: The carbonate system in hypersaline lakes: the Laguna Negra case (Puna of Catamarca, Argentina). In: Farias ME (ed) Microbial ecosystems in Puna environments: biofilms, microbial mats microbialites and endoevaporites. Springer, New York

Gomez FJ, Kah LC, Bartley J, Astini RA (2014) Microbialites in a high-altitude Andean Lake: multiple controls in carbonate precipitation and lamina accretion. PALAIOS 29:233–249

Gomez FJ, Mlewski C, Boidi FJ, Farías ME, Gérard E (2018) Calcium carbonate precipitation in diatom-rich microbial mats: the Laguna Negra Hypersaline Lake, Catamarca, Argentina. J Sediment Res 88(6):727–742. https://doi.org/10.2110/jsr.2018.37

González-Muñoz MT, Rodriguez-Navarro C, Martínez-Ruiz F, Arias JM, Merroun ML, Rodriguez-Gallego M (2010) Bacterial biomineralization: new insights from Myxococcus-induced mineral precipitation. In: Pedley HM, Rogerson M (eds) Tufas and speleothems: unravelling the microbial and physical controls, vol 336. Geological Society; Special Publications, London, pp 31–50

Harris JK, Caporaso JG, Walker JJ, Spear JR, Gold NJ, Robertson CE, Hugenholtz P, Goodrich J, McDonald D, Knights D, Marshall P, Tufo H, Knight R, Pace NR (2013) Phylogenetic stratigraphy in the Guerrero Negro hypersaline microbial mat. ISME J 7:50–60. https://doi.org/10.1038/ismej.2012.79

Hawkins C, Angheluta L, Hammer Ø (2014) Hydrodynamic shadowing effect during precipitation of dendrites in channel flow. Phys Rev E 89:022402

Hawkins C, Angheluta L, Hammer Ø, Jamtveit B (2013) Precipitation dendrites in channel flow. EPL 102:54001. https://doi.org/10.1209/0295-5075/102/54001

Imhoff JF (2014) International Committee on Systematics of Prokaryotes Subcommittee on the taxonomy of phototrophic bacteria: minutes of the closed online meeting. Int J Syst Evol Microbiol 64:3910–3912. https://doi.org/10.1099/ijs.0.068908-0

Ivanova N, Rohde C, Munk C, Nolan M, Lucas S, Del Rio TG et al (2011) Complete genome sequence of Truepera radiovictrix type strain (RQ-24 T). Stand Genomic Sci 4:91–99. https://doi.org/10.4056/sigs.1563919

Jimenez-López C, Jroundi F, Rodriguez-Gallego M, Arias JM, González- Muñoz MT (2007) Biomineralization induced by Myxobacteria. In: Méndez-Vilas A (ed) Communicating current research and educational topics and trends in applied microbiology, vol 1. Formatex, Badajoz, pp 143–154

Karami A, Sarshar M, Ranjbar R, Sorourizanjani R (2014) The phylum Spirochaetaceae. In: Rosenberg E, Delong EF, Lory S, Stackebrandt E, Thompson F (eds) The prokaryotes. Springer, Berlin, Heidelberg, pp 915–929

Luo H, Moran MA (2014) Evolutionary ecology of the marine Roseobacter clade. Microbiol Mol Biol Rev 78:573–587. https://doi.org/10.1128/MMBR.00020-14

Makarova KS, Omelchenko MV, Gaidamakova EK, Matrosova VY, Vasilenko A et al (2007) Deinococcus geothermalis: the pool of extreme radiation resistance genes shrinks. PLoS One 26:e955

McIlroy SJ, Nielsen PH (2014) The family Saprospiraceae. In: Rosenberg E, DeLong EF, Lory S, Stackebrandt E, Thompson F (eds) The prokaryotes. Springer, Berlin, Heidelberg, pp 863–889

Mlewski EC, Pisapia C, Gomez F, Lecourt L, Soto Rueda E, Benzerara K, Ménez B, Borensztajn S, Jamme F, Réfrégiers M, Gérard E (2018) Characterization of pustular mats and related Rivularia-rich laminations in oncoids from the Laguna Negra Lake (Argentina). Front Microbiol 9:996. https://doi.org/10.3389/fmicb.2018.00996

Nedashkovskaya O, Kim S, Kyun Han S, Lysenko AM, Rohde M, Rhee MS et al (2004) Maribacter gen. nov., a new member of the family Flavobacteriaceae, isolated from marine habitats, containing the species Maribacter sedimenticola sp. nov., Maribacter aquivivus sp. nov., Maribacter orientalis sp. nov. and Maribacter ulvicola sp. nov. Int J Syst Evol Microbiol 54:1017–1023. https://doi.org/10.1099/ijs.0.02849-0

Nelson WC, Stegen JC (2015) The reduced genomes of Parcubacteria (OD1) contain signatures of a symbiotic lifestyle. Front Microbiol 6:713

Orphan VJ, Jahnke LL, Embaye T, Turk KA, Pernthaler A, Summons RE et al (2008) Characterization and spatial distribution of methanogens and methanogenic biosignatures in hypersaline microbial mats of Baja California. Geobiology 6:376–393. https://doi.org/10.1111/j.1472-4669.2008.00166.x

Overmann J, van Gemerden H (2000) Microbial interactions involving sulfur bacteria: implications for the ecology and evolution of bacterial communities. FEMS Microbiol Rev 24:591–599. https://doi.org/10.1111/j.1574-6976.2000.tb00560.x

Pagani I, Chertkov O, Lapidus A, Lucas S, Del Rio TG, Tice H et al (2011) Complete genome sequence of Marivirga tractuosa type strain (H-43). Stand Genomic Sci 4:154–162. https://doi.org/10.4056/sigs.1623941

Park S, Akira Y, Kogure K (2014) The family Rhodothermaceae. In: Rosenberg E, Delong EF, Lory S, Stackebrandt E, Thompson F (eds) The prokaryotes. Springer, Berlin, Heidelberg, pp 849–856

Rascovan N, Maldonado J, Vazquez MP, Farías ME (2016) Metagenomic study of red biofilms from Diamante Lake reveals ancient arsenic bioenergetics in haloarchaea. ISME J 10:299–309

Rasuk MC, Fernandez AB, Kurth D, Contreras M, Novoa F, Poiré D, Farías ME (2015) Bacterial diversity in microbial mats and sediments from the Atacama Desert. Microb Ecol 71:44–56. https://doi.org/10.1007/s00248-015-0649-9

Rasuk MC, Fernández AB, Kurth D, Contreras M, Novoa F, Poiré D, Farías ME (2016) Bacterial diversity in microbial mats and sediments from the Atacama Desert. Microb Ecol 71:44–56. https://doi.org/10.1007/s00248-015-0649-9

Rasuk MC, Kurth D, Flores MR, Contreras M, Novoa F, Poire D, Farías ME (2014) Microbial characterization of microbial ecosystems associated to evaporites domes of gypsum in Salar de Llamara in Atacama Desert. Microb Ecol 68:483–494

Reitner J, Paul J, Arp G, Hause-Reitner D (1996) Lake Thetis domal microbialites. Acomplex framework of calcified biofilms and organomicrites (Cervantes, Western Australia), in global and regional controls on biogenic sedimentation. In: Reitner J, Neuweiler F, Gunkel F (eds) Reef evolution research reports, vol SB2. Göttinger Arb. Geol. Paläont, Sonderband, Göttingen, pp 85–89

Riding R (2008) Abiogenic, microbial and hybrid authigenic carbonate crusts: components of Precambrian stromatolites. Geologia Croatica 61(2–3):73–103

Schink B (2002) Synergistic interactions in the microbial world. Antonie Van Leeuwenhoek 81:257–261. https://doi.org/10.1023/A:1020579004534

Scholten JCM, Joye SB, Hollibaugh JT, Murrell JC (2005) Molecular analysis of the sulfate reducing and archaeal community in a meromictic soda lake (Mono Lake, California) by targeting 16S rRNA, mcrA, apsA, and dsrAB genes. Microb Ecol 50:29–39. https://doi.org/10.1007/s00248-004-0085-8

Shalygin S, Pietrasiak N, Gomez F, Mlewski C, Gerard E, Johansen JR (2018) Rivularia halophila sp. nov. (Nostocales, Cyanobacteria): the first species of Rivularia described with the modern polyphasic approach. Eur J Phycol 53:537–548. https://doi.org/10.1080/09670262.2018.1479887

Talbot MR (1990) A review of paleohydrological interpretation of carbon and oxygen isotopic ratios in primary lacustrine carbonates. Chem Geol (Isot Geosci Sect) 80:261–279

Teske A, Stahl DA (2002) Microbial mats and biofilms: evolution, structure and function of fixed microbial communities. In: Staley JT, Reysenbach AL (eds) Biodiversity of microbial life. Wiley-Liss, Inc., New York, pp 49–100

Toneatti DM, Albarracín VH, Flores MR, Polerecky L, Farías ME (2017) Stratified bacterial diversity along physico-chemical gradients in high-altitude modern stromatolites. Front Microbiol 8:646. https://doi.org/10.3389/fmicb.2017.00646

Valero-Garces BL, Delgado-Huertas A, Ratto N, Navas A (1999) Large 13C enrichment in primary carbonate from Andean Altiplano lakes, northwest Argentina. Earth Planet Sci Lett 171:253–266

Valero-Garces BL, Delgado-Huertas A, Ratto N, Navas A, Edwards L (2000) Paleohydrology of Andean saline lakes from sedimentological and isotopic records, Northwestern Argentina. J Paleolimnol 24:343–359

Wong HL, Smith DL, Visscher PT, Burns BP (2015) Niche differentiation of bacterial communities at a millimetre scale in Shark Bay microbial mats. Sci Rep 5:15607. https://doi.org/10.1038/srep15607

Zhu T, Dittrich M (2016) Carbonate precipitation through microbial activities in natural environment, and their potential in biotechnology: a review. Front Bioeng Biotechnol 4:4. https://doi.org/10.3389/fbioe.2016.00004

Chapter 14
Extreme Microbiology at Laguna Socompa: A High-Altitude Andean Lake (3570 m a.s.l.) in Salta, Argentina

Virginia Helena Albarracín, Fátima Silvina Galván, and María Eugenia Farías

14.1 Introduction

"High-Altitude Andean Lakes" (HAAL) are unexplored ecosystems of shallow lakes and salt flats at altitudes between 3000 and 6000 m located at The Central Andes region. These pristine environments are a rich source of poly-extremophilic microbes that adapt to various physical and chemical stresses. For instance, large daily ambient thermal amplitude, alkalinity, high concentrations of arsenic (up to 200 ppm) and dissolved salts (Escudero et al. 2007; Farías et al. 2011a; Dib et al. 2008; Escalante et al. 2009; Ordoñez et al. 2009; Albarracín et al. 2015, 2016). Nonetheless, the main factor limiting life at the HAAL is the high solar irradiance. The values of UV-B flux at noon reach 10 W m^{-2} with a monthly average of daily insolation of 6.6 kWh m^{-2} d^{-1}; these parameters are among the highest in the world (Albarracín et al. 2015, 2016; Portero et al. 2019; Escudero et al. 2007; Cabrol et al.

V. H. Albarracín (✉)
Laboratorio de Investigaciones Microbiológicas de Lagunas Andinas (LIMLA), Planta Piloto de Procesos Industriales y Microbiológicos (PROIMI), CCT, CONICET, Tucumán, Argentina

Laboratorio de Microbiología Ultraestructural y Molecular, Centro Integral de Microscopía Electrónica (CIME, CONICET, UNT), Tucumán, Argentina

Facultad de Ciencias Naturales e Instituto Miguel Lillo, Universidad Nacional de Tucumán, Tucumán, Argentina
e-mail: cime@tucuman-conicet.gov.ar

F. S. Galván
Laboratorio de Investigaciones Microbiológicas de Lagunas Andinas (LIMLA), Planta Piloto de Procesos Industriales y Microbiológicos (PROIMI), CCT, CONICET, Tucumán, Argentina

Laboratorio de Microbiología Ultraestructural y Molecular, Centro Integral de Microscopía Electrónica (CIME, CONICET, UNT), Tucumán, Argentina

M. E. Farías
LIMLA-PROIMI-CONICET, Tucuman, Argentina

M. E. Farías (ed.), *Microbial Ecosystems in Central Andes Extreme Environments*, https://doi.org/10.1007/978-3-030-36192-1_14

2014). The intense UV radiation is due to the low latitude and high altitude at HAAL, thin ozone layer, and a clear sky (Portero et al. 2019).

UV radiation is subdivided into ultraviolet A (UV-A) (315–400 nm), UV-B (280–315 nm), and UV-C (<280 nm) (Comité Internationale de l'Eclairage, CIE 2011). As the solar UV-C spectrum is filtered by the atmospheric gases O_2, H_2O, CO_2, and mainly O_3, the UV range affecting life on Earth falls only between 280 to 400 nm. Of the UV spectrum that reaches the earth, 95% is UV-A; it causes cell damage indirectly by producing photooxidation of compounds and reactive oxygen species (Hernández et al. 2007; Portero et al. 2019). UV-B represents less than 5% of incident solar radiation caused oxidative damage, and UV-B is also able to excite and ionize biologically relevant molecules, such as lipids, proteins, and DNA. DNA damage produced by UV-B irradiation generated dimers or adducts formation between adjacent pyrimidines bases on one DNA strand (Mitchell and Karentz 1993; Portero et al. 2019). The dimerization products called cyclobutane pyrimidine dimers (CPDs) constitute 70–80% of total photoproducts, while the adducts accounting for the rest and are termed 6–4 [pyrimidine-2′-one] pyrimidines or (6–4) photoproducts ((6–4)PPs) photoproducts ((6–4)PPs) (Mitchell and Karentz 1993; Friedberg et al. 2005).

UV irradiation and the effect on microbial diversity and ecological impact was studied in Laguna Socompa. It is considered a hypersaline and moderate alkaline lake and is located at 3570 m a.s.l. at the base of the active Socompa volcano. The microbialites thriving on its shore are the highest on Earth (Farías et al. 2013), which exhibit an atypical microbial community with abundant representatives of *Deinococcus–Thermus*, Rhodobacteraceae, Desulfobacterales, and Spirochaetes. Their metagenome showed a high proportion of sequences depicting less than 80% identity to the best hit in the NCBI database, an indication of novel distant lineages (Farías et al. 2013). Several bacterial and archaeal strains from Socompa stromatolites were isolated, physiologically characterized, and subjected to genome pyrosequencing (Farías et al. 2011b; Ordoñez et al. 2013; Gorriti et al. 2014; Albarracín et al. 2016). Screening of the genomes of these strains, i.e., *Exiguobacterium* sp. S17, *Sphingomonas* sp. S17, *Nesterenkonia* sp. Act20, and *Salinivibrio* spp., yielded genes putatively encoding essential traits for survival under multiple extreme environmental conditions, e.g., high levels of UV radiation, elevated salinity, and the presence of poisonous arsenic concentrations (Farías et al. 2011b; Belfiore et al. 2013; Ordoñez et al. 2013, 2015; Gorriti et al. 2014; Albarracín et al. 2016). Additionally, a new species of *Salinivibrio* was recently proposed in this microbial ecosystem: *Salinivibrio socompensis* (Galisteo et al. 2019).

Thus, the study of microbes associated with Socompa stromatolites and their environment opens a window to the past by providing a modern analog to their Precambrian counterparts. On the other hand, they serve as models for further understanding of adaptation and the function of proteins under extremely harsh conditions, which offers a biotechnological interest.

In this chapter, we compare the microbial diversity of the different microbial niches found in Laguna Socompa (3570 m), i.e., lake water, sediments, stromatolites, and soil surrounding the lake.

14.2 The Environment

In a Tertiary and Cuaternary volcanic and volcaniclastic geological setting, Laguna Socompa (Fig. 14.1) is located at the northern part of Salta Province in Argentina near the border with Chile (S 24°35′34″ W 68°12′42″) at the base of the still active Socompa volcano (6031 m a.s.l.). With incipient access roads and far from any urban population, the lake is isolated and suffers from arid climate and desert conditions: daily temperatures range from 20 to −10 °C in summer and 10 to −40 °C in winter (Farías et al. 2011a, 2013).

The lake is shallow (60 cm) and presents high arsenic content (32 mg/L). A salinity gradient (from 3 mS in the west side to 125 mS in the east side) was likewise reported (Farías et al. 2011a). Stromatolites are located around the border of the lagoon, on the east side. During the summer, they are partially exposed depending on the tide and hydrological regime, but they are completely submerged during

Fig. 14.1 Laguna Socompa. (**a**) General Landscape showing the lake and the stromatolites at the shore. (**b**) Closer view of the stromatolites domes. (**c**) Perpendicular cut of the top 5 cm of the stromatolites showing the different colored layers

winter and spring seasons. Hydrothermal water is introduced to Laguna Socompa as part of the modern Andean volcanic system causing the water surrounding stromatolites to be relatively warm (20–24 °C), alkaline, rich in dissolved ions such as Na^+, K^+, Mg^{2+}, Ca^{2+}, Cl^-, and SO_4^{-2}, rich in organic carbon, nitrate, phosphate, silicate, and iron (Farías et al. 2013; Albarracín et al. 2015; Rasuk et al. 2017). In fact, this scenario is deemed to resemble some of the conditions that prevailed anciently on Earth and Mars, rendering models for evolutionary studies.

14.3 Microbial Diversity

Microorganisms that colonize extreme environments are called extremophiles, a group that includes representatives of all three domains (Bacteria, Archaea, and Eukarya). They are categorized into subgroups according to the environmental specificity of their surroundings, i.e., psychrophilic, thermophilic, halophilic, alkalophilic, and acidophilic (Albarracín et al. 2016). In their natural habitat, many extremophiles tolerate several extremes combined and are better termed "polyextremophiles" to indicate their ability to cope with several concurrent stressors. This definition also corresponds to Socompa's microbes since they simultaneously deal with the stress produced by extraordinary UV exposure, hipersalinity, high arsenic content, and alkaline pH values (pH 8–10).

Laguna Socompa (LS) and surrounding areas harbor a distinct and varied microbiodiversity thriving in most niches: as dispersed cells in shallow waters (plankton), sediments (benthos), or forming cooperative benthonic structures such as microbial mats and microbialites (Albarracín et al. 2015). Moreover, the dry, hypersaline, and alkaline soil surrounding the lake also showed an interesting microbiome. Diversity is highly dependent on the microecosystem considered, which is briefly reviewed in the next sections.

14.3.1 Microbes from the Lake Water/Sediments

HAAL are usually shallow lakes (10–400 cm) where benthic communities share members with planktonic communities, as winds might often stir up the sediment (Cabrol et al. 2007). A remarkable aspect of HAAL microbiodiversity is that, in general, bacterial communities dominate planktonic microecosystems—with Archaea represented to a much lower extent (Escudero et al. 2007; Demergasso et al. 2010; Dorador et al. 2013; Rascovan et al. 2016). In Laguna Socompa, halophilic Euryarchaeota were detected from water and sediment samples by culture-dependent and independent methods (Maldonado et al. 2018). DGGE bands sequencing identified the non-culturable community as belonging to the class Halobacteria, Methanobacteria, Methanococci, and Methanomicrobia with representative genus *Haloterrigena*, *Natrinema*, *Natrialba*, *Methanobacterium*,

Methanotorris, and *Methanococcus* (Maldonado et al. 2018). Nevertheless, only members of the class Halobacteria were isolated: *Natrinema*, *Halorubrum*, and *Natronorubrum*. All isolates were considered neutrophil with pH 8 being the optimum value; they also grew between 10% and 30% NaCl, a typical feature of the Halobacteriaceae family (Oren 2006). In accordance with their isolation origin, all strains were able to tolerate a total UV-B dose of up to 66.11 kJ m^{-2}. The pigmented strains displayed a higher tolerance to UV-B than the non-pigmented strains, suggesting a protective mechanism of carotenoid pigments against UVR damage (Moeller et al. 2005; Chattopadhyay 2008).

Bacteria were also present in Laguna Socompa water; *Salinivibrio* sp. S35, recently reported as *Salinivibrio socompensis* (Galisteo et al. 2019), which is gram-negative with motile, non-sporulating curved rods (Gorriti et al. 2014). The genome was sequenced and compared with other members of the same genus. Among all genes, the ones annotated as coding for cobalamine biosynthesis, Tra elements and Type II/IV secretion systems were considered to have an ecological interest. These genes may enable S35 to interact with dominant diatoms of the Socompa Lake through exudates of cobalamin, a vitamin that is necessary for the development of such microalgae communities (Haines and Guillard 1974; Farías et al. 2013; Gorriti et al. 2014). This is possible as diatoms represented the main algal group on LS's phytoplankton (Farías et al. 2011b); the genera *Nitzschia* and *Navicula* were reported as the most abundant. Cyanobacteria such as chrysophycea, euglenophyta, and chlorophyceae were also found, but in lower numbers and diversity (Farías et al. 2011b). Macroscopic eukayotes (e.g., crustaceans or insects) were not reported for this lake; otherwise, they would graze on the microbial mats and would not allow the formation of laminated structures such as the stromatolites (Farías et al. 2011a).

14.3.2 Stromatolites

Microbial mats are benthic communities growing on a solid substrate (e.g., sand, rock or sediments) as vertical layers of functional groups of microbes embedded in a mineral–organic matrix, most often exopolysaccharides, silicates, and carbonates (Kremer et al. 2008; Dupraz et al. 2009; Glunk et al. 2011; Albarracín et al. 2016). When these organic sedimentary structures present various degrees of mineral precipitation and lithification they are called microbialites (Baumgartner et al. 2009; Dupraz et al. 2009; Glunk et al. 2011; Albarracín et al. 2016). Typically, vertical stratification on mats/microbialites originates from the physicochemical gradients and their indigenous microbial activity. In phototrophic microbial mats, cyanobacteria and phototrophic eukaryotes harvest light and fix CO_2. The food chain of the microbialite and of the surrounding ecosystem is based on the organic matter and extracellular polymers (EPS) of these primary producers. In deeper layers, the system becomes anoxic and is fueled by microbial degradation of carbon reserves, i.e., fermenting and producing low-molecular organic acids and alcohols. Methanogenic and sulfate-reducing bacteria oxidize these byproducts and produce methane and sulfide. This last compound is also a sulfur-oxidizing bacteria subtract (Farías et al. 2013; Dupraz et al. 2009; Baumgartner et al. 2009; Decho et al. 2005; Visscher and Stolz 2005).

In recent years, microbial mat communities with different degrees of lithification at HAAL and the Atacama Desert area have been characterized and proposed as early Earth models. For instance, microbial mats formed mainly by *Proteobacteria* and *Bacteroidetes* were recorded in Laguna Llamara and Laguna Cejar in Chile (Rasuk et al. 2016). Also, in the Salar de Llamara of the Atacama Desert, dome-shaped biohermos, which present a stratified distribution of microbial communities, were described (Pueyo et al. 2001; Rasuk et al. 2014). Llamara endoevaporitic dome diversity changed in different seasons (summer and winter) and at the top and bottom layers. *Alpha* and *Gamma Proteobacteria* were dominant in summer and winter samples followed by *Bacteroidetes*. *Planctomycetes* were third in abundance (Rasuk et al. 2014). In the bottom layer, substantial differences were found between samples collected in winter and summer: while *Proteobacteria* were predominant in summer, they diminished in winter (26%); in contrast, *Bacteroidetes* were under-represented in summer and were predominant in winter. *Verrucomicrobia* (*Spartobacteria*) became very abundant only in the summer bottom layer. Minority groups were more abundant during winter enlarging the biodiversity (*Planctomycetes*, *OP8*, *OP1*, *OD1*, *Lentisphaerae*, *Firmicutes*, *Gemmatimonadetes*, *NKB19*, *Spirochaetas*, etc.). As in other high Andean microbial ecosystems, percentages of detected *Cyanobacteria* were low (2% in both seasons) (Rasuk et al. 2014). Likewise, the Diamante Lake, in the caldera of the Galán volcano, Antofagasta de la Sierra, Catamarca (4600 m a.s.l.), showed a special organo-sedimentary ecosystem at the bottom of calcareous rocks, i.e., a red biofilm of coccoid cells embedded in extracellular polymeric substances (EPS) and associated with gaylussite crystals. Metagenomics analyses showed it was composed almost exclusively of haloarchaea (94% is classified as Archaea and 6% as Bacteria). All Archaea sequences were assigned to the family Halobacteriaceae and nearly 20 different haloarchaea genera were identified as *Halorubrum* (55%), *Natronomonas* (14%), *Halonotius* (4%), *Halohasta* (2%), *Natronococcus* (2%), and *Halorhabdus* (1%) (Rascovan et al. 2016).

In turn, Laguna Socompa's microbialites were reported as columnar round dome-shaped, laminar mats and fit perfectly to the term "modern stromatolites" (Farías et al. 2013). Fossil stromatolites are considered as the earliest evidence of life on Earth, with geological records dating back 3.5 billion years (Burne and Moore 1987; Schopf 2006; Allwood et al. 2006). Modern stromatolites only exist today at some places, and the most thoroughly studied ones are located at the hyper-saline region of Hamelin Pool, Shark Bay in Western Australia (Goh et al. 2009), and at the shallow subtidal regions at the margin of Exuma Soundin the Bahamas (Foster et al. 2009). The Socompa stromatolites (Fig. 14.2) are rich in silica due to diatomeas frustules and calcium carbonate (aragonite). The main diatomeas found at the stromatolites included *Cymbella* sp., *Navicula* sp., *Hantzschia* sp., *Nitzschia* sp., *Synedra* sp., *Surirella* sp., *Rhopalodia* sp., and *Pinnularia* sp. The analysis of bacterial 16S pyrotags showed Proteobacteria as the most abundant phylum (34% of

Fig. 14.2 Inside the stromatolite. Scanning electron microscopy showed the microbes, mainly cyanobacteria and diatoms, associated with aragonite crystals in the first layers of the stromatolites. Bar: 10 μm

all sequences). Other main phyla were Spirochetes (8%), *Deinococcus–Thermus* (7%), Bacteroidetes (6%), Firmicutes (5%), Cyanobacteria (3%, dominated by the genus Microcoleus), and Chloroflexi (1%). One percent of all sequences correspond to the 16S rRNA gene of Bacillaryophyta (diatoms) chloroplast (Toneatti et al. 2017).

Scompa's stromatolites were found submerged along the southern shore, near several hydrothermal inputs (26 °C) and thereby protected from the direct incidence of high solar radiation levels during the summer. In the winter, the water level decreased and, consequently, top layers of the stromatolites were directly exposed to UV-A/B light affecting its structure and microbial distribution; this phenomenon was also observed in experimental conditions at the lab. When the stromatolite is exposed to direct UV, a white-pinkish crust dominated by *Deinococci–Thermus* microbes covered the surface followed by a second layer (ca 0.5–1.5 mm thickness) of cyanobacteria. This may allow cyanobacteria to minimize exposure to deleterious UV radiation while staying close to the surface to perform photosynthesis. When the mats are submerged in the lake or exposed to darkness in the lab, the top layer became green, indicating a phototaxis mechanism for cyanobacteria (Farías et al. 2013). A detailed metagenomic analyses of the different layers of the stromatolite showed how microbial communities shift from a UV-high/oxic world to an IR-low/anoxic/high H_2S environment within the first 7 mm. These drastic changes of conditions force stratification and metabolic specialization of the bacterial community, thus modulating the chemical faces of the Socompa stromatolites. The oxic zone was dominated by *Deinococcus* sp. at the top surface (0.3 mm), followed by a second layer of *Coleofasciculus* sp. (0.3–2 mm). At middle layers, there were anoxygenic phototrophic Alphaproteobacteria, along with an increasing diversity of phyla including Bacteroidetes and Spirochaetes. Deeper layers (5–7 mm) were mostly occupied by sulfate reducers of Deltaproteobacteria, Bacteroidetes, and Firmicutes, next to a high diversity and equitable community of rare, unclassified, and candidate phyla.

Apart from the UV stress, the microbial community has to deal with high concentration of arsenic as a selective pressure. The arsenic resistance mechanism developed by these microbial communities consisted mainly of the reduction and extrusion of As(V) by diverse Acr3 efflux pumps, associated with ArsC (thioredoxin type) reductases, which was also observed in strains that were isolated from Socompa. In the stromatolites, a complete set of arsenic metabolism genes was unraveled (Kurth et al. 2017), including ArsM methyltransferases from the most varied phyla, as well as respiratory AioA oxidases, ArrA reductases, and even novel ArxA oxidases; therefore, the community is able to carry out a complete arsenic cycle, even using it as a source of energy, allowing microorganisms to grow and adapt to high concentration of arsenic.

Isolation programs of bacteria from the stromatolites also yielded novel strains; Rasuk et al. (2017) characterized four rare actinobacteria; according to 16S ADNr sequencing, strain ST80 (Accession Number KJ187450) showed 100% identity to *Micromonospora maritima* D10-9-5, while strain ST82 (Accession Number KJ187452) presented 98.46% identity to *Blastococcus jejuensis* KST3-10. Likewise, strain ST84 (Accession Number KJ187454) had 99.92% identity to *Micrococcus yunnanensis* YIM 65004 and strain ST85 99.85% identity to *Streptomyces pratensis* ch24 (Accession Number KJ187455). This is the first report of the presence of *Micromonospora, Blastococcus, Micrococcus,* and *Streptomyces* strains in stromatolites. Interestingly, the *Blastococcus* genus belonging to the Geodermatophilaceae family comprises species considered model microbes to study the ability to thrive on or within stones (Urzi et al. 2001). For instance, *Blastococcus saxobsidens* strains were isolated from calcareous stones (Urzi et al. 2004) and genome-sequenced together with *Modestobacter marinus* and *Geodermatophilus obscurus* (Sghaier et al. 2016). These microbes also have the ability to resist adverse environmental conditions such as ultraviolet light, ionizing radiation, desiccation, and heavy metals (Montero-Calasanz et al. 2014; Gtari et al. 2012). This resistance to environmental hazards represents a trait of Terrabacteria, a phylogenetic group that colonized land 3.05–2.78 Ga and that comprises Actinobacteria, Firmicutes, Cyanobacteria, Chloroflexi, and *Deinococcus-Thermus*.

Another gram-positive strain *Exiguobacterium* sp. S17 (Fig. 14.3) was isolated from the stromatolite. The genus *Exiguobacterium* is one of the most widespread and representative genera in the HAAL, being detected by direct (pure culture isolation) and indirect (DGGE) techniques (Ordoñez et al. 2009) and also present in other lakes. Representatives of this genus were detected in a wide range of habitats, including cold and hot environments (Vishnivetskaya et al. 2006; Rodrigues and Tiedje 2008). This fact confers substantial interest to the genus as a potential model system for the investigation of attributes that may correlate with adaptation and evolution of organisms to diverse thermal regimes (Vishnivetskaya et al. 2009). Several studies have shown that some strains of *Exiguobacterium* possess unique

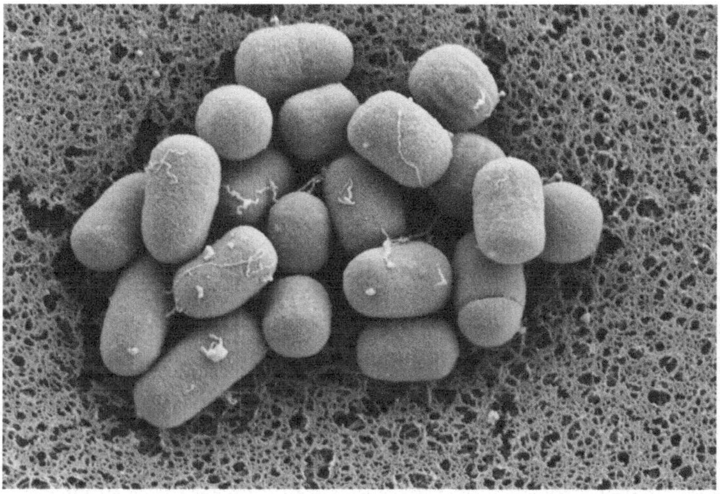

Fig. 14.3 *Exiguobacterium* sp. S17. A gram-positive strain isolated from the stromatolites and observed by scanning electron microscopy. Bar: 500 nm

properties of interest for applications in biotechnology, bioremediation, industry, and agriculture (Pattanapipitpaisal et al. 2002; López et al. 2005; Okeke et al. 2007).

As most HAAL strains, S17 presents a multi-resistance profile against arsenic, UV-B, and high salinity. This bacteria was further studied for its remarkable resistance and molecular mechanisms to cope with arsenic salts in laboratory conditions. Indeed, *Exiguobacterium* sp. S17 demonstrated to be much more resistant than other *Exiguobacterium* strains as it was able to grow in media amended with concentrations up to 100 mM of arsenate and 5 mM arsenite. Arsenic-induced molecular mechanisms were studied in S17 by a proteomic approach: 25 proteins were described as differentially expressed under As(V) or As(III) exposition; all (100%) of these proteins were expressed or over-expressed in the presence of As(V), while 32% and 48% of these were detected in the As(III)-supplemented medium with As(III) and/or in the control, respectively. The over-expression of proteins in arsenic-grown strains was related to oxidative stress response (e.g., SOD). The over-expression of SOD in *Exiguobacterium* sp. S17 under As exposition is thus considered to be directly involved in the cellular response to reduce the alterations, such as DNA damage and protein oxidation, caused by the presence of this metalloid. In the presence of arsenate, heat-shock proteins (Hsps) were over-expressed, e.g., DnaK chaperone, GroEL, and cpn10. The Hsps have an important function against environmental stress conferring tolerance to high temperatures, high salt, and heavy metal concentrations (Tkâĉová and Angeloviĉová 2012; Ron 2006; Kiriyama et al. 2001; Tedengren et al. 1999; Parsell and Lindquist 1993). So too, one universal stress protein, UspA was expressed; conversely, the Hsps proteins, GroEL, Cpn10, and UspA, were not detected in the presence of arsenite, probably because the cell removes the arsenite and the defense mechanism in the presence of this metalloid may not be activated. Ordoñez (2012) reported that the genome of *Exiguobacterium* sp. S17 carries a copy of an ArsB gene encoding a protein that acts as an efflux pump of cytoplasmic arsenite. In addition, an As(III) resistance protein (ARC3) that acts as a flow pump was also reported. The combined work of both efflux pumps (arsB and ACR3) might explain the enhanced tolerance of this strain to As(III) (Ordoñez et al. 2015). Two proteins involved in transport of arsenic were detected only in the presence of As(V): ABC transporter that binds and uses the energy of ATP to transport substances across membranes (Couoh Uicab et al. 2010) and a protein member of the family of toxic anion resistance proteins where it could be assumed that cells promote the elimination of toxic anionic arsenite as a mechanism of arsenic resistance. Sigma 54 modulation protein/ribosomal protein S30EA, ribosomal protein S2, ribosome recycling factor, prolyl-tRNA synthetase, and elongation factor TS are all over-expressed in the presence of As(V). All of these proteins are involved in transcriptional and translational processes. Baker-Austin et al. (2007) suggested that the amount of ribosomal sub-units increases due to the requirement of increased protein biosynthesis to combat the cellular stress caused by the presence of metalloids. Screening of response against arsenic stress by 2DE also yielded proteins that belong to the general metabolism pathways (carbohydrate and amino acid metabolism) in the presence of As(V). The results could suggest that in *Exiguobacterium* sp. S17 significantly upregulates the increment of energy-providing metabolic pathways to overcome the cellular stress caused by the exposure to arsenic (Belfiore et al. 2013).

Biofilm formation was also studied in S17. The formation of biofilm was evident in S17 cells grown in media supplemented with different doses of arsenic (50, 100, 150, 200, 250, and 350 mM for arsenate (As[V])), being able to grow at all tested concentrations. It was observed that arsenic had strong effects on cell adhesion directly influencing the number of attached cells per sample and the probability of cell–cell aggregation and, therefore, on the biofilm structure. Anderson and Cook (2004) conducted a comparable study on *Exiguobacterium* sp. WK6 and explain this phenomenon as a result of a detoxifying arsenate to arsenite process in which the external pH becomes alkaline, allowing the bacteria to grow for extended periods of time. As discussed by Ordoñez et al., this fact should be explained by the promotion of a more active metabolism as suggested by the over expression of glycolysis proteins in the presence of As[V] (Belfiore et al. 2013; Ordoñez et al. 2015; Zannier et al. 2019).

Gram-negative bacteria were also presented in the stromatolites. *Salinivibrio* sp. S10B and S34 were isolated from the top and the medium pink layer, respectively (Gorriti et al. 2014). These obligate halophiles were able to grow with up to 15% NaCl in the culture media. Moreover, they grew in medium supplemented with 200 mM of arsenate and 2.5 mM arsenite and resisted a UV-B radiation dose as high as 19 kJ m^{-2}. Accordingly, their genome analyses revealed genes with a potential role in DNA repair such as genes for RecBCD helicase/nuclease and UvrABC endonuclease holoenzymes involved in the recombination repair. The ssDNA binding protein RecA acts centrally in SOS response (i.e., Rec repair system) and other homologous-recombination-based DNA repair, initiating the exchange of strands between two recombining DNA molecules. RecX modulates RecA activity by direct protein–protein interaction, blocking the extension of RecA filaments to avoid aberrant DNA transactions. Gene homologs of deoxyribodipyrimidine photolyase, and one gene coding for a transcriptional regulator of the Mer family, associated with photolyases were also found. Photolyases are enzymes that catalyze the light-dependent monomerization (300–600 nm) of cyclobutyl pyrimidine dimers, which are formed between adjacent bases on the same DNA strand upon exposure to UV radiation.

S10B presented several ORFs with similarity to phage genes flanked by IS elements, tRNAs, and integrases genes, indicating an important role of horizontal genetic transfer in genome evolution in the Socompa Lake. In turn, S34 presented a gene coding for a capsular polysaccharide synthesis enzyme. These enzymes may be responsible for biofilm formation, a strategy also used by the S17 strain for survival during periods of nutrient scarcity and/or environmental fluctuations. It may help S34 to establish favorable interactions with other bacteria and to build the stromatolites matrix. Interestingly, genes identified with clustered regularly interspaced short palindromic repeats elements (CRISPR) were also found in the S34 genome indicating the existence of adaptive immunity for defense against the invasion of foreign genetic elements.

14.3.3 Soil Microbes

Dry soil surrounding the lake and in the vicinity was also subjected to microbial isolation programs and biodiversity studies. Rasuk et al. (2017) reported the isolation of 29 actinobacteria strains from LS soil; most of the strains belong to the genus *Streptomyces*, including a novel strain Act15 with low similarity to a type strain of *Streptomyces pratensis* ch24T (JQ806215). The rest of the isolates corresponded to the so-called "rare actinobacteria": *Microbacterium, Nesterenkonia, Kocuria, and Arthrobacter.* The actinobacteria were characterized as poly-extremophiles for their multiple resistance profile toward UV, NaCl, alkalinity, and arsenic. Moreover, they showed potential for antimicrobial compounds production as all the isolates exhibited antagonistic activities against each other and against at least one of the control bacteria including *E. coli*, *Bacillus*, *Enterococcus faecalis*, *Staphylococcus aureus,* and two yeasts (*Rhodotorula* sp.).

14.4 Conclusions and Future Prospects

High solar irradiation and other extremes challenged microbial communities from High-Altitude Andean Lakes, in many forms, to select resistance/adaptation phenotypes (Albarracín et al. 2012, 2014, 2016). Albeit these conditions, a rich biodiversity was reported for all HAAL, and especially in Laguna Socompa where microbiota diversity is not only evident in terms of phylogenetic distribution of microbes but also as a consequence of the existence of differential micro-niches where microbial communities can be specialized and evolved.

The Socompa's stromatolites are a clear example of adaptation and specialization; this associative behavior pursues a common benefit to microbial survival. Moreover, the physicochemical gradient within the stromatolites explains the spatial compartmentalization of diverse taxonomic groups in a "cooperative" microbial society and the concomitant functional specialization of each layer. As microbialites thriving under a natural high-UV radiation regime and in an arsenic-rich environment, they represent a modern analog of their counterparts to study the process of microbialite formation in a close proxy environment to early Earth. The study of specific metabolisms of As, UV-resistance, and other extremes provides an open window to better understand the ecology, biogeochemistry, and evolution of Precambrian analogs.

Understanding the limits for life precedes the applications of extremophiles and their biomolecules in industrial processes. Microbial strains isolated from Laguna Socompa are probed to tolerate multiple harsh conditions and, thus, are excellent models for studying adaptive responses and mechanisms involved in UV, salt, and arsenic resistance in environmental microbes. On-going mining of available genomes of model Socompa polyextremophilic microbes and several metagenomes of microbialites is revealing that these microbes possess enzymatic and metabolic potential to be of biotechnological use, e.g., antibiotic production, anti-age products, solar creams, and bioremediation.

The uniqueness of the Socompa environment, especially the stromatolites, with regard to their broad unknown microbial diversity demands active protection policies and continuous monitoring to avoid the huge impact of nearby and current mining activities menacing the preservation of these microbial treasures. This contribution aims to give arguments to this novel goal.

Acknowledgements The authors acknowledge the generous financial support from the PICT V Bicentenario 2010 1788, 3825, PICT 2013 2991, 730 Projects (FONCyT, Argentina) and PIUNT G603 Projects. V.H.A. was supported by a Marie Curie FP7-People-2010-IIF EXTREMOPHIL(273831) in Germany and its return phase in Argentina (PIIFR-GA-2010-910831-EXTREMOPHIL). M.E.F. and V.H.A. are researchers from the National Research Council (CONICET) in Argentina. S.G. is recipient of a fellowship from CONICET. Electron micrographs used in this study were taken at the Research Center for Electron Microscopy and Core Facility (CIME) belonging to UNT and CCT, CONICET, Tucuman. Environmental Secretary Salta province Goverment. Tolar Grande Community.

References

Albarracín VH, Gärtner W, Farias ME (2016) Forged under the sun: life and art of extremophiles from Andean lakes. Photochem Photobiol 92:14–28

Albarracín VH, Kurth D, Ordoñez OF, Belfiore C, Luccini E, Salum GM, Piacentini RD, Farías ME (2015) High-up: a remote reservoir of microbial extremophiles in central Andean wetlands. Front Microbiol 6:1404

Albarracín VH, Pathak GP, Douki T, Cadet J, Borsarelli CD, Gärtner W, Farías ME (2012) Extremophilic Acinetobacter strains from high-altitude lakes in Argentinean Puna: remarkable UV-B resistance and efficient DNA damage repair. Orig Life Evol Biosph 42(2–3):201–221

Albarracín VH, Simon J, Pathak GP, Valle L, Douki T, Cadet J, Borsarelli CD, Farias ME, Gärtner W (2014) First characterisation of a CPD-class I photolyase from a UV-resistant extremophile isolated from high-altitude Andean lakes. Photochem Photobiol Sci 13(5):739–750

Allwood AC, Walter MR, Kamber BS, Marshall CP, Burch IW (2006) Stromatolite reef from the Early Archaean era of Australia. Nature 441:714–718

Anderson CR, Cook GM (2004) Isolation and characterization of arsenate-reducing bacteria from arsenic-contaminated sites in New Zealand. Curr Microbiol 48(5):341–347

Baker-Austin G, Dopson M, Wexler M, Sawers RG, Stemmler A, Rosen BP, Bond PL (2007) Extreme arsenic resistance by the acidophilic archaeon 'Ferroplasma acidarmanus' Fer1. Extremophiles 11:425–434

Baumgartner LK, Spear JR, Buckley DH, Pace NR, Reid RP, Dupraz C, Visscher PT (2009) Microbial diversity in modern marine stromatolites, Highborne Cay, Bahamas. Environ Microbiol 11:2710–2719

Belfiore C, Ordoñez OF, Farias ME (2013) Proteomic approach of adaptive response to arsenic stress in Exiguobacterium sp.S17, an extremophile strain isolated from a high-altitude Andean Lake stromatolite. Extremophiles 17(3):421–431

Burne R, Moore L (1987) Microbialites: organosedimentary deposits of benthic microbial communities. PALAIOS 2:241–254

Cabrol NA, Feister U, Hader DP, Piazena H, Grin EA, Klein A (2014) Record solar UV irradiance in the tropical Andes. Front Environ Sci 2:19

Cabrol NA, Wettergreen D, Warren-Rhodes K, Grin EA, Moersch J, Diaz GC et al (2007) Life in the Atacama: searching for life with rovers (science overview). J Geophys Res Biogeosci 112(G4):G04S02

Chattopadhyay MK (2008) Cryotolerance in bacteria: interlink with adaptation to other stress factors. Trends Microbiol 16:455

Comité Internationale de l'Eclairage (2011) Recommendations on minimum levels of solar UV exposure. Technical report CIE 201:2011. CIE, Vienna. ISBN: 978-3-902842-39-8

Couoh Uicab YL, Canto Canche BB, Islas Flores I (2010) Review of characteristics of ABC transporters involved in fungal pathogenesis. Tecnociencia Chihuahua 4(2):87–96

Decho AW, Visscher PT, Reid RP (2005) Production and cycling of natural microbial exopolymers (EPS) within a marine stromatolite. Palaeogeogr Palaeoclimatol Palaeoecol 219:71–86

Demergasso C, Dorador C, Meneses D, Blamey J, Cabrol N, Escudero L, Chong G (2010) Prokaryotic diversity pattern in high-altitude ecosystems of the Chilean Altiplano. J Geophys Res Biogeosci 115:G00D09

Dib J, Motok J, Zenoff VF, Ordoñez O, Farías ME (2008) Occurrence of resistance to antibiotics, UV-B, and arsenic in bacteria isolated from extreme environments in high-altitude (above 4400 m) Andean wetlands. Curr Microbiol 56:510–517

Dorador C, Vila I, Witzel K-P, Imhoff JF (2013) Bacterial and archaeal diversity in high altitude wetlands of the Chilean Altiplano. Fundam Appl Limnol 182:135–159

Dupraz C, Reid RP, Braissant O, Decho AW, Norman RS, Visscher PT (2009) Processes of carbonate precipitation in modern microbial mats. Earth Sci Rev 96:141–162

Escalante G, Campos VL, Valenzuela C, Yañez J, Zaror C, Mondaca MA (2009) Arsenic resistant bacteria isolated from arsenic contaminated river in the Atacama Desert (Chile). Bull Environ Contam Toxicol 83:657–661

Escudero L, Chong G, Demergasso C (2007) Investigating microbial diversity and UV radiation impact at the high-altitude Lake Aguas Calientes, Chile. In: Instruments, methods, and missions for astrobiology X; 66940Z

Farías ME, Poiré DG, Arrouy MJ, Albarracín VH (2011a) Modern stromatolite ecosystems at alkaline and hypersaline high-altitude lakes in the Argentinean Puna. In: Tewari V, Seckbach J (eds) STROMATOLITES: interaction of microbes with sediments cellular origin, life in extreme habitats and astrobiology. Springer Netherlands, Dordrecht, pp 427–441

Farías ME, Rascovan N, Toneatti DM, Albarracín VH, Flores MR, Poire DG, Collavino MM, Aguilar OM, Vazquez MP, Polerecky L (2013) The discovery of stromatolites developing at 3570 m above sea level in a high-altitude volcanic lake Socompa, Argentinean Andes. PLoS One 8:e53497

Farías ME, Revale S, Mancini E, Ordoñez O, Turjanski A, Cortez N et al (2011b) Genome sequence of Sphingomonas sp. S17, isolated fromanalkaline, hyperarsenic, and hypersaline volcano-associated lake at high altitude in the Argentinean Puna. J Bacteriol 193(14):3686–3687

Foster JS, Green SJ, Ahrendt SR, Golubic S, Reid RP, Hetherington KL, Bebout L (2009) Molecular and morphological characterization of cyanobacterial diversity in the stromatolites of Highborne Cay, Bahamas. ISME J 3(5):573

Friedberg EC, Walker GC, Siede W, Wood RD (eds) (2005) DNA repair and mutagenesis. American Society for Microbiology Press, Washington, DC

Galisteo C, Sánchez-Porro C, de la Haba RR, López-Hermoso C, Fernández AB, Farias ME, Ventosa A (2019) Characterization of *Salinivibrio socompensis* sp. nov., a new halophilic bacterium isolated from the high-altitude Hypersaline Lake Socompa, Argentina. Microorganisms 7:241

Glunk C, Dupraz C, Braissant O, Gallagher KL, Verrecchia EP, Visscher PT (2011) Microbially mediated carbonate precipitation in a hypersaline lake, Big Pond (Eleuthera, Bahamas). Sedimentology 58:720–736

Goh F, Allen MA, Leuko S, Kawaguchi T, Decho AW, Burns BP, Neilan BA (2009) Determining the specific microbial populations and their spatial distribution within the stromatolite ecosystem of Shark Bay. ISME J 3(4):383

Gorriti MF, Dias GM, Chimetto LA, Trindade-Silva AR, Silva BS, Mesquita MA, Gregoracci GB, Farias ME, Thompson C (2014) Genomic and phenotypic attributes of novel salinivibrios from stromatolites, sediment and water from a high altitude lake. BMC Genomics 15(1):473

Gtari M, Essoussi I, Maaoui R, Sghaier H, Boujmil R, Gury J (2012) Contrasted resistance of stone-dwelling Geodermatophilaceae species to stresses known to give rise to reactive oxygen species. FEMS Microbiol Ecol 80:566–577

Haines KC, Guillard RR (1974) Growth of vitamin b12-requiring marine diatoms in mixed laboratory cultures with vitamin b12-producing marine bacteria 1 2. J Phycol 10(3):245–252

Hernández KL, Quiñones RA, Daneri G, Farias ME, Helbling EW (2007) Solar UV radiation modulates daily production and DNA damage of marine bacterioplankton from a productive upwelling zone (36°S), Chile. J Exp Mar Biol Ecol 343:82–95

Kiriyama MT, Oka M, Takehana M, Kobayashi S (2001) Expression of a small heat shock protein 27 (HSP27) in mouse skin tumors induced by UVB-irradiation. Biol Pharm Bull 24:197–120

Kremer B, Kazmierczak J, Stal LJ (2008) Calcium carbonate precipitation in cyanobacterial mats from sandy tidal flats of the North Sea. Geobiology 6:46–56

Kurth D, Amadio A, Ordoñez OF, Albarracín VH, Gärtner W, Farías ME (2017) Arsenic metabolism in high altitude modern stromatolites revealed by metagenomic analysis. Sci Rep 7(1):1024

López L, Pozo C, Rodelas B, Calvo C, Juarez B, Martinez-Toledo MV, Gonzalez-Lopez J (2005) Identification of bacteria isolated from an oligotrophic lake with pesticide removal capacities. Ecotoxicology 14(3):299–312

Maldonado MJ, Albarracín VH, Lara JA, Ferrero MA, Farías ME (2018) Culture-dependent and-independent methods reveal dominance of halophilic Euryarchaeota in high-altitude Andean lakes. Aquat Microb Ecol 81(2):171–188

Mitchell DL, Karentz D (1993) The induction and repair of DNA photodamage in the environment. In: Young AR, Bjorn LO, Moan J, Nultsch W (eds) Environmental UV photobiology. Plenum Press, New York, pp 345–377

Moeller RE, Gilroy S, Williamson CE, Grad G, Sommaruga R (2005) Dietary acquisition of photoprotective compounds (mycosporine-like amino acids, carotenoids) and acclimation to ultraviolet radiation in a freshwater copepod. Limnol Oceanogr 50:427–439

Montero-Calasanz MDC, Hofner B, Göker M, Rohde M, Spröer C, Hezbri K, Klenk HP (2014) Geodermatophilus poikilotrophi sp. nov.: a multitolerant actinomycete isolated from dolomitic marble. Biomed Res Int 80:566–577

Okeke BC, Laymon J, Oji C, Crenshaw S (2007) Rapid bioreduction of hexavalent chromium in water by Exiguobacterium sp. GS1. In: ASM general 107th meeting. ASM Press, Toronto, p Q-199

Ordoñez OF (2012) Factores ambientales extremos en ecosistemas microbianos de humedales altoandinos: mecanismos de adaptacion. Ph.D. thesis from Universidad Nacional de Tucumán, Tucumán, Argentina

Ordoñez OF, Flores MR, Dib JR, Paz A, Farías ME (2009) Extremophile culture collection from Andean lakes: extreme pristine environments that host a wide diversity of microorganisms with tolerance to UV radiation. Microb Ecol 58(3):461–473

Ordoñez OF, Lanzarotti E, Kurth D, Cortez N, Farías ME, Turjanski AG (2015) Genome comparison of two Exiguobacterium strains from high altitude andean lakes with different arsenic resistance: identification and 3D modeling of the Acr3 efflux pump. Front Environ Sci 3:1–12

Ordoñez OF, Lanzarotti E, Kurth D, Gorriti MF, Revale S, Cortez N, Vazquez MP, Farias ME, Turjanski AG (2013) Draft genome sequence of the polyextremophilic Exiguobacterium sp. strain S17, isolated from Hyperarsenic Lakes in the Argentinian Puna. Genome Announc 1:e01059–e01013

Oren A (2006) The order Halobacteriales. In: Dworkin M, Falkow S, Rosenberg E, Schleifer KH, Stackebrandt E (eds) The prokaryotes. Springer, New York, pp 113–164

Parsell DA, Lindquist S (1993) The function of heat-shock proteins in stress tolerance: degradation and reactivation of damaged proteins. Annu Rev Genet 27:437–496

Pattanapipitpaisal P, Mabbett AN, Finlay JA, Beswick AJ, PatersonBeedle M, Essa A, Wright J, Tolley MR, Badar U, Ahmed N, Hobman JL, Brown NL, MAcaskie LE (2002) Reduction of Cr(VI) and bioaccumulation of chromium by gram positive and gram negative microorganisms not previously exposed to CR-stress. Environ Technol 23:731–745

Portero LR, Alonso-Reyes DG, Zannier F, Vazquez MP, Farías ME, Gärtner W, Albarracín VH (2019) Photolyases and cryptochromes in UV-resistant bacteria from high-altitude Andean lakes. Photochem Photobiol 95(1):315–330

Pueyo JJ, Chong G, Jensen A (2001) Neogene evaporites in desert volcanic environments: Atacama Desert, northern Chile. Sedimentology 48(6):1411–1431

Rascovan N, Maldonado J, Vazquez MP, Farías ME (2016) Metagenomic study of red biofilms from Diamante Lake reveals ancient arsenic bioenergetics in haloarchaea. ISME J 10(2):299

Rasuk MC, Fernández AB, Kurth D, Contreras M, Novoa F, Poiré D, Farías ME (2016) Bacterial diversity in microbial mats and sediments from the Atacama Desert. Microb Ecol 71(1):44–56

Rasuk MC, Ferrer GM, Kurth D, Portero LR, Farías ME, Albarracín VH (2017) UV-resistant Actinobacteria from high-altitude Andean lakes: isolation, characterization and antagonistic activities. Photochem Photobiol 93(3):865–880

Rasuk MC, Kurth D, Flores MR, Contreras M, Novoa F, Poire D, Farias ME (2014) Microbial characterization of microbial ecosystems associated to evaporites domes of gypsum in Salar de Llamara in Atacama desert. Microb Ecol 68(3):483–494

Rodrigues DF, Tiedje JM (2008) Coping with our cold planet. Appl Environ Microbiol 74(6):1677–1686

Ron EZ (2006) Bacterial stress response. In: Dworkin M, Falkow S, Resenberg E, Schleifer KH, Stackebrandt E (eds) The prokaryotes. Springer, Singapore, pp 1012–1027
Schopf JW (2006) Fossil evidence of Archaean life. Philos Trans R Soc Lond Ser B Biol Sci 361:869–885
Sghaier H, Hezbri K, Ghodhbane-Gtari F, Pujic P, Sen A, Daffonchio D (2016) Stone-dwelling actinobacteria Blastococcus saxobsidens, Modestobacter marinus and Geodermatophilus obscurus proteogenomes. ISME J 10:21–29
Tedengren M, Olsson B, Bradley B, Zhou LZ (1999) Heavy metal uptake, physiological response and survival of the blue mussel (Mytilus edulis) from marine and brackish waters in relation to the induction of heat-shock protein 70. Hydrobiologia 393:261–269
Tkáĉová J, Angeloviĉová M (2012) Heat shock proteins (Hsps): a review. Anim Sci Biotechnol 45(1):349–353
Toneatti DM, Albarracín VH, Flores MR, Polerecky L, Farías ME (2017) Stratified bacterial diversity along physico-chemical gradients in high-altitude modern stromatolites. Front Microbiol 8:646
Urzi C, Brusetti L, Salamone P, Sorlini C, Stackebrandt E, Daffonchio D (2001) Biodiversity of Geodermatophilaceae isolated from altered stones and monuments in the Mediterranean basin. Environ Microbiol 3:471–479
Urzi C, Salamone P, Schumann P, Rohde M, Stackebrandt E (2004) Blastococcus saxobsidens sp. nov., and emended descriptions of the genus Blastococcus Ahrens and Moll 1970 and Blastococcus aggregatus Ahrens and Moll 1970. Int J Syst Evol Microbiol 54:253–259
Vishnivetskaya TA, Kathariou S, Tiedje JM (2009) The Exiguobacterium genus: biodiversity and biogeography. Extremophiles 13(3):541–555
Vishnivetskaya TA, Petrova MA, Urbance J, Ponder M, Moyer CL, Gilichinsky DA, Tiedje JM (2006) Bacterial community in ancient Siberian permafrost as characterized by culture and culture-independent methods. Astrobiology 6(3):400–414
Visscher PT, Stolz JF (2005) Microbial mats as bioreactors: populations, processes, and products. Palaeogeogr Palaeoclimatol Palaeoecol 219:87
Zannier F, Portero LR, Ordoñez OF, Martinez L, Farias ME, Albarracín VH (2019) Poly-extremophilic bacteria from high altitude Andean lakes: arsenic resistance profiles and biofilm production. Biomed Res Int 2019:1. https://doi.org/10.1155/2019/1231975

Chapter 15
Mats and Microbialites from Laguna La Brava

María Cecilia Rasuk, Pieter T. Visscher, Manuel Contreras Leiva, and María Eugenia Farías

15.1 Introduction

The Salar de Atacama is a tectonic intermontane basin filled with Tertiary to Quaternary clastic and evaporitic deposits of continental origin. The hydrogeological setting of the salar is complex, mostly receiving groundwater input but also surface water, predominantly from the east (Risacher et al. 2003). The dominant input into Laguna La Brava is upwelling of sulfidic groundwater containing leached volcanic material, including lithium, sodium, and arsenic. A series of lakes have formed in the lowest region of the Atacama basin, including Laguna de Piedra, Tebenquiche, Chaxas, Burro Muerto, and La Brava (Risacher and Alonso 1996). The environmental conditions in which these lakes have formed are characterized by (1) high ultraviolet (UV) radiation (Cabrera and Pizarro 1992), due to less light scatter; (2) extreme diel temperature fluctuations typical of desert environments; (3) net evaporation, producing hypersaline water; (4) extremely low relative humidity (Cáceres et al. 2007); and (5) high arsenic and lithium concentrations in the water, due to volcanic events (Lara et al. 2012). The conditions in Laguna La Brava are ideal for supporting the development of microbial mats and microbialites

M. C. Rasuk
Laboratorio de Investigaciones Microbiológicas de Lagunas Andinas (LIMLA), Planta Piloto de Procesos Industriales Microbiológicos (PROIMI), Centros Científico Tecnológicos (CCT), Consejo Nacional de Investigaciones Científicas y Técnicas (CONICET), Tucumán, Argentina

P. T. Visscher
Department of Marine Sciences, University of Connecticut, Storrs, CT, USA

Department of Geoscience, University of Connecticut, Storrs, CT, USA

M. C. Leiva
Centro de Ecología Aplicada (CEA), Ñuñoa, Santiago, Chile

M. E. Farías (✉)
LIMLA-PROIMI-CONICET, Tucuman, Argentina

M. E. Farías (ed.), *Microbial Ecosystems in Central Andes Extreme Environments*, https://doi.org/10.1007/978-3-030-36192-1_15

(Dupraz et al. 2011). It is a shallow hypersaline lake surrounded by the thick gypsum crust of the salar, with a seasonally fluctuating water level, a salinity gradient (with around 119 g/L in the area of the mats and 72 g/L in the area of the microbialites), and alkaline conditions (pH 8).

15.2 Benthic Microbial Ecosystems in Laguna La Brava

The gradients in the dissolved chemicals, water depth, salinity, and exposure to strong evaporation support a number of different microbial ecosystems in Laguna La Brava. In total, six different types of sedimentary ecosystems have been described (Fig. 15.1), the most abundant of which are nonlithifying microbial mats (Fig. 15.1c). These mats are located along the western and southern shoreline of the lake, typically submerged in 3–10 cm of water during the wet season and partially exposed to the air during the dry season. They are abundant on the southwestern shoreline and have semispheroidal morphologies covered by pink or black biofilms (Fig. 15.1a, b, d). They grow upward until they reach the water/air interface, after which some expand laterally, forming platforms (Fig. 15.1a). The western shore of the lake harbors partially lithified structures, which are called rhizome-associated concretions; they are found where *Distichlis spicata* (Gramineae) grows abundantly on a carbonate substrate and consists of a laminated microbial subsurface community, which resembles the microbialite platforms described above. Oncolites, or concentric microbialite deposits, are distributed along the northern and northwestern shoreline, where they can be found in dense patches. Their internal structure is laminated, and their size ranges from several millimeters to around 10 cm in diameter.

15.2.1 Sediment Geochemistry

Some of the microbial mats and microbialites have been investigated in detail, using needle-type microelectrodes. The results showed various degrees of active photosynthesis and sulfide production. The distribution of O_2 and sulfide with depth was determined during the peak of photosynthesis, which occurred in the middle of the day (12:00–14:00). The intensity of the photosynthetically active radiation (PAR) was 1410–1620 μE m^{-2} s^{-1} at the surface of the mat and 1850–2550 μE m^{-2} s^{-1} at the surface of a pink microbialite and a black one, the latter being submerged in shallower water. An increasing oxygen concentration with depth was found in all three mat types; the highest maximum value (>200% in the microbialites versus ~140% in the microbial mat) and the steepest profile were observed in black microbialites, suggesting that O_2 production and consumption are greater in microbialites than in nonlithified microbial mats. The sulfide profiles confirmed greater activity in microbialites, as documented by the highest H_2S concentration at a 10 mm depth and steeper profiles. Similar differences between profiles in microbialites and nonlithifying mats were documented for other hypersaline microbialite systems

Fig. 15.1 Microbial ecosystems in Laguna La Brava. **a** Microbialite platforms. **b** Rhizome-associated concretions. **c** Submerged microbialites. **d** Microbialite section. **e** Oncolite section. **f** Microbial mat section

(e.g., Dupraz et al. (2004, 2009)). This broadens the evidence supporting the concept that steeper geochemical gradients associated with higher metabolic rates in the entire community support lithification of microbial mats. High concentrations of polysulfides (S_x^{2-}; ~190 μM sulfane S) and thiosulfate ($S_2O_3^{2-}$; ~112 μM) were measured in both nonlithifying mats and microbialites (pink and black) than during the daytime (Farias et al. 2017), arguing for an important role of the sulfur cycle in benthic microbial ecosystems. Of note, the concentrations and the rates of change (i.e., the turnover) of both polysulfides and thiosulfate were 2–3 times higher in the mats than in the microbialites, suggesting that a substantial part of the oxygen produced by cyanobacteria is used in cycling of sulfur (Visscher and Van Gemerden 1993a, b; Van Gemerden 1993); consequently, the production is greater than the respiration (i.e., $P/R > 1$, Pinckney and Reid 1997), favoring carbonate precipitation (Dupraz and Visscher 2005).

The mineralogy in all of the samples revealed that they precipitated more aragonite. Concerning halite, it was identified as a major mineral in early studies, but later investigations conclusively showed that this was an artifact (Sancho-Tomas et al. 2018; Farias et al. 2017).

15.3 Microbial Diversity and Alternative Pathways for Carbon Fixation

The microbial ecosystems of La Brava revealed an unusual community in which Euryarchaeota, Crenarchaeota, Acetothermia, Firmicutes, and Planctomycetes were the most abundant groups (Figs. 15.2, 15.3). Surprisingly, cyanobacterial sequences, which are typically a major component and the main primary producer in microbial mats and microbialites (Van Gemerden 1993; Pinckney and Reid 1997), were relatively insignificant or absent.

The diversity in the microbial mats at La Brava showed an unusual contribution from the archaeal domain: 43% of the total diversity was made up by members of this domain (see Farias et al. (2017) in comparison with Ley et al. (2006) and Baumgartner et al. (2009a, b)), with 29% belonging to the Euryarchaeota and 14% to the Crenarchaeota (Fig. 15.3). The bacterial phyla in the microbial mat included Planctomycetes (14%), Firmicutes (11%), and Acetothermia (6%) as the most abundant taxa. In contrast, the microbialite diversity was dominated by the phylum Planctomycetes (35%), followed by Euryarchaeota (17%) and Proteobacteria (9%) (Fig. 15.2a), taking into account only the functional groups belonging to the bacterial domain, represented in all of the layers. The dominant groups were aerobic heterotrophs, fermenters, anaerobic heterotrophs, and anoxygenic phototrophs, with the most representatives. It is important to highlight that the minor metabolic group that was represented was oxygenic phototrophs (Fig. 15.2b).

Analyses of the archaeal functional groups in the mat sample showed that aerobic heterotrophic representatives from this domain dominated the upper two layers and methanogens were found only in the deepest layer. Marine benthic group B

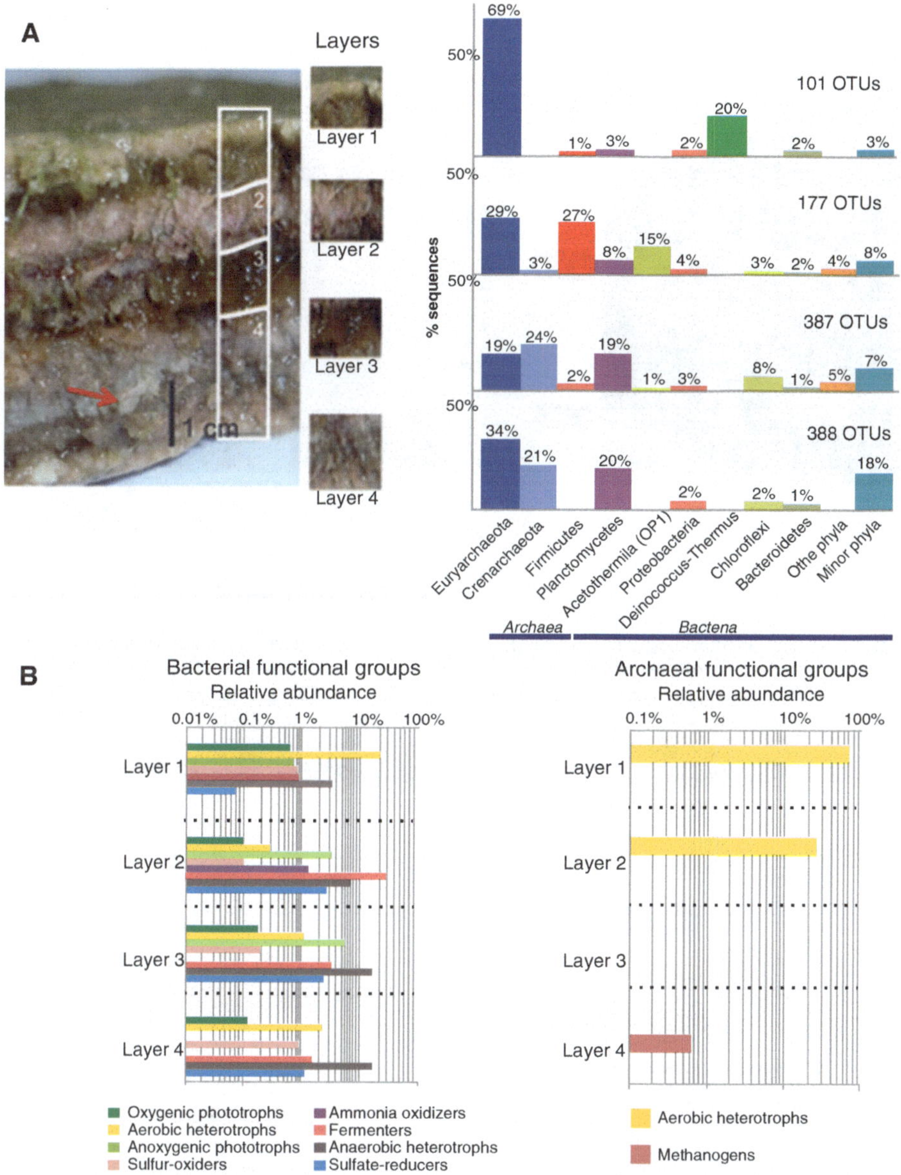

Fig. 15.2 Prokaryotic diversity in microbial mats (Farias et al. 2017). **a** Phylum abundance based on bacterial 16S rRNA gene sequences. **b** Percentage of sequences belonging to Bacteria and Archaea metabolic groups (log scale)

and D (MBGB) phylotypes, which are major contributors to crenarchaeal diversity, were present in all layers and decreased with depth; the functional roles of these groups remain unknown (Fig. 15.2b). In the microbialite sample, a distribution similar to that of the bacterial metabolic groups in the mat sample was found.

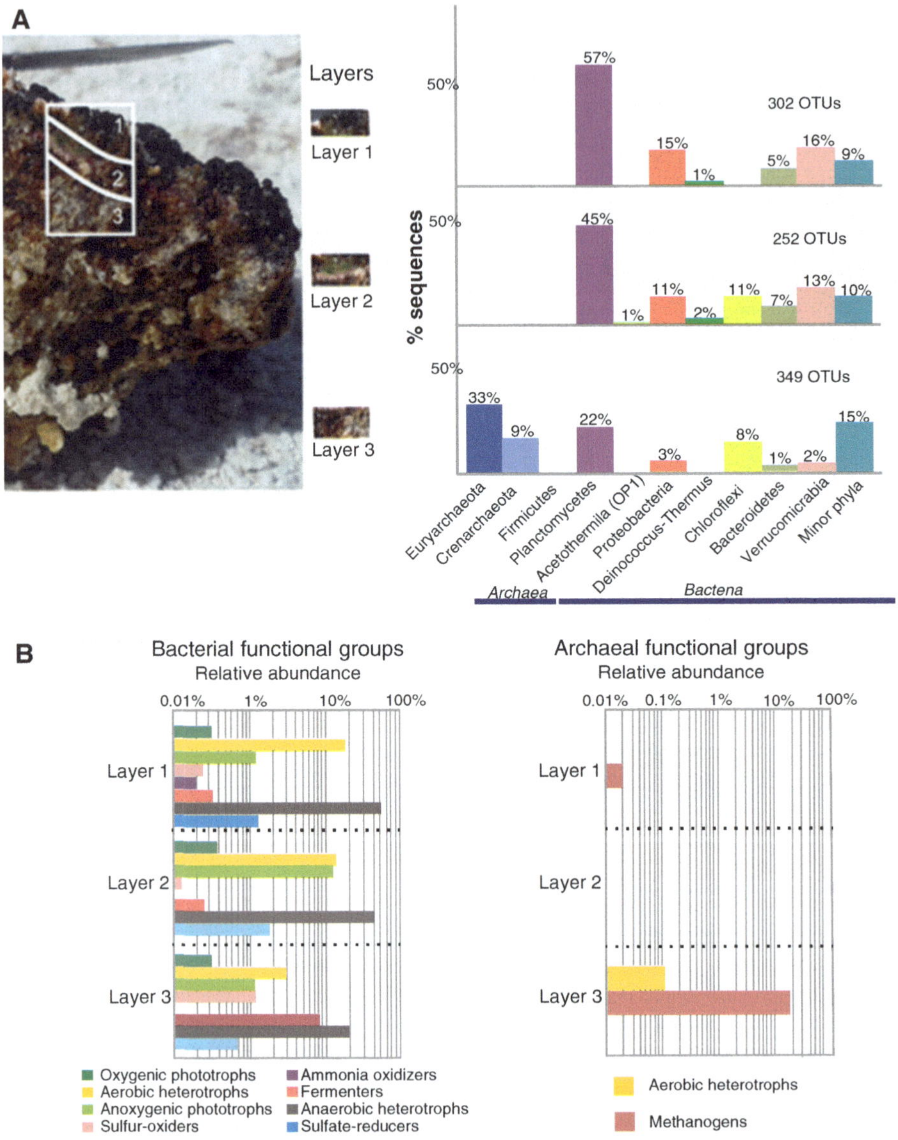

Fig. 15.3 Prokaryotic diversity in black microbialites (Farias et al. 2017). **a** Phylum abundance based on bacterial 16S rRNA gene sequences. **b** Percentage of sequences belonging to Bacteria and Archaea metabolic groups (log scale)

The diversity in La Brava samples reported by Farias et al. (2017) included Euryarchaeota, Crenarchaeota, Firmicutes, Planctomycetes, and Acetothermia as the major phyla; in contrast, a previous study had shown that Bacteroidetes and Proteobacteria constituted the major phyla in the bacterial diversity of both microbial mats and microbialites (Farías et al. 2014).

As outlined above, typically, the main carbon-fixing organisms in microbial mats are cyanobacteria (Visscher et al. 1991; Van Gemerden 1993; Pinckney and Reid 1997). In the benthic microbial ecosystems studied in Laguna La Brava, cyanobacteria were always low in abundance both in the microbialites (2–4% of the total diversity) and in the microbial mats (<1%). Although chlorophyll *a* (Chl*a*) was found in the water column of the lake (≤1 μg/L) (Farías et al. 2014), investigations of the microbial mats and microbialites at La Brava showed almost no Chl*a* levels.

The maximum oxygen values at the subsurface of the microbial mats and microbialites ranged from 140% to 200% as discussed above—values considerably lower than those observed in other hypersaline mats (400–600% of air saturation), most of which were at sea level (Dupraz et al. 2004, 2009; Glunk et al. 2011; Tkavc et al. 2011; Casillas-Martinez et al. 2005a, b; Visscher et al. 2003) and thus exposed to lower levels of UV radiation.

Possibly, a few cyanobacterial phyla highly adapted to the extreme conditions at La Brava were present in low diversity but in large numbers and with a low specific Chl*a* content (Farías et al. 2014). Alternatively, diatoms, which were readily observed by scanning electron microscopy, could have contributed to the O_2 production.

Anoxygenic phototrophic sequences were well represented in the mat sample and microbialite sample from La Brava. Furthermore, the presence of sulfide, thiosulfate, and polysulfides suggests the use of thiosulfate and/or polysulfide by the anoxygenic phototrophs. These findings agree with others in similar high-altitude systems such as Socompa (Farías et al. 2013), Llamara (Rasuk et al. 2014), and Tebenquiche (Fernandez et al. 2016), and also in geographically different microbial mats or microbialite systems such as Shark Bay (Wong et al. 2015; Ruvindy et al. 2016) and Guerrero Negro (Ley et al. 2006), all of which are systems where anoxygenic phototrophs are abundant.

The low abundance of phototrophs implies that in the microbial ecosystems of La Brava, particularly the microbial mats, a large fraction of the carbon appears to be fixed by alternative pathways. Perhaps little-known pathways and/or poorly characterized phylogenetic groups such as Planctomycetes, Firmicutes, Acetothermia, Euryarchaeota, and Crenarchaeota play an important autotrophic role.

15.4 Arsenic Metabolism in Laguna La Brava, an Early-Earth Counterpart

As described in other chapters of this book, potential analogs of Precambrian ecosystems can be found today in remote areas such as the hypersaline lakes in the Altiplano Andino. In addition, La Brava receives drained groundwater containing leached volcanic material (Corenthal et al. 2016; Boutt et al. 2016), in which high concentrations of arsenic and sulfide are found. Hence, the microbial metabolism in the mats is possibly driven by anoxygenic photosynthesis using reduced sulfur and arsenic compounds. Anaerobic processes could include fermentation, methanogenesis, sulfate reduction, and likely arsenate reduction. The widespread volcanism and geothermal activity characteristic of the Precambrian environment would present considerably more arsenic accumulation on the Earth's surface than there is today (Cabral and Beaudoin 2007; Witt-Eickschen et al. 2009). However, very few biogeochemical data supporting this claim have been put forward so far. Recently, Sforna et al. (2014) reported the discovery of 2.7-billion-year-old microbial stromatolites from Tumbiana, in which fossilized microorganisms and biogenic minerals were associated with high concentrations of As. On the basis of hierarchical length-scale information on the element distribution and interelement correlations, Sforna and colleagues argued that Tumbiana stromatolites were formed by microbial mats in an anoxic, salt-saturated lake setting and, importantly, that As(III) oxidation occurred there by anoxygenic photosynthesis and/or denitrification. As a modern counterpart of this Precambrian system, living microbial mats from Laguna Brava were investigated, using an array of conventional geochemical techniques (Sancho-Tomás et al 2018). This experimental approach allowed us to unravel the relationship between the microbial mat activity, the occurrence of minerals, the arsenic speciation, the distribution of major and trace elements, and their relationship with the mineralogy and exopolymeric substances (EPS). In that way, it was shown that As was not linked to Ca or Si, and it was only moderately related to Fe, resulting from sorption onto an iron (oxy)hydroxide mineral. In that way, a number of features that support the biological cycling of arsenic have been documented. These include the incorporation of arsenic into EPS, the identification of organic-rich globules containing abundant arsenic with no other trace metals, and the heterogeneous distribution of As(III) and As(V), either together or separately and at different scales. Although there is a lack of accurate information on sulfur distribution and speciation, the high concentrations of both sulfur and arsenic in the water column support the notion of a coupled arsenic–sulfur cycle.

15.5 Concluding Remarks

From the microbiological point of view, Laguna La Brava is one of the most extraordinary environments in the Central Andes wetlands. This is mainly because of the unusual microbial diversity, which includes mainly Planctomycetes, Firmicutes, Acetothermia, Euryarchaeota, and Crenarchaeota, with low representation of phototrophs; this has an implication for possible alternative carbon-fixing pathways. In addition, the potential biogeochemical role of arsenic in these systems would recreate Precambrian microbial metabolism. All of these features occur in one of the locations of major environmental impact on the planet, due to mining activities, with active implications for scientists laying the foundations to preserve these primitive ecosystems.

Acknowledgements The authors acknowledge the generous financial support provided by Proyectos de Investigación Científica y Tecnológica (PICT) V Bicentenario 2010 1788, 3825, and Centro de Ecología Aplicada (CEA), and acknowledge Luis Ahumada and Marco Contreras for field logistics.

References

Baumgartner LK, Dupraz C, Buckley DH et al (2009a) Microbial species richness and metabolic activities in hypersaline microbial mats: insight into biosignature formation through lithification. Astrobiology 9:861–874. https://doi.org/10.1089/ast.2008.0329

Baumgartner LK, Spear JR, Buckley DH et al (2009b) Microbial diversity in modern marine stromatolites, Highborne Cay, Bahamas. Environ Microbiol 11:2710–2719. https://doi.org/10.1111/j.1462-2920.2009.01998.x

Boutt DF, Hynek SA, Munk LA, Corenthal LG (2016) Rapid recharge of fresh water to the halite-hosted brine aquifer of Salar de Atacama, Chile. Hydrol Process 30:4720–4740. https://doi.org/10.1002/hyp.10994

Cabral AR, Beaudoin G (2007) Volcanic red-bed copper mineralisation related to submarine basalt alteration, Mont Alexandre, Quebec Appalachians, Canada. Miner Deposita 42:901–912. https://doi.org/10.1007/s00126-007-0141-7

Cabrera D, Pizarro G (1992) Changes in chlorophyll a concentration, copepod abundance and UV and PAR penetration in the water column during the ozone depletion in Antarctic Lake Kitiesh. Arch Hydrobiol 43:123–134

Cáceres L, Gómez-Silva B, Garró X et al (2007) Relative humidity patterns and fog water precipitation in the Atacama Desert and biological implications. J Geophys Res Biogeosciences 112. https://doi.org/10.1029/2006JG000344

Casillas-Martinez L, Gonzalez ML, Fuentes-Figueroa Z, Castro CM, Nieves-Mendez D, Hernandez C, Ramirez W, Sytsma RE, Perez-Jimenez J, Visscher PT (2005a) Community structure, geochemical characteristics and mineralogy of a hypersaline microbial Mat, Cabo Rojo, PR. Geomicrobiol J 22(6):269–281

Casillas-Martinez L, Gonzalez ML, Rivera M, Fuentes Z, Nieves D, Hernandez C, RiosVelazquez C, Ramirez W, Visscher PT (2005b) Interrelations among communities, physiological structure and resulting mineralogy in the hypersaline mats of the Cabo Rojo salterns. Geomicrobiol J 22:269–281

Corenthal LG, Boutt DF, Hynek SA, Munk LA (2016) Regional groundwater flow and accumulation of a massive evaporite deposit at the margin of the Chilean Altiplano. Geophys Res Lett 43:8017–8025. https://doi.org/10.1002/2016GL070076

Dupraz C, Visscher PT (2005) Microbial lithification in marine stromatolites and hypersaline mats. Trends Microbiol 13(9):429–438

Dupraz C, Visscher PT, Baumgartner LK, Reid RP (2004) Microbe-mineral interactions: early carbonate precipitation in a hypersaline lake (Eleuthera Island, Bahamas). Sedimentology 51(4):745–765

Dupraz C, Reid RP, Braissant O, Decho AW, Norman RS, Visscher PT (2009) Processes of carbonate precipitation in modern microbial mats. Earth-Sci Rev 96(3):141–162

Dupraz C, Reid RP, Visscher PT (2011) Modern microbialites. In: Reitner J, Thiel V (eds) Encyclopedia of geobiology. Springer, Berlin, pp 617–634

Farías ME, Rascovan N, Toneatti DM et al (2013) The discovery of stromatolites developing at 3570 m above sea level in a high-altitude volcanic lake Socompa, Argentinean Andes. PLoS One 8:e53497. https://doi.org/10.1371/journal.pone.0053497

Farías ME, Contreras M, Rasuk MC et al (2014) Characterization of bacterial diversity associated with microbial mats, gypsum evaporites and carbonate microbialites in thalassic wetlands: Tebenquiche and La Brava, Salar de Atacama, Chile. Extremophiles 18:311–329. https://doi.org/10.1007/s00792-013-0617-6

Farias ME, Rasuk MC, Gallagher KL et al (2017) Prokaryotic diversity and biogeochemical characteristics of benthic microbial ecosystems at La Brava, a hypersaline lake at Salar de Atacama, Chile. PLoS One 12:e0186867. https://doi.org/10.1371/journal.pone.0186867

Fernandez AB, Rasuk MC, Visscher PT et al (2016) Microbial diversity in sediment ecosystems (evaporites domes, microbial mats, and crusts) of hypersaline Laguna Tebenquiche, Salar de Atacama, Chile. Front Microbiol 7:1284. https://doi.org/10.3389/fmicb.2016.01284

Glunk C, Dupraz C, Braissant O, Gallagher KL, Verrecchia EP, Visscher PT (2011) Microbially mediated carbonate precipitation in a hypersaline lake, Big Pond (Eleuthera, Bahamas). Sedimentology 58(3):720–736

Lara J, Escudero González L, Ferrero M et al (2012) Enrichment of arsenic transforming and resistant heterotrophic bacteria from sediments of two salt lakes in Northern Chile. Extremophiles 16:523–538. https://doi.org/10.1007/s00792-012-0452-1

Ley RE, Harris JK, Wilcox J et al (2006) Unexpected diversity and complexity of the Guerrero Negro hypersaline microbial mat. Appl Environ Microbiol 72:3685–3695. https://doi.org/10.1128/AEM.72.5.3685-3695.2006

Pinckney JL, Reid RP (1997) Productivity and community composition of stromatolitic microbial mats in the Exuma Cays. Bahamas Facies 36(204207):111116

Rasuk MC, Kurth D, Flores MR et al (2014) Microbial characterization of microbial ecosystems associated to evaporites domes of gypsum in Salar de Llamara in Atacama desert. Microb Ecol 68:483–494. https://doi.org/10.1007/s00248-014-0431-4

Risacher F, Alonso H (1996) Geoquimica del Salar de Atacama, parte 2: evolucion de las aguas. Andean Geol 23:123–134

Risacher F, Alonso H, Salazar C (2003) The origin of brines and salts in Chilean salars: a hydrochemical review. Earth-Sci Rev 63:249–293. https://doi.org/10.1016/S0012-8252(03)00037-0

Ruvindy R, White RAIII, Neilan BA, Burns BP (2016) Unravelling core microbial metabolisms in the hypersaline microbial mats of Shark Bay using high-throughput metagenomics. ISME J 10:183–196. https://doi.org/10.1038/ismej.2015.87

Sancho-Tomás M, Somogyi A, Medjoubi K, Bergamaschi A, Visscher PT, Van Driessche AES, Gérard E, Farias ME, Contreras M, Philippot P (2018) Distribution, redox state and (bio) geochemical implications of arsenic in present day microbialites of Laguna Brava, Salar de Atacama. Chem Geol 490:13–21

Sforna MC, Philippot P, Somogyi A et al (2014) Evidence for arsenic metabolism and cycling by microorganisms 2.7 billion years ago. Nat Geosci 7:811–815. https://doi.org/10.1038/ngeo2276

Tkavc R, Gostinčar C, Turk M, Visscher PT, Oren A, Gunde-Cimerman N (2011) Bacterial communities in the 'petola' microbial mat from the Sečovlje salterns (Slovenia). FEMS Microbiol Ecol 75(1):48–62

Van Gemerden H (1993) Microbial mats: a joint venture. Mar Geol 113:3–25

Visscher PT, van Gemerden H (1993a) Sulfur cycling in laminated marine microbial ecosystems. In: Oremland RS (ed) Biogeochemistry of global change. Springer, Boston, MA. https://doi.org/10.1007/978-1-4615-2812-8_37

Visscher PT, van Gemerden H (1993b) Sulfur cycling in laminated marine ecosystems. In: Oremland RS (ed) Biogeochemistry of global change: radiatively active trace gases. Chapman and Hall, New York, pp 672–693

Visscher PT, Quist P, van Gemerden H (1991) Methylated sulfur compounds in microbial mats: in situ concentrations and metabolism by a colorless sulfur bacterium. Appl Environ Microbiol 57:1758–1763

Witt-Eickschen G, Palme H, O'Neill HSC, Allen CM (2009) The geochemistry of the volatile trace elements As, Cd, Ga, In and Sn in the Earth's mantle: new evidence from in situ analyses of mantle xenoliths. Geochim Cosmochim Acta 73:1755–1778. https://doi.org/10.1016/j.gca.2008.12.013

Wong HL, Smith D-L, Visscher PT et al (2015) Niche differentiation of bacterial communities at a millimeter scale in Shark Bay microbial mats. Sci Rep 5:15607. https://doi.org/10.1038/srep15607

Chapter 16
The Carbonate System in Hypersaline Lakes: The Case of Laguna Negra (in the Puna Region of Catamarca, Argentina)

Fernando Javier Gomez, Flavia Jaquelina Boidi, Estela Cecilia Mlewski, and Emmanuelle Gérard

16.1 Introduction

Lacustrine sedimentary carbonates are an excellent record of environmental and climate conditions during carbonate deposition (Alonzo-Zarza and Tanner 2010a, b; Frantz et al. 2014; Petryshyn et al. 2016). Physicochemical processes and microbiological activity can trigger carbonate precipitation and dissolution by controlling the pH, alkalinity, and availability of ions and carbon species (for example, Ca^{+2} and CO_3^{-2}) (Visscher and Stolz 2005; Arp et al. 2001), and this have an impact on calcite and aragonite chemical equilibrium (Morse et al. 2007). In addition, extreme changes in environmental factors, such as salinity and temperature (which are particularly important in Laguna Negra and other systems in the Puna region), have a significant influence by controlling the equilibrium constants involved in the reactions of the carbonate system (Saas and Ben-Yaakov 1977; Weiss 1974) (for reviews, see Zeebe and Wolf-Gladrow 2001; Mucci 1983). Here, available hydrochemical data from Laguna Negra are compiled (from Gomez et al. 2014), as well as data from other lakes in the Chilean, Bolivian, and Argentinian region to make some general comparisons. The compiled data and references can be seen in Table 1, which is a file presented as supporting online material in GitHub (https://github.com/FernandoJGomez/Gomez-et-al.-2019_Elsevier_Book_Puna), together with the Jupyter notebooks that provide most of the plots for this contribution.

Laguna Negra is a shallow (≤2 m) hypersaline lake located 4200 m above sea level. The high-altitude, hot, and dry climate (with a strongly negative water balance)

F. J. Gomez (✉) · F. J. Boidi · E. C. Mlewski
Centro de Investigaciones en Ciencias de la Tierra (CICTERRA), Consejo Nacional de Investigaciones Científicas y Técnicas (CONICET)–Universidad Nacional de Córdoba (UNC), Córdoba, Argentina

E. Gérard
Institut de Physique du Globe de Paris, Université de Paris, Centre national de la recherche scientifique (CNRS), Unité mixte de recherche (UMR) 7154, Paris, France

M. E. Farías (ed.), *Microbial Ecosystems in Central Andes Extreme Environments*, https://doi.org/10.1007/978-3-030-36192-1_16

favors the precipitation of evaporites (halite, gypsum, calcite, etc.) and produces favorable conditions for the development of microbial mats, where abundant carbonate precipitation takes place, particularly in the mixing zone between groundwater springs and the main lake waters. For details of the microbial biodiversity, as well as the textural and geochemical aspects of these carbonates, see Gomez et al. (2014, 2018), Mlewski et al. (2018), Buongiorno et al. (2018), Shalygin et al. (2018), and Boidi et al. (2019, chapter 14 this volume). Beeler et al. (2020) provide further details regarding the Laguna Negra hydrochemistry and water isotope signature. Here, we focus on discussing some relevant aspects regarding the carbonate system in Laguna Negra and similar lakes in the region.

The carbonate system parameters were calculated by using published information on these hypersaline systems. References that provided the necessary data to calculate the carbonate system parameters (cf. Zeebe and Wolf-Gladrow 2001) and calcite–aragonite solubility (Mucci 1983) were selected. Given that some data recorded either salinity or conductivity, to estimate the salinity from the available conductivity data, a polynomy was fitted to those data where both salinity and conductivity were known. Then this was used to calculate the salinity when only the conductivity was known. Weiss (1974) was used for Henry's constant (K_H). Given that the experimental determinations of the acidity constants (K_1 and K_2), as published by Roy et al. (1993), were focused on seawater and were not parameterized for high salinity, Saas and Ben-Yaakov (1977) was used because these were determined for salinities of up to ~300 parts per trillion (ppt) (see also Beeler et al. 2020). It is worth mentioning that most of the compiled studies (supplementary Table 1) did not publish full water chemistry that would allow use of a more proper speciation model to calculate cation and anion activities (for example, Pitzer 1971) and mineral saturation states (Harvie et al. 1984). Our approach here seems reasonable given our main goal, which is just to make some general comparisons of the carbonate systems in Laguna Negra and other lakes in the region. Besides, the effect of high ionic strength on the activities of the carbonate system ions is implicit in the Saas and Ben-Yaakov (1977) experimental determination of K_1 and K_2. Solubility product constants for calcite and aragonite (K_{spc} and K_{spa}, respectively) were calculated by using the data from Mucci (1983), although it is necessary to be aware that these were parameterized only for salinities of up to 45 ppt (almost an order of magnitude lower). The Python Jupyter notebooks that perform the calculations and plots can be downloaded from the GitHub website, together with Table 1 (https://github.com/FernandoJGomez/Gomez-et-al.-2019_Elsevier_Book_Puna).

16.2 The Laguna Negra Carbonate System: Comparison with Other Lakes in the Puna

Figure 16.1 shows dissolved inorganic carbon (DIC) and alkalinity (Alk) data from Gomez et al. (2014), recalculated by using Saas and Ben-Yaakov (1977) and Mucci (1983). As expected, the alkalinity and DIC show a strong positive covariation and these variables also positively correlate with salinity (Fig. 16.1) (see also Beeler

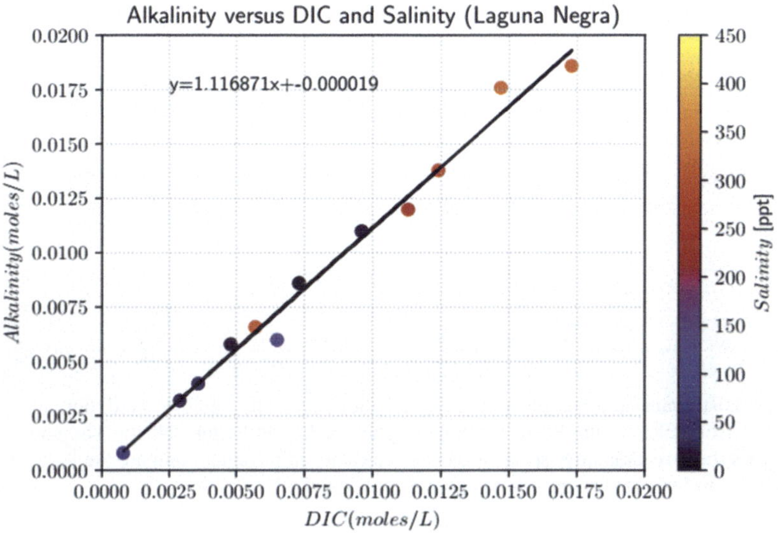

Fig. 16.1 Scatter plot of alkalinity (Alk) and dissolved inorganic carbon (DIC) values in Laguna Negra (recalculated from Gomez et al. 2014). The *color bar* shows how salinity correlates with these variables. The equation is the trend line adjusted to the data.

et al. 2020) for additional data and a deeper analysis using DIC stable isotopes). Fitting a trend line to a scatter plot of the Laguna Negra data shows that the Alk:DIC ratio is 1.12:1. Calcium carbonate precipitation consumes alkalinity and DIC in a 2:1 ratio (Zeebe and Wolf-Gladrow 2001). The Laguna Negra Alk:DIC ratio of 1.12:1 suggests that, besides carbonate precipitation, more DIC is also lost through other processes (a smaller slope in the trend line). Most probably, this is due to CO_2 degassing, which also consumes DIC, and strong degassing has been suggested to explain covariation and extreme enrichments in carbon and oxygen stable isotopes (Gomez et al. 2014; Buongiorno et al. 2018; Beeler et al. 2020). Beeler et al. (2020) did more detailed work, which also recorded a similar Alk:DIC ratio value (1.27:1) and showed, by mass balance carbon isotope modeling, that 60% of DIC loss was due to carbonate precipitation and the remaining loss was due to other processes, where CO_2 degassing was suggested (Beeler et al. 2020).

As observed in Fig. 16.2, most DIC and alkalinity values for the lakes are ≤0.025 moles/L and ≤0.05 moles/L, respectively, with a median below 0.01 moles/L. The same is observed for Laguna Negra. Both (DIC and alkalinity) are enriched (~1 ORDER of magnitude higher) when compared with seawater (typically 0.0021 for DIC and 0.00236 moles/L (Zeebe and Wolf-Gladrow 2001). The values are also higher than those in some hypersaline systems such as the Dead Sea (Barkan et al. 2001). The median DIC and alkalinity values in Laguna Negra are similar to those in all of the lakes (Fig. 16.2). Given that many of these lakes are also enriched in calcium (Ca^{+2}) (Table 1), it is expected that many of them are also saturated in carbonate minerals (Fig. 16.7).

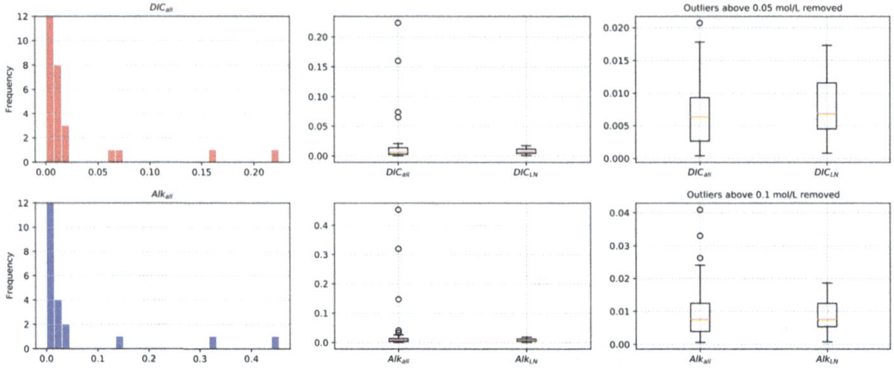

Fig. 16.2 Histograms and box plots of compiled alkalinity (Alk) and dissolved inorganic carbon (DIC) data from lakes in the Puna–Altiplano region, including Laguna Negra. The *panel on the right* shows the box plots after some outliers were removed to better compare the Laguna Negra data (Alk_{LN} and DIC_{LN}) with the whole data set (Alk_{All} and DIC_{All})

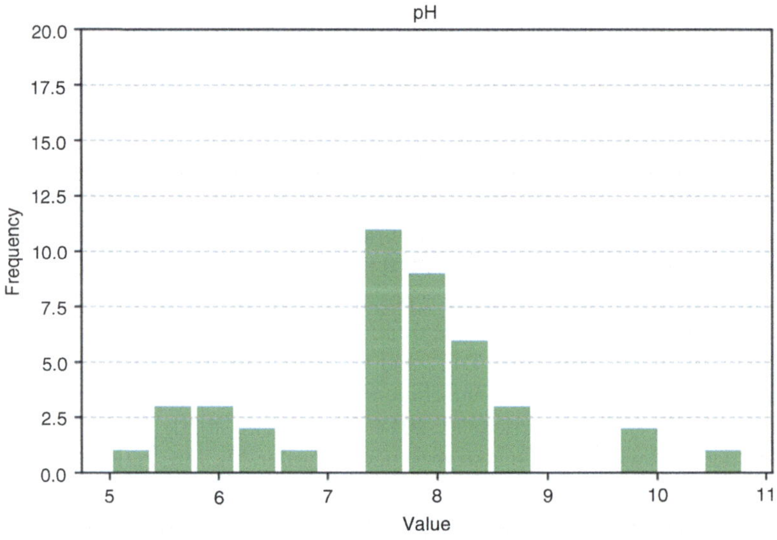

Fig. 16.3 Compiled pH data from the Puna–Altiplano lakes

16.3 pH and Salinity of Some Puna–Altiplano Lakes

When analyzing saline lakes of the Puna–Altiplano region, it can be seen that the pH has a bimodal distribution, with pH values of around 5.5–6 and 7.5–8, with just a few alkaline lakes having a pH above 9.5 (Fig. 16.3). The salinity is also highly variable because of the net balance between evaporation, mineral precipitation, and

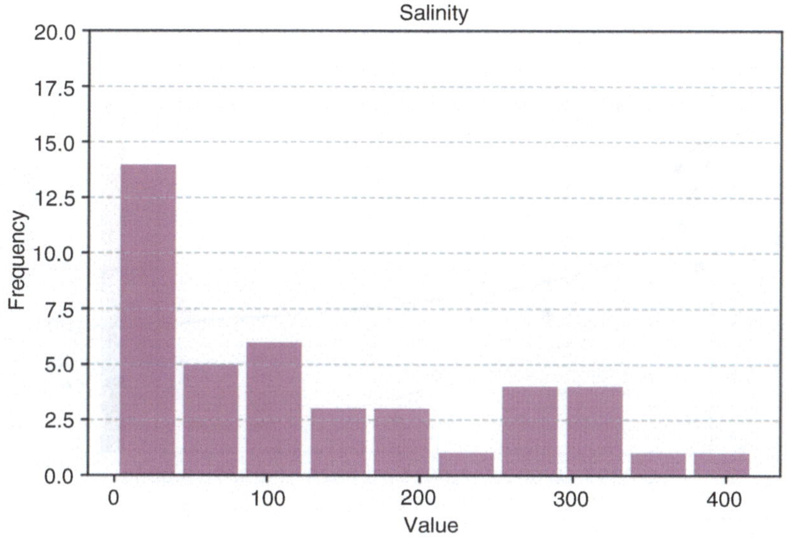

Fig. 16.4 Compiled salinity data from the Puna–Altiplano lakes

freshwater input, with some lakes reaching 400 ppt (Fig. 16.4). Laguna Negra shows a lower pH in the main lake (pH 5.6) and a higher pH in the inlet, as well as in the mixing zone waters (pH up to 7.8). The salinity is also variable in Laguna Negra, where the inlet waters (9 ppt) and main lake waters (325 ppt) are the two end members (Table 1). Even when a relatively low pH has been recorded in the main lake, it has been shown to be variable, since a pH above 7 has also been recorded (data not shown, and Beeler et al. 2020). These changes in the pH could be related to increased CO_2 input into the main lake area or could also be due to carbonate precipitation events, which can also increase CO_2 (Barkan et al. 2001; Golan et al. 2017), thus lowering the pH through the following reaction:

$$Ca^{+2} + 2HCO_3^- \leftrightarrow CaCO_3 + CO_2 + H_2O \tag{16.1}$$

According to the Le Châtelier principle, when degassing occurs, reaction (16.1) proceeds to the right to restore the equilibrium, and carbonate precipitation takes place, also producing CO_2. Even when for each mole of carbonate precipitated, CO_2 should increase by 1 mole as well, it does not actually happen, because of buffering, since CO_2 is transformed into bicarbonate (HCO_3^-) (Zeebe and Wolf-Gladrow 2001).

16.4 CO_2 Degassing and pH

Measured pH and calculated CO_2 covariates obtained (by use of the pH and alkalinity) in a nonlinear way and a decreasing logarithmic function can be fitted to these data (Fig. 16.5). The salinity also shows a trend with these variables: as salinity

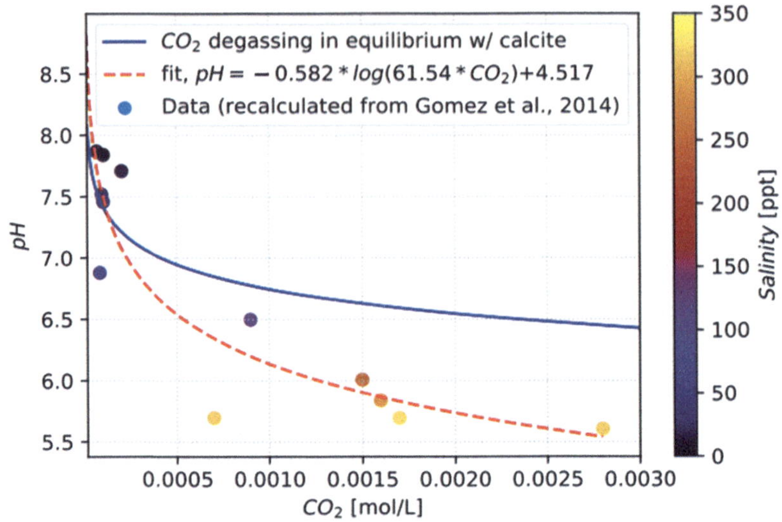

Fig. 16.5 CO_2 versus pH data from Laguna Negra (*colored dots* according to salinity). The *red dashed trend line* represents an exponential function fitted to the data, and the *blue continuous line* shows values calculated according to Eqs. (16.3) and (16.4)

increases, CO_2 and the pH decrease (Fig. 16.5). When CO_2 is lost through degassing to the atmosphere (or is consumed by photosynthesis and other metabolic processes), the pH increases, increasing the carbonate ion concentration (CO_3^{-2}), and, if enough calcium (Ca^{+2}) is available to overcome carbonate saturation, this triggers carbonate precipitation according to calcite equilibrium (Eqs. 16.1 and 16.2).

$$Ca^{+2} + CO_3^{-2} \leftrightarrow CaCO_3 \quad (16.2)$$

If equilibrium with calcite is assumed, it can be seen how the pH may change as CO_2 degassing occurs (Fig. 16.5). This can be shown by modeling the pH versus CO_2 in equilibrium with calcite (through Eqs. 16.3 and 16.4, from Kano (2019); see the derivation in the supporting online material). The model follows the same trend as the data (Fig. 16.5), and differences in the pH may occur because even when CO_2 degassing increases the pH, carbonate precipitation also produces CO_2 and lowers the pH. Thus, the net balance of these driving processes will define the final pH and CO_2 concentration during the degassing-precipitation trend (Fig. 16.5).

$$H^+ = \left(K_2^{1/3} \left(K_H K_1 \right)^{2/3} / 2K_{sp}^{1/3} \right) pCO_2^{2/3} \quad (16.3)$$

$$pH = -\log\left[H^+ \right] \quad (16.4)$$

16.5 Calcite and Aragonite Saturation and Processes Triggering Carbonate Precipitation

The main problem when calculating the saturation state of carbonate minerals in hypersaline lakes is the lack of experimental data on the solubility product (K_{sp}) of calcite and aragonite at high ionic strength and salinities. The classic work of Mucci (1983) focused on seawater and salinities of up to 45 ppt. Saas and Ben-Yaakov (1977) determined K_1 and K_2 for salinities of up to 309 ppt (at 30 °C) and suggested a K_{sp} value for calcite of $6.7 * 10^{-4}$. This is about 1000 times higher than typical values for seawater (Saas and Ben-Yaakov 1977), and this would lower the saturation state for carbonate minerals. With these lower values for K_{sp} and the saturation state of calcite, it is implicit that, at high ionic strength, the activities of carbonate and calcium ions will be lesser, given ion interactions and pairing that decrease ion activities. With the aim of illustrating the effect of salinity and using available data, Fig. 16.6 shows saturation states for calcite (Ω_{cal}) versus DIC and salinity to show how, as salinity increases, the saturation state decreases (using K_1 and K_2 from Saas and Ben-Yaakov (1977) and the K_{sp} of calcite from Mucci (1983)). As a reference, the white dot in Fig. 16.6 represents a sample of the groundwater springs with salinity of 9.1 and DIC of 0.0173 moles/L (sample IW-7 from Laguna Negra; Gomez et al. 2014), which is oversaturated.

As observed in Fig. 16.7, the saturation index (S.I. = $\log(\Omega)$, where S.I. > 0 and S.I. < 0 indicates over- and undersaturation of carbonate minerals, respectively) suggests that the Laguna Negra lake waters are undersaturated (given the relatively low pH of 5.7) and that most mixing zone and inlet waters are oversaturated with calcite. When the saturation state was calculated with a pH of 7, as has also been observed (Beeler et al. 2020), the main lake waters became slightly oversaturated, so the main lake may fluctuate between oversaturated and undersaturated states.

Strong evaporation, as is observed in Laguna Negra, increases the saturation state and CO_2 degassing, also increasing the pH and thus triggering carbonate precipitation. As salinity increases, carbonate precipitation takes place and thus the pH becomes lower, as well as the saturation index (Fig. 16.7). Beeler et al. (2020) documented that the pH and DIC spatially change from the groundwater springs to the mixing zones and reaching the main lake, and this has been interpreted as being related to calcite precipitation. It is the net balance of these complex interacting processes that controls the saturation state and mineral precipitation in Laguna Negra.

As shown in Fig. 16.7 (left panel), most lakes are oversaturated with calcite and aragonite, and, although the Laguna Negra samples are variable, they show similar degrees of saturation when compared with other lakes (Fig. 16.7, right panel).

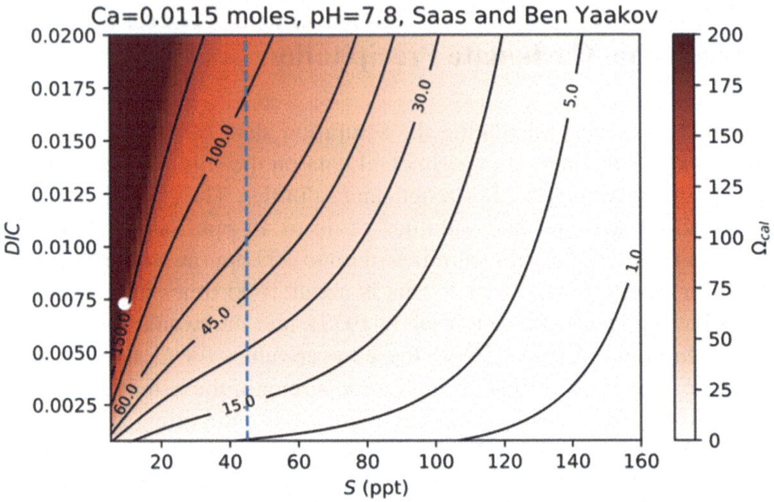

Fig. 16.6 Omega versus dissolved inorganic carbon (DIC) and salinity at pH 7.84 and Ca = 0.0115 moles, using the acidity constants (K_1 and K_2) from Saas and Ben-Yaakov (1977) and the solubility product constant for calcite (K_{spc}) from Mucci (1983). Note that at constant DIC and pH values, when salinity increases, the saturation state for calcite (Ω_{cal}) decreases. The *vertical blue dotted line* shows the salinity range up to which the Mucci (1983) K_{spc} is valid

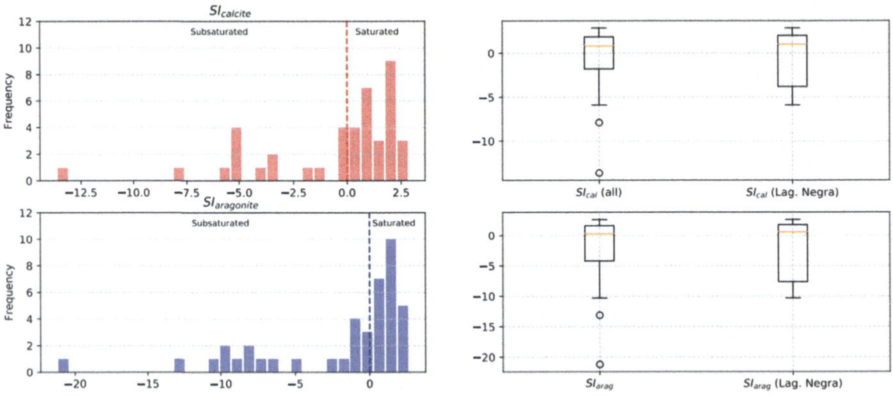

Fig. 16.7 Saturation index (S.I.) histograms and box plots for calcite and aragonite for the whole data set (including Laguna Negra) and for Laguna Negra itself. The *vertical red and blue dotted lines* show the equilibrium line where S.I. = 0

16.6 The Mg/Ca Molar Ratio of Laguna Negra and Other Puna–Altiplano Lakes

The Mg/Ca molar ratio for the Laguna Negra waters fluctuates between ~0.08 and ~2, like those for more than half (62%; Fig. 16.8) of the lakes in the Chilean–Argentinian–Bolivian Puna–Altiplano area. Given that, as shown by experimental

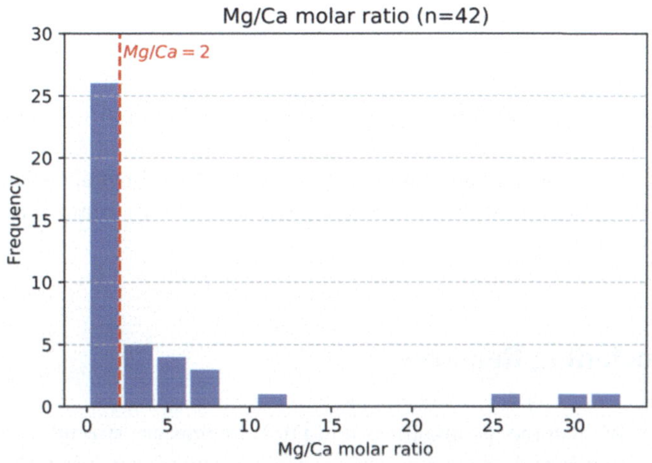

Fig. 16.8 Mg/Ca molar ratio in lakes in the Puna–Altiplano region

and thermodynamic models (see Sun et al. 2015 and references therein), Mg/Ca ratios below 2 are considered important for calcite precipitation, it is expected that calcite should be the predominant calcium carbonate mineral species (if oversaturation is reached). The other 38% is above this boundary (Fig. 16.8) and some of them even show Mg/Ca molar ratios that are extremely high (between 25–35), such as Diamante Lake (in Argentina) and Chaxa and Tebenquiche Lakes (in Chile; there, gypsum and halite predominate, and aragonite is subordinated). Thus, aragonite is expected to be the main carbonate mineral in these lakes. Even when mineralogy data are scarce, it has been observed that in Diamante and La Brava Lakes, aragonite precipitation prevails (M. E. Farías, personal communication, 2019). In hypersaline lakes, the Mg/Ca molar ratio will be controlled by the local geology (weathering of the available rocks and minerals), hydrothermal input (in some cases), and the evolution of the lake during evaporation, so this ratio, as opposed to that in seawater, is expected to be highly variable and thus the carbonate mineralogy will be as well. Further studies integrating hydrochemistry, evaporite mineralogy, and local geology would be necessary to have a better understanding of the main controls on calcium carbonate mineralogy in hypersaline lakes.

It is worth mentioning that the aragonite/calcite ratio is controlled not only by the Mg/Ca molar ratio but also by many other processes, such as the precipitation rate, temperature, salinity, and solution composition (see Morse et al. 2007; Jones 2017 for a recent review). In Laguna Negra, only two samples were close to the value at which aragonite should precipitate, one coming from the groundwater and the other from the mixing zone. In Laguna Negra, calcite is the predominant mineral, as observed in the main lake, in the mixing zone sediments, in the oncoids (Gomez et al. 2014; Buongiorno et al. 2018), and within the microbial mats. It is worth mentioning that aragonite traces have been recorded within some laminae in the oncoids (Buongiorno et al. 2018). More interestingly, aragonite has recently been recorded during a local diatom bloom event (Gomez et al. 2018), associated with cyanobacteria such as *Rivularia* (Mlewski et al. 2018), in areas where the Mg/Ca ratio is clearly below 2. This suggests the presence of other controls on carbonate polymorph selection, probably related to the kinetic influence of microbial metabolism—in this case, photosynthesis, given the direct association with cyanobacteria and diatoms (Gomez et al. 2018; Boidi et al. 2019). In addition, *Rivularia* has been observed to concentrate calcium on the cell surface (Mlewski et al. 2018) and could be locally modifying the Mg/Ca ratio on the sheats and thus controlling carbonate polymorph selection.

16.7 Concluding Remarks

In hypersaline environments, alkalinity and DIC changes are strongly controlled by a set of physicochemical processes such as evaporation and strong CO_2 degassing. In Laguna Negra, CO_2 degassing, as well as carbonate precipitation, have been shown to be a main control on the pH, alkalinity, and DIC evolution of lake waters. In addition, water mixing also contributes to changes in these variables and can be particularly important in increasing saturation states, thus promoting carbonate precipitation as observed in the mixing zone in Laguna Negra. By changing chemical equilibrium, microbial activity can also modify alkalinity and DIC. Photosynthesis consumes CO_2, lowering DIC and thus changing the pH, and consequently triggering carbonate precipitation. Anoxic degradation of organic matter increases alkalinity, also promoting oversaturation of carbonate minerals. Exopolymeric substances and other organic matrices can also serve as nucleation sites for mineral precipitation. Thus, the net balance of these combined microbiological and physicochemical processes will define the textural and chemical signatures preserved in these sedimentary carbonates.

The Mg/Ca molar ratio is known to influence the carbonate polymorph that may precipitate, and this has been observed in several lakes in the Puna–Altiplano area, where a high Mg/Ca molar ratio favors aragonite precipitation. Examples are Diamante Lake (in Argentina) and Tebenquiche and Chaxa Lakes (in Chile), where abundant aragonite has been recorded. Although the prevalence of aragonite at high Mg/Ca ratios is supported by thermodynamics, several processes control carbonate polymorph selection. For example, aragonite precipitation has been observed in Laguna Negra, where the Mg/Ca molar ration is low, and is related to the photosynthetic activity of diatoms and cyanobacteria, suggesting kinetic control of mineral polymorphism.

There is a need to gain better understanding of the carbonate system in hypersaline lakes, particularly given that some of the parameters and equilibrium constants that are useful for calculation of carbon species concentrations have not been determined for high-ionic-strength solutions such as those that are typically observed in the Puna–Altiplano region. Improving this understanding can help us to better understand geomicrobiological aspects and the environmental record of carbonate precipitation in modern and ancient hypersaline systems.

References

Alonzo-Zarza AM, Tanner LH (2010a) Carbonates in continental settings: facies, environments and processes. In: Developments in sedimentology, vol 61. Elsevier, Amsterdam, 378p

Alonzo-Zarza AM, Tanner LH (2010b) Carbonates in continental settings: geochemistry, diagenesis and applications. In: Developments in sedimentology, vol 62. Elsevier, Amsterdam, 319p

Arp G, Reimer A, Reitner J (2001) Photosynthesis-induced biofilm calcification and calcium concentrations in the Phanerozoic oceans. Science 292:1701–1704

Barkan E, Luz B, Lazar B (2001) Dynamics of the carbon dioxide system in the Dead Sea. Geochimi Comsochim Acta 65(3):355–368

Beeler S, Gomez FJ, Bradley A (2020) Controls of extreme isotopic enrichment in modern microbialites and associated abiogenic carbonates. Geochim Cosmochim Acta 269:136–149

Boidi FJ, Mlewski C, Gomez FJ, Gerard E (2019) Chapter 13: Physicochemical and microbiological processes involved in carbonate precipitation in hypersaline lakes: a review of the Laguna Negra case Puna of Catamarca, Argentina. In: Farias ME (ed) Microbial ecosystems in Puna environments: biofilms, microbial mats microbialites and endoevaporites. Springer, New York

Buongiorno J, Gomez FJ, Fike D, Kah LC (2018) Mineralized microbialites as archives of environmental evolution, Laguna Negra, Catamarca Province, Argentina. Geobiology 1:1–24

Dupraz C, Visscher PT (2005) Microbial lithification in marine stromatolites and hypersaline mats. Trends Microbiol 13:429–438. https://doi.org/10.1016/j.tim.2005.07.008

Frantz CM, Petryshyn VA, Marenco PJ, Tripati A, Berelson WM, Corsetti FA (2014) Dramatic local environmental change during the Early Eocene Climatic Optimum detected using high resolution chemical analyses of Green River Formation stromatolites. Palaeogeogr Palaeoclimatol Palaeoecol 405:1–15

Golan R, Lazar B, Wurgaft E, Lensky N, Ganor J, Gavrieli I (2017) Continuous CO_2 escape from the hypersaline Dead Sea caused by aragonite precipitation. Geochim Cosmochim Acta 207:43. https://doi.org/10.1016/j.gca.2017.02.031

Gomez FJ, Kah LC, Bartley J, Astini RA (2014) Microbialites in a high-altitude Andean Lake: multiple controls in carbonate precipitation and lamina accretion. Palaios 29:233–249

Gomez FJ, Mlewski C, Boidi FJ, Farías ME, Gérard E (2018) Calcium carbonate precipitation in diatom-rich microbial mats: the Laguna Negra Hypersaline Lake, Catamarca, Argentina. J Sediment Res 88(6):727–742. https://doi.org/10.2110/jsr.2018.37

Harvie CE, Moller N, Weare JH (1984) The prediction of mineral solubilities in natural waters: the Na-K-Mg-Ca-H-CIS04-0H-HC03-C03-CO2H20 system to high ionic strengths at 25°C. Geochim Cosmochim Acta 48:723–751

Jones B (2017) Review of aragonite and calcite crystal morphogenesis in thermal springs. Sediment Geol 354:9–23

Mlewski EC, Pisapia C, Gomez F, Lecourt L, Soto Rueda E, Benzerara K, Ménez B, Borensztajn S, Jamme F, Réfrégiers M, Gérard E (2018) Characterization of pustular mats and related rivularia-rich laminations in oncoids from the Laguna Negra Lake (Argentina). Front Microbiol 9:996. https://doi.org/10.3389/fmicb.2018.00996

Morse JW, Arvidson RS, Lüttge A (2007) Calcium carbonate formation and dissolution. Chem Rev 107:342–381

Mucci A (1983) The solubility of calcite and aragonite in seawater at various salinities, temperatures and one atmosphere of total pressure. Am J Sci 283:780–799

Petryshyn VA, Juarez Rivera M, Agić H, Frantz CM, Corsetti FA, Tripati AE (2016) Stromatolites in Walker Lake (Nevada, Great Basin, USA) record climate and lake level changes ~35,000 years ago. Palaeogeogr Palaeoclimatol Palaeoecol 451:140–151

Pitzer KS (1971) Thermodynamics of electrolytes. I. Theoretical basis and general equations. J Phys Chem 77:268

Roy RN, Vogel KM, Porter-Moore C, Pearson T, Good CE, Millero FJ, Campbell DM (1993) The dissociation constants of carbonic acid in seawater at salinities 5 to 45 and temperatures 0 to 45 Celsius. Mar Chem 44:249–267

Saas E, Ben-Yaakov S (1977) The carbonate system in hypersaline solutions: Dead Sea Brines. Mar Chem 5:183–199

Shalygin S, Pietrasiak N, Gomez F, Mlewski C, Gerard E, Johansen JR (2018) Rivularia halophila sp. nov. (Nostocales, Cyanobacteria): the first species of Rivularia described with the modern polyphasic approach. Eur J Phycol 53:537–548. https://doi.org/10.1080/09670262.2018.1479887

Sun I, Saivenkataraman J, Chen W, Persson KA, Gerbran C (2015) Nucleation of metastable aragonite $CaCO_3$ in seawater. Proc Natl Acad Sci U S A 112:3199–3204

Visscher P, Stolz J (2005) Microbial mats as bioreactors: populations, processes and products. Palaeogeogr Palaeoclimatol Palaeoecol 219:97–100

Weiss RF (1974) Carbon dioxide in water and seawater: the solubility of a non-ideal gas. Mar Chem 2:203–215

Zeebe R, Wolf-Gladrow D (2001) CO_2 in seawater: equilibrium, kinetics, isotopes, Elsevier oceanographical series. Elsevier, Amsterdam, 346p

Part VI
Generalities and Particularities of Microbial Ecosystems at the Central the Andes

Chapter 17
Integral Prospection of Andean Microbial Ecosystem Project

María Eugenia Farías, Patricio G. Villafañe, and Agustina I. Lencina

17.1 Introduction: Particular Characteristics of Andean Microbial Ecosystems

When microbial ecosystems first started to be reported 10 years ago, nobody really had a notion of the relevance they would have in the Central Andean region. Consequently, the heritage of the microbialites reported in El Peinado, Laguna Negra, Laguna Pozo Bravo, Laguna La Brava, etc. promises to position the Andes as reservoirs of the most relevant modern microbialites on the planet (Table 17.1, Fig. 17.1). Furthermore, the number of different ecosystems is worth paying close attention to, as it gives rise to questions such as: What favors the development of these ecosystems? What are the conditions that influence the precipitation of a carbonaceous or a gypsum system at such a short distance and under similar environmental conditions, such as at the Atacama salt flat? Why are oncolites distributed so? Until now, it has been possible only to survey the systems and to carry out more in-depth studies in some of them to try to achieve their preservation. Throughout the prospection

M. E. Farías (✉)
LIMLA-PROIMI-CONICET, Tucuman, Argentina

P. G. Villafañe
Laboratorio de Investigaciones Microbiológicas de Lagunas Andinas (LIMLA), Planta Piloto de Procesos Industriales Microbiológicos (PROIMI), Centros Científico Tecnológicos (CCT), Consejo Nacional de Investigaciones Científicas y Técnicas (CONICET), Tucumán, Argentina

A. I. Lencina
Laboratorio de Investigaciones Microbiológicas de Lagunas Andinas (LIMLA), Centro de Investigaciones y Transferencia Catamarca (CITCa), Centros Científico Tecnológicos (CCT), Consejo Nacional de Investigaciones Científicas y Técnicas (CONICET), San Fernando del Valle de Catamarca, Catamarca, Argentina

M. E. Farías (ed.), *Microbial Ecosystems in Central Andes Extreme Environments*, https://doi.org/10.1007/978-3-030-36192-1_17

Table 17.1 Prospected environments and their global positions, chemical characteristics, kinds of Andean microbial ecosystems (AMEs) reported, mineralogy, and references

Region	Lagoon, lake, or wetland	Global position	Altitude (m asl)	Maximal conductivity (mS/cm)	pH	Temperature (°C)	O_2 (%)	Arsenic (mgL^{-1})	Ecosystem	Mineralogy	References
Argentina	Laguna Socompa	24°31′34.96″S 68°12′24.26″W	3650	165	8.6	20	6.92	18	Microbialites, evaporitic mats, oncolites	Aragonite, thenardite	Farías et al. (2013); Kurth et al. (2017)
	Socompa volcanic fumarole	24°31′43.29″S 68°11′53.76″W	3650	ND	6.5	26	NF	0.05	Biofilms	ND	Lynch et al. (2012)
	Ojo de Tolar	24°37′23.51″S 67°22′14.35″W	3515	201.7	6.5	14	2.87	0.59	Evaporitic mats	Gypsum, halite	Farías et al. (2011)
	Llullaillaco volcanic fumarole	24°47′24.88″S 68°18′38.19″W	6381	ND	8.5	ND	NF	ND	Biofilms	ND	Schröder (2000)
	Laguna Diamante	26°0′49.75″S 67°2′10.08″W	4595	217.3	11	14	1.02	354	Microbialites, biofilms	Carbonate	Rascovan et al. (2016); Ordóñez et al. (2018)
	Diamante hydrothermal vent	26°1′57.91″S 66°59′9.98″W	4828	ND	6.5	85	ND	0.05	Microbialites	ND	Rascovan et al. (2016)
	Laguna Carachipampa	26°27′2.29″S 67°30′39.38″W	3018	87.31	7.8	31.3	3.8	ND	Microbial mats, oncolites, microbialites	Carbonate	Informe Mineria Catamarca
	Complejo Ojos de Campo	25°30′49.63″S 67°34′39.20″W	4081	256.1	8.5	18	5.18	18	Microbialites, microbial mats	Carbonate	Informe Mineria Catamarca
	Laguna Pozo Bravo	25°30′48.55″S 67°34′37.18″W	3338	148	7.8	14	17.1	15.5	Microbial mats	Carbonate	Informe Mineria Catamarca
	Las Quinoas	25°52′15.50″S 67°54′25.56″W	3338	256.1	8.5	18	5.18	18	Microbialites, microbial mats	Carbonate	Informe Mineria Catamarca
	Bosques del salar, Laguna Verde	25°28′45.73″S 67°33′17.72″W	3343	222	7.44	11.4	7.1	ND	Endoevaporites, microbial mats	Gypsum, halite	This report
	Humedal Rojo	25°28′43.44″S 67°33′23.04″W	3343	ND	ND	ND	ND	ND	Microbial mats	Halite, calcite	This report
	El Peinado	26°30′46.16″S 68°5′50.09″W	3748	16.28	7.9	22.5	5.3	ND	Microbialites, oncolites, microbial mats	Carbonate	This report
	Tres Quebradas	27°22′20.15″S 68°40′37.43″W	3100	13.24	8.23	33.4	36.3	ND	Oncolites, microbial mats	Carbonate	This report
	Laguna Negra	27°38′34.82″S 68°33′53.00″W	4101	103.8	7.5	15	1.32	3	Microbial mats, microbialites, oncolites	Carbonate	Gomez et al. (2014)
	Laguna Vilama	22°35′S 66°55′W	4600	117	7.1	ND	ND	11.8	Microbial mats	ND	Farías et al. (2009)

Chile	Laguna La Brava	23°43′30.79″S 68°14′29.38″W	2300	187.6	8.63	32.4	3.02	20.64	Microbial mats, microbialites, oncolites, *Rhizobium*-associated mats	Halite, gypsum, carbonate	Farías et al. (2014, 2017)
	Laguna Tebenquiche	23°3′35.58″S 68°13′0.98″W	2340	108.7	8.6	21.5	2.2	2.3	Microbial mats, endoevaporites, *Rhizobium*-associated mats	Halite, gypsum, carbonate	Farías et al. (2014); Thiel et al. (2010); Fernandez et al. (2016)
	Laguna Cejar	23°43′30.79″S 68°14′29.38″W	2300	8.2	5.5	ND	ND	ND	Microbial mats, endoevaporites	Halite, gypsum	Rasuk et al. (2015); Stivaletta et al. (2011)
	Laguna Baltinache	23°8′7.47″S 68°15′31.13″W	2315	ND	ND	ND	ND	ND	Endoevaporites	Halite	This report
	Laguna Chaxas	29°24.9′S 68°10′W	2300	149.9	7.59	28.5	109.7	2.27	Microbial mats, microbialites	Carbonate	Thiel et al. (2010)
	Laguna Barros Negros	23°21.7′S 68°9.1′W	2300	175.9	7.61	30.8	47.8	ND	Endoevaporites, microbial mats	Halite, gypsum	This report
	Laguna Salada Saladita	23°41.1′S 68°9.66′W	2300	26.5	8.05	18.5	104.1	2.6–17.86	Microbial mats	ND	This report
	Salar Aguas de Quelana	23°24.9′S 68°5.77′W	2300	116.0	6.6	26.6	111.4	1.3–18.56	Microbial mats, oncolites	ND	This report
	Salar de Llamara	21°0′2.15″S 68°23′9.72″W	4157	7.6	1.2	20.5	3.45	9.1	Endoevaporites	Halite, gypsum	Demergasso et al. (2003); Rasuk et al. (2014, 2015)
		20°40′46.88″S 68°42′18.43″W	3745	7.6	0.7	36.5	ND	ND	Microbial mats	Halite, gypsum	
	Salar de Ascotán	23°29′55.38″S 67°41′40.96″W	4330	122.9	9.37	42	3.5–19.7	0.005–82	Evaporitic mats	ND	This report
	Salar de Punta Negra	23°1′59.40″S 67°17′9.98″W	4320	45	9.70	21	0.4–11.8	0.0004–1.4	Microbial mats	ND	This report
	Salar de Tara	20°19′3.57″S 68°51′51.55″W	3790	169.6	8.14	22.7	ND	0.132–39.8	Microbial mats	ND	This report
	Salar de Huasco	23°30′24.27″S 67°34′43.44″W	4210	109.4	9.21	18.1	0.5–11	0.0002–47.0	Microbial mats	ND	Dorador et al. (2008)
	Salar de Aguas Calientes	23°30′24.27″S 67°34′43.44″W	4515	131.5	7.63	52	ND	0.58–55.0	Microbial mats	ND	This report
	Salar de Carcote	21°22′57.64″S 68°21′35.80″W	4157	200	9.01	24.8	0.5–11.5	0.005–9.0	Microbial mats, endoevaporites	Gypsum, halite	This report
	Campo de Géiseres El Tatio	22°20′5.56″S 66°59′40.74″W	4200	16.9 and 23	7.2	86	ND	ND	Microbial mats	ND	Summers Engel et al. (2012)

(continued)

Table 17.1 (continued)

Region	Lagoon, lake, or wetland	Global position	Altitude (m asl)	Maximal conductivity (mS/cm)	pH	Temperature (°C)	O_2 (%)	Arsenic (mgL^{-1})	Ecosystem	Mineralogy	References
Bolivia	Campo de Géiseres Sol de Mañana	22°26.12′S 67°45.47′W	4859	ND	ND	86	ND	34.5	Biofilms	ND	This report
	Laguna Verde	22°47.27′S 67°49.24′W	4333	67.4	8.6	8	ND	ND	Evaporitic mats	ND	Fleming and Prufert-Bebout (2010)
	Laguna Blanca	22°46.89′S 67°47.62′W	4328	42.4	7.9	5	ND	ND	Microbial mats	ND	Fleming and Prufert-Bebout (2010)
	Laguna Hedionda	22°26.92′S 67°23.09′W	4553	35.9	10.5	5	ND	ND	Microbial mats	ND	This report
	Laguna Mama Khumu	22°15.88′S 67°05.25′W	4483	17.5	8.7	1	ND	ND	Microbial mats	ND	This report
	Salar de Uyuni	20°17.74′S 67°22.69′W	3655	88.7	7.3	28	ND	ND	Endoevaporites	Halite	This report
	Laguna Capina	21°58.99′S 67°36.84′W	4441	22.5	9.22	1	ND	ND	Microbial mats	ND	This report
	Laguna Busch	21°58.99′S 67°36.84′W	4441	22.5	9.22	1	ND	ND	Microbial mats	ND	This report
	Laguna Pastos Grandes	21°44.05′S 67°57.28′W	4509	99.7	7.0	28	ND	34.5	Microbial mats	ND	This report
	Laguna Honda	21°37′S 68°03.5′W	4124	78.4	8.21	13.5	ND	21.9	Microbial mats	ND	This report
	Laguna Loromayu	22°23.82′S 67°12.65′W	4670	54.2	8.64	10.6	ND	21.8	Microbial mats	ND	This report
	Salar, San Pedro de Quemes	20°52.70′S 67°52.50′W	3670	ND	ND	ND	ND	ND	Endoevaporites	ND	This report

The temperature, conductivity, O_2, and pH data are considered variable, since they may describe seasonal variations throughout the year
asl above sea level, *ND* not detected, *NF* Not found

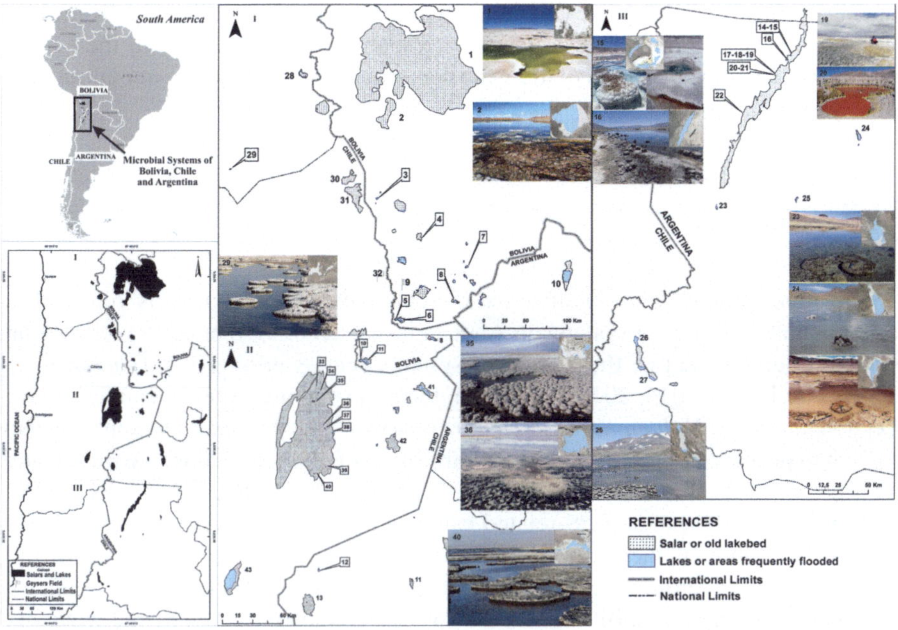

Fig. 17.1 Maps of Bolivia, Chile, and Argentina, showing the locations of salars and lakes with microbial systems. **Bolivia**: *1* Salar de Uyuni, *2* Salar San Pedro de Quemes, *3* Laguna Hedionda, *4* Laguna Capina, *5* Laguna Verde, *6* Laguna Blanca, *7* Laguna Mama Khumu, *8* Laguna Busch, *9* Campo de Géiseres. **Argentina**: *10* Laguna Vilama, *11* Ojos de Tolar, *12* Salar de Socompa, *13* Salar Llullaillaco, *14* Humedal Rojo, *15* Laguna Verde, *16* Laguna Pozo Bravo, *17* Ojo Luz, *18* Ojo María, *19* Ojo Valentina, *20* Ojo Azul, *21* Ojo Naranja, 22 Las Quinoas, *23* Laguna El Peinado, *24* Laguna Diamante, *25* Laguna Carachipampa, *26* Laguna Tres Quebradas, *27* Laguna Negra. **Chile**: *28* Salar de Huasco, *29* Laguna Llamara, *30* Salar de Carcote, *31* Salar de Ascotán, *32* Campo de Géiseres El Tatio, *33* Laguna Baltinache, *34* Laguna Cejar, *35* Laguna Tebenquiche, *36* Laguna Chaxas, *37* Laguna Barros Negros, *38* Salar Aguas de Quelana, *39* Laguna Salada, *40* Laguna La Brava, *41* Salar de Tara, *42* Salar de Aguas Calientes, *43* Laguna Punta Negra

of the Andean microbial ecosystems (AMEs), some shared characteristics have been found from the geological, physical, and chemical points of view [(1) active volcanic incidence: all of the microbial ecosystems that have been found are in some way connected to areas where active volcanoes are present; (2) underground water input; (3) mixed zones with different salinities: underground low-conductivity water and salar thalassic water; (4) two kinds of microbialites found: oncolites (at Laguna Negra, Tres Quebradas, Las Quinoas, etc.) or domes with thrombolites at the bottom and stromatolites at the top surface (at La Brava, Pozo Bravo, Ojos Bravos, and El Peinado)] and from the biological point of view [(5) predominance of diatoms, the main component in all studied systems; (6) predominance of anaerobic over aerobic photosynthetic microorganisms; (7) microbial rhodopsin as the main system for producing adenosine triphosphate (ATP); (8) arsenic resistance and bioenergetic mechanisms; and (9) predominance of Carbon fixation pathways other than the Calvin cycle]. The biological aspects of these are being studied thoroughly in our lab and are briefly discussed below.

17.2 Diatoms

Associations among microbialites and diatoms have been extensively reported in microbialite and mat ecosystems in Andean lakes, denoting that eukaryotes are an important component of these Andean systems (Farías et al. 2013; Gomez et al. 2014; Rasuk et al. 2014, 2016a, b). So far, the main diatom groups recognized have been *Achnanthes*, *Halamphora*, *Navicula*, *Surirella*, and *Striatula*. Diatoms are also known to produce copious amounts of extracellular polymers (EPSs) (Awramik and Riding 1988; Brinkmann 2007; Stal and Défarge 2005; Underwood et al. 2010; Urbani et al. 2012); thus, EPS-related calcification of diatoms is also possible. In this sense, although they are predominant in microbial mats, their role is still unknown. Thus, diatoms might also have a role as primary EPS producers in microbial mats. Another possibility is that they are accumulated as dust by the wind and accreted by the EPS that covers the systems. This could be the case in Laguna Socompa microbialites, a system surrounded by diatomite sedimentary rocks, which are constantly weathered by the wind. The Socompa microbialites are located in an area where the wind deposits all of the eroded material, including fossilized diatoms from diatomite rocks. In addition, copious diatom occurrence in microbialites might ensue from in situ flourishing during the winter (as these systems are submerged only during this season), from deposition of fossilized diatoms that come from weathered diatomite deposits around the lagoon, or as a sum of all of the above (Farías et al. 2013).

In addition to EPS production, diatoms could be involved in carbonate precipitation by forming bacteria–diatom aggregates with precipitation of nanoglobular carbonates. These microformations have been largely observed in Scanning Electron Microscopy (SEM) in Laguna Negra, Laguna Socompa, Laguna Tebenquiche, and La Brava microbial mats, and they have also been recorded inside carbonate microbialites and gypsum evaporites (Farías et al. 2017; Fernandez et al. 2016; Rasuk et al. 2014, 2016a, b; Gomez et al. 2018).

17.3 Microbial Diversity and Carbon Fixation Pathways

The microbial diversity associated with each microbial ecosystem also presents great variation where, in general, Proteobacteria, Bacteroidetes, and Firmicutes dominate most mat systems. However, examples of some peculiarities worth highlighting are the phylum Thermi (*Deinococcus radiodurans*), which is predominant on the top surfaces of carbonatic systems (in the Socompa, Tebenquiche, and La

Brava mats), probably because they are the most ultraviolet (UV) resistant ones and are able to dominate such an exposed area. Nevertheless, in gypsum systems (in the Llamara and Tebenquiche gypsum domes) they do not dominate the top layers, probably because the gypsum mineral is able to reflect UV radiation, providing a more protected niche where UV radiation is not a selecting factor. What is more, the surfaces of the Socompa system become green as a result of cyanobacterial dominance when UV radiation is removed (Farías et al. 2013). Another particularity of these systems is the small proportion of Cyanobacteria, supported by small amounts of chlorophyll *a* in pigment analyses (Farías et al. 2013, 2014; Rasuk et al. 2014), and the small proportion of Calvin cycle genes reported in metagenomic analyses (Kurth et al. 2017; Lynch et al. 2012; Rascovan et al. 2016).

The hypothesis developing around the small proportion of Cyanobacteria in Andean systems is that the extreme conditions of elevated UV radiation incidence, arsenic, conductivity, and pH are likely to prevent cyanobacterial development. As a result, anoxygenic photosynthesizers might complement primary production, complementing that performed by diatoms. Purple sulfur and nonsulfur bacteria from the classes Alphaproteobacteria and Gammaproteobacteria are involved in this metabolism and are present in most of the studied environments. Usually, members of the Rhodospirillaceae and Rhodobacteraceae families are abundant. Green sulfur and nonsulfur bacteria (Chlorobi and Chloroflexi) are also present but less abundant. Pigment analyses of AME layers of microbialites, evaporites, and mats in the Socompa, Llamara, Tebenquiche, and La Brava ecosystems have demonstrated the existence of bacteriochlorophyll pigments, supporting the contribution of anoxygenic photosynthetic bacteria (Farías and Contreras 2013; Farías et al. 2014; Rasuk et al. 2014).

Carbon fixation in phototrophs goes through the Calvin–Benson cycle, reductive tricarboxylic acid (rTCA) cycle, and 3-hydroxypropionate (3HP) bi-cycle (Tang et al. 2011). The abundance of carbon fixation pathways as alternatives to Calvin–Benson in AME environments supports the role of anoxygenic phototrophs or even chemolithotrophs in primary production, as has been observed in other systems (Gomez-Saez et al. 2017). Members of the Verrucomicrobia, Planctomycetes, Crenarchaeota, Spirochaetes, and OP1 phyla present in AME carbon fixation could also be involved in alternative carbon fixation pathways (Berg 2011). In that sense, in metagenomic studies from Socompa, analysis of the abundance of key enzymes from each of the known carbon fixation pathways shows that the most abundant is ATP citrate lyase (which represents the rTCA cycle), followed by ribulose-1,5-bisphosphate carboxylase/oxygenase (RuBisCO) (for the Calvin–Benson cycle), and the bifunctional enzyme carbon monoxide dehydrogenase/acetyl coenzyme A (CODH/acetyl-CoA) synthase (for the reductive acetyl-CoA pathway (Wood–Ljungdahl pathway)) (Kurth ct al. 2017). This pattern was also observed in a similar study of La Brava and Tebenquiche microbial mats. In Llamara evaporites, a large number of genes related to alternative carbon fixation pathways, such as the ancient reductive acetyl-CoA, were revealed. CODH/acetyl-CoA synthase (a key enzyme of the pathway) was even more strongly represented than RubisCO from the classic Calvin–Benson cycle (Gutiérrez-Preciado et al. 2018). Laguna Diamante biofilms dominated by Archaea revealed only an incomplete rTCA pathway (Rascovan et al. 2016). Consistent with the observations in AMEs, a functional analysis carried out on the Shark Bay stromatolite metagenome also showed a significant abundance of alternative carbon fixation pathways, such as rTCA and 3-hydroxypropionate/4-hydroxybutyrate (3HP/4HB) cycles. Although it has not been demonstrated yet, it is presumed that these pathways would have potential relevance in the carbon cycle and precipitation process in this microbial system (Ruvindy et al. 2016). Moreover, analysis with carbon isotopes in Tumbiana Formation stromatolites demonstrated that the fixation of atmospheric CO_2 in these structures did not involve the Calvin–Benson cycle, suggesting that primitive organisms may have used the reductive acetyl-CoA pathway to carry out this process (Slotznick and Fischer 2016).

17.4 Archaea in Andean Microbial Ecosystems

Archaea have also been detected as an important component of microbial diversity in Andean environments in general (Maldonado et al. 2018). In AMEs, multiple diversity studies have been performed, amplifying the V4 hypervariable region of the 16S ribosomal RNA gene. A high abundance of Archaea was detected in sedimentary ecosystems in Laguna Tebenquiche and Laguna La Brava, using RK primers (F515 and R806) designed to be universal for a broad range of archaeal and bacterial taxa (Farías et al. 2017; Fernandez et al. 2016). However, Archaea were almost absent in a previous study carried out in Laguna Tebenquiche, using F563-R802 universal primers, which are reliable for amplifying bacteria but not Archaea (Farías et al. 2013, 2014; Rasuk et al. 2014, 2016a; Stivaletta et al. 2010). A similar situation occurred with red biofilms from Laguna Diamante, showing a predominance of Firmicutes when F563-R802 universal primers were used and dominance of Halobacteria, with 94% of 16S rDNA gene sequences when the taxonomic composition was analyzed on the basis of metagenomic shotgun sequencing of total DNA (Rascovan et al. 2016).

What is more, in Tebenquiche and La Brava, AME analyses layer by layer showed that Euryarchaeota were mainly represented by aerobic heterotrophs belonging to *Halobacterium* in the upper layers and sulfate reducers belonging to Thermoplasmatales in the deeper layers. In addition, Crenarchaeota—represented by the miscellaneous marine benthic group B and D (MBGB)—were reported for the first time in mat systems dominating anoxygenic layers at La Brava and Tebenquiche (Fernandez et al. 2016; Farías et al. 2017).

Thus, the incidence of Archaea in the communities should be carefully considered in Andean AMEs studies, not only in biodiversity studies but also in functional analyses, since in Andean AMEs, nothing is as it seems. That is what we found in the entire communities of Laguna Diamante microbialites dominated by *Halorubrum*—hitherto considered anaerobic heterotrophs—which showed a possible dominant role in arsenic respiration and bacteriorhodopsin production, from the moment when metagenomic analyses revealed a surprisingly high abundance of genes used for arsenite oxidation (*aioBA*) and respiratory arsenate reduction (*arrCBA*). These results suggested that Haloarchaea might use arsenic compounds as bioenergetic substrates (Rascovan et al. 2016), and the results in isolated *Halorubrum* from the Laguna Diamante system reinforced that theory (Ordoñez et al. 2018).

Methanogenic Archaea were also detected in sampling of microbial mats in the El Tatio geyser field, which indicated that methanogenic Archaea are dominated by microorganisms within the genera *Methanospirillum* and *Methanobrevibacter*. These microbes are associated with a number of unclassified Archaea in the class Thermoplasmata, the order Halobacteriales, and unclassified Crenarchaeota. In addition, the preliminary results include an unclassified Thaumarchaeota clone—a member of the recently proposed third archaeal phylum, Thaumarchaeota. Nonspecific microbial mat samples from a nonethanogenic site included only crenarchaeal clones within the Desulfurococcales order of Thermoprotei (Franks et al. 2008).

17.5 Microbial Rhodopsin Systems

Microbial rhodopsins are considered the simplest energy-harvesting photoreceptors, consisting of a single membrane-embedded protein covalently binding a retinal (vitamin A aldehyde) chromophore. These proteins are able to absorb and transduce light into energy, performing a variety of functions (Ernst et al. 2014). Microbial rhodopsin systems also represent an important characteristic of AMEs, as they have been detected in metagenomes (Kurth and Farías 2016; Rascovan et al. 2016; Rasuk et al. 2016a, b). Most of these functions are found in isolated genomes, as well as in AME metagenomic sequences. However, microbial rhodopsins have wide diversity—many of them having unassigned functions—and this is reflected in AME microbial rhodopsins, which include several sequences with unknown functions. Interestingly, AME rhodopsins are taxonomically related to phyla with low relative abundance in the metagenomes.

To complement the results found in metagenome analyses, bacterial and archaeal genomes isolated from those AMEs were mined for microbial rhodopsin systems. As a result, xanthorhodopsin and proteorhodopsin were detected in *Salinivibrio* and *Exiguobacterium* genomes, respectively, both isolated from Socompa stromatolites. These microbial rhodopsin systems have never been reported before in these phylogenetic groups. They were heterologously expressed in *Escherichia coli BL21*, and their functional capacity was determined (Albarracín et al. 2016). The putative xanthorhodopsin found in *Salinibacter* S34, isolated from Socompa stromatolites by Gorriti et al. (2014), was named xanthorhodopsin 34 (XR34). To confirm whether XR34 encodes a functional photoactive protein, the recombinant protein was successfully expressed in *E. coli BL21* cells. The biochemical steady state, together with the spectral properties of XR34, was determined, showing its functionality. It has been suggested that proteorhodopsins play a key role in carbon cycles and energy flux in aquatic microbial systems (Beja et al. 2001; Bohorquez et al. 2012; de la Torre et al. 2003). It has been suggested that in marine microbial ecosystems, where environmental conditions are variable and periods of starvation occur, organisms that express proteorhodopsin could have an adaptive advantage. In this scenario, the photoproteins would serve to harvest extra energy by phototrophy (Albarracín et al. 2016; Bohorquez et al. 2012; Gómez-Consarnau et al. 2010). This could also be the case in the AMEs that present proteorhodopsins and develop under extreme conditions.

17.6 The Prospection

The prospected ecosystems and their geographic data (e.g., data on their altitude, global position, physicochemical parameters of the water, and AMEs present) are listed in Table 17.1. Below, a brief description of these systems is presented.

17.6.1 Chile: The Atacama Desert

In Chile, most of the studied sites are within the Atacama Desert, related to its interest as a Mars analog model (Azua-Bustos et al. 2012). Here, a wide range of hypersaline environments with different characteristics are associated with salt deposits (Risacher et al. 2003). The extreme conditions, along with the input of groundwater, promote the development of either carbonaceous or gypsum precipitation systems (Farías et al. 2014). The best-known locations, reported in several publications, include Laguna Cejar, Laguna Brava, Laguna Tebenquiche, and Salar de Llamara. All of them harbor several types of AMEs, and some of their main features are described in the following sections. Among the locations reported to contain AMEs in this work (Table 17.1, Fig. 17.1), we highlight the Salar de Aguas Calientes and Salar de Carcote. In Aguas Calientes, microbial mats are located in an area irrigated by a warm stream (27 °C) coming from the north, close to a *bofedal* (high-altitude peatland). The bottom layers of the mat show yellow lines, which are probably due to arsenic sulfide. Yellow minerals have also been observed at the Salar de Coposa and Salar de Tara, and previously at the Salar de Ascotán (Demergasso et al. 2007). The Salar de Carcote presents a series of wetlands in the western region, surrounded by small wet patches and pools, which harbor microbial mats, endoevaporite ecosystems, and a great diversity of invertebrates at their larval stages. Finally, the most recently found ecosystems (found in December 2017), which have just begun to be studied, include black carbonatic microbialites in Laguna Chaxas, gypsum domes inhabited by endoevaporitic microbial communities in Laguna Barros Negros, and abundant development of microbial mats in Aguas de Quelana and Salada Saladita.

17.6.2 Argentine: The Puna

In Argentina, besides the already mentioned AMEs in Laguna Socompa and Laguna Diamante, another well studied site is Laguna Negra (Table 17.1, Fig. 17.1 [location 27]). AMEs can be found all over the Andean area; however, recent work in Catamarca Province has revealed several new locations. The most recently discovered, with studies in progress, include microbialites at Laguna Tres Quebradas and evaporitic domes with typical three-color layers (yellow, green, and purple) at Laguna Verde and Humedal Rojo (Salar de Antofalla). At this latter site, because of the magnitude of the evaporitic ecosystems inhabited by green layers, we are starting to introduce the concept of *bosques del salar* (salar forests) on the basis that large areas of the salars are alive and deserve to be protected like forests, since they could function as carbon sinks in completely deserted areas. Other locations in Catamarca where diversity has been analyzed include:

- Laguna Verde (salar forest): Evaporitic domes inhabited typically by three color pigments (yellow, green, and purple). These ecosystems have only recently been reported, so there is no relevant information available yet (Fig. 17.1 [location 15]).
- Laguna Pozo Bravo: The microbialites of Pozo Bravo are stromatolite-thrombolite-like types. Three different morphological external types have been recognized: domic, discoidal, and tabular shapes, with heights from 0.05 to 0.7 m. Tabular shapes are predominant on the eastern margin, constituting a 500 m–long continuous belt. Meanwhile, on the western margin, the three shapes are present with a predominance of domic and discoidal shapes. Also, on this same margin, microbial mats are more abundant (Fig. 17.1 [location 16]).
- Puquios Ojos de Campo: This is a complex of eight small pools (around 40 m in diameter) close to each other, with very different characteristics. Ojo Naranja is the most striking pool because of its red color, and it has the highest conductivity. The parameters listed in the table pertain to this pool. Ojo Naranja, Ojo Valentina, and Ojo María harbor microbial mats, while Ojo Luz presents a narrow microbialite belt across its west margin (Fig. 17.1 [locations 17–21]).
- Las Quinoas: The microbialites of the Las Quinoas wetland run for 500 m, forming a belt between the vega to the west and the salt flat to the east. Oncolite-type microbialites are present in the small channels, which go from the vega into the salar. Their diameter varies between 1 and 12 cm on average, and they have rounded/subrounded and discoidal external morphologies. In addition, microbial mats are found, which present lithification in the dry season. DNA extraction from this system has proved to be difficult, but the efforts continue (Fig. 17.1 [location 22]).
- Hydrothermal vent located in the Galán Volcano crater: A microbial mat has developed around a fissure from which geothermally heated (about 80 °C) water issues, running down to Laguna Diamante. In comparison with other AMEs, the microbial diversity is unusual, dominated by Chloroflexi, Bacteroidetes, and Armatimonadales (Fig. 17.1 [location 24]).

- Laguna Farallon: Located at the southwest margin of the Galán Volcano crater, in a different drainage basin from Laguna Diamante, this freshwater lake harbors microbial mats with abundant Betaproteobacteria.
- Laguna Carachipampa: Microbial mat systems appear 2 km along the eastern lake margin, although it was in small hydrothermal pools near the lake that the main microbial systems were registered. Associated directly with the small hydrothermal pools, there are flat oncolites and microbial buildups with a flat top. Other types are large carbonate "pavements" with internal lamination (Fig. 17.1 [location 25]).
- Laguna El Peinado: Microbialites extend about 2 km along the western margin and have been described as being like stromatolites, forming small mounds up to 20 cm thick, with a patchy distribution along the emergent shore of Laguna El Peinado (Valero-Garcés et al. 2001). So far, El Peinado microbialite system is the largest one registered at the Puna in Catamarca Province (Fig. 17.1 [location 23]).
- Salar de la Laguna Verde: This salt flat is located in western Catamarca Province and includes several lakes where AMEs have developed:
  - Laguna Tres Quebradas: Microbialites are located in the Salado River delta fan, which receives groundwater flows from the east margin. These microbialites are mainly formed by calcite, and they occupy an area of 14,000 m^2. Two main external morphologies can be distinguished: round/subrounded and discoidal types. On the basis of their internal structure, they have been distinguished as oncolitic and stromatolitic microbialites (Fig. 17.1 [location 26]).
  - Laguna Negra: The microbialites and associated carbonate crusts of Laguna Negra are intimately associated with the microbial mat system and display a diverse set of morphologies at both the macroscopic and microscopic scales. The morphologies include millimeter- to centimeter-thick laminated crusts, which form patchy to laterally extensive pavements, and centimeter- to decimeter-scale concentrically laminated disks, spheres, and flattened domes, which are referred to as oncoids. In addition to these discrete morphologies, oncoids also occur as complex composite structures. Oncoids are broadly distributed in Laguna Negra as both subaqueous and subaerially exposed surface accumulations, and as discrete bodies within the upper meter of sediment (Gomez et al. 2014) (Fig. 17.1 [location 27]).

17.6.3 Bolivia: The Altiplano

On the Bolivian Altiplano, a 2015 campaign to the Sud Lípez region in the Departamento de Potosi explored more than 20 saline lakes. Microbial mats were found in Laguna Hedionda and Laguna Mama Khumu. In Laguna Hedionda a high concentration of sulfates (530 mg L^{-1}) may be one of the causes of the lake's characteristic hydrogen sulfide odor. The name Laguna Hedionda means "Stinking Lake" in Spanish. Large amounts of exopolysaccharides accumulating on the lake's shores may have prevented DNA extraction. Laguna Mama Khumu is characterized mainly by its ulexite reserves and other less concentrated minerals. Another iconic place in the Bolivian Andean region is the Salar de Uyuni, which is probably one of the largest salt flats in the world, with an area of about 10,000 km^2. Microbiological studies at this site are very recent (dC Rubin et al. 2017; Haferburg et al. 2017).

Acknowledgments The authors acknowledge the generous financial support provided by Proyectos de Investigación Científica y Tecnológica (PICT) V Bicentenario 2010 1788, 3825, PIO-CITCA CONICET. To the guides: Luis Ahumada and Marco Contreras for field logistics and collaboration. To the original Kolla communities from Antofalla, Tolar Grande, El Peñon and Antofagasta de la Sierra (Argentine), and Atacama (Chile) for their hospitality during during our extenses field trips and their commitment in preserving the natural and cultural heritage of the Andes.

The authors want to thank and acknowledge Dr. Mauricio Guzman for his contribution to the Bolivian survey and Dr. Tatiana Stepanenko and Dr. Jamie Foster for their expertise, guidance and collaboration with the manuscript editing.

References

Albarracín VH, Kraiselburd I, Bamann C, Wood PG, Bamberg E, Farías ME et al (2016) Functional green-tuned proteorhodopsin from modern stromatolites. PLoS One 11:1–18. https://doi.org/10.1371/journal.pone.0154962

Awramik SM, Riding R (1988) Role of algal eukaryotes in subtidal columnar stromatolite formation. Proc Natl Acad Sci U S A 85:1327–1329. https://doi.org/10.1073/pnas.85.5.1327

Azua-Bustos A, Urrejola C, Vicuña R (2012) Life at the dry edge: microorganisms of the Atacama Desert. FEBS Lett 586:2939–2945. https://doi.org/10.1016/j.febslet.2012.07.025

Beja O, Spudich EN, Spudich JL, Leclerc M, DeLong EF (2001) Proteorhodopsin phototrophy in the ocean. Nature 411:786–789

Belluscio A (2009) A high window in the past. Nat Dig 6:34–36. https://doi.org/10.1038/ndigest.2009.091134

Belluscio A (2010) Hostile volcanic lake teems with life. Nat News. https://doi.org/10.1038/news.2010.161

Berg IA (2011) Ecological aspects of the distribution of different autotrophic CO_2 fixation pathways. Appl Environ Microbiol 77:1925–1936. https://doi.org/10.1128/AEM.02473-10

Bohorquez LC, Ruiz-Pérez CA, Zambrano MM (2012) Proteorhodopsin-like genes present in thermoacidophilic high-mountain microbial communities. Appl Environ Microbiol 78:7813–7817. https://doi.org/10.1128/AEM.01683-12

Brinkmann N (2007) Assessing the diversity of pennate benthic diatoms in calcifying biofilms of hard water creeks. In: Kusber W-H, Jahn R (eds) Proceedings of the First Central European Diatom Meeting, Berlin, 23–25 Mar 2007, pp 11–14. https://doi.org/10.3372/cediatom.102

dC Rubin SS, Marín I, Gómez MJ, Morales EA, Zekker I, San Martín-Uriz P et al (2017) Prokaryotic diversity and community composition in the Salar de Uyuni, a large scale, chaotropic salt flat. Environ Microbiol 19:3745–3754. https://doi.org/10.1111/1462-2920.13876

de la Torre JR, Christianson LM, Beja O, Suzuki MT, Karl DM, Heidelberg J et al (2003) Proteorhodopsin genes are distributed among divergent marine bacterial taxa. Proc Natl Acad Sci U S A 100:12830–12835. https://doi.org/10.1073/pnas.2133554100

Demergasso CS, Guillermo CD, Lorena EG, Mur JJP, Pedrós-Alió C (2007) Microbial precipitation of arsenic sulfides in Andean salt flats. Geomicrobiol J 24:111–123. https://doi.org/10.1080/01490450701266605

Demergasso C, Chong G, Galleguillos P, Escudero L, Martínez-Alonso M, Esteve I (2003) Microbial mats from the Llamará salt flat, Northern Chile. Rev Chil Hist Nat 76:485–499.
Dorador C, Busekow A, Vila I, Imhoff JF, Witzel KP (2008) Molecular analysis of enrichment cultures of ammonia oxidizers from the Salar de Huasco, a high altitude saline wetland in northern Chile. Extremophiles 12:405–414.
Ernst OP, Lodowski DT, Elstner M, Hegemann P, Leonid S, Kandori H (2014) Microbial and animal rhodopsins: structures, functions and molecular mechanisms. Chem Rev 114:126–163. https://doi.org/10.1021/cr4003769
Farías ME, Fernández-Zenoff V, Flores R, Ordóñez O, Estévez C (2009) Impact of solar radiation on bacterioplankton in Laguna Vilama, a hypersaline Andean lake (4650 m). Journal of Geophysical Research: Biogeosciences, 114(G2). https://doi.org/10.1029/2008jg000784
Farías ME, Poiré DG, Arrouy MJ, Albarracin VH (2011) Modern stromatolite ecosystems at alkaline and hypersaline high-altitude lakes in the Argentinean Puna STROMATOLITES: Interaction of Microbes with Sediments (pp. 427–441): Springer.
Farias ME, Rascovan N, Toneatti DM, Albarracín VH, Flores MR, Poiré DG, et al (2013) The Discovery of Stromatolites Developing at 3570 m above Sea Level in a High-Altitude Volcanic Lake Socompa, Argentinean Andes. PLoS One, 8. https://doi.org/10.1371/journal.pone.0053497
Farías ME, Contreras M (2013) Ecosistemas microbianos. ¿Nuevo patrimonio de la humanidad? Rev Bitacora Ecológicas 1
Farías ME, Contreras M, Rasuk MC, Flores MR, Poiré DG, Novoa F et al (2014) Characterization of bacterial diversity associated with microbial mats, gypsum evaporites and carbonate microbialites in thalassic wetlands: Tebenquiche and La Brava, Salar de Atacama, Chile. Extremophiles 18:311–329
Farías ME, Rasuk MC, Gallagher KL, Contreras M, Kurth D, Fernandez AB et al (2017) Prokaryotic diversity and biogeochemical characteristics of benthic microbial ecosystems at La Brava, a hypersaline lake at Salar de Atacama, Chile. PLoS One 12, e0186867. https://doi.org/10.1371/journal.pone.0186867
Fernandez AB, Rasuk MC, Visscher PT, Contreras M, Novoa F, Poire DG et al (2016) Microbial diversity in sediment ecosystems (evaporites domes, microbial mats, and crusts) of hypersaline Laguna Tebenquiche, Salar de Atacama, Chile. Front Microbiol 7:1284. https://doi.org/10.3389/fmicb.2016.01284
Fleming ED, Prufert-Bebout L (2010) Characterization of cyanobacterial communities from high-elevation lakes in the Bolivian Andes. J. Geophys. Res 115, https://doi.org/10.1029/2008jg000817
Franks M, Bennett P, Omelon C, Engel A (2008) Diversity of archaeal consortia in an arsenic-rich hydrothermal system. Paper presented at the American Geophysical Union Fall Meeting, San Francisco, 15–19 Dec 2008
Gomez FJ, Kah LC, Bartley JK, Astini RA (2014) Microbialites in a high-altitude Andean lake: multiple controls on carbonate precipitation and lamina accretion. Palaios 29:233–249. https://doi.org/10.2110/palo.2013.049
Gomez FJ, Mlewski C, Boidi FJ, Farías ME, Gérard E (2018) Calcium carbonate precipitation within diatom-rich microbial mats: the Laguna Negra hypersaline lake, Catamarca (Argentina). J Sediment Res 88:727–742
Gómez-Consarnau L, Akram N, Lindell K, Pedersen A, Neutze R, Milton DL et al (2010) Proteorhodopsin phototrophy promotes survival of marine bacteria during starvation. PLoS Biol 8:2–11. https://doi.org/10.1371/journal.pbio.1000358
Gomez-Saez GV, Pop Ristova P, Sievert SM, Elvert M, Hinrichs K-U, Bühring SI (2017) Relative importance of chemoautotrophy for primary production in a light exposed marine shallow hydrothermal system. Front Microbiol 8:702
Gorriti MF, Dias GM, Chimetto LA, Trindade-Silva AE, Silva BS, Mesquita MMA et al (2014) Genomic and phenotypic attributes of novel salinivibrios from stromatolites, sediment and water from a high altitude lake. BMC Genomics, 15, 473. https://doi.org/10.1186/1471-2164-15-473
Gutiérrez-Preciado A et al (2018) Functional shifts in microbial mats recapitulate early Earth metabolic transitions. Nat Ecol Evol 2(11):1700–1708. https://doi.org/10.1038/s41559-018-0683-3

Haferburg G, Gröning JAD, Schmidt N, Kummer NA, Erquicia JC, Schlömann M (2017) Microbial diversity of the hypersaline and lithium-rich Salar de Uyuni, Bolivia. Microbiol Res 199:19–28. https://doi.org/10.1016/j.micres.2017.02.007

Kurth D, Amadio A, Ordoñez OF, Albarracín VH, Gärtner W, Farías ME (2017) Arsenic metabolism in high altitude modern stromatolites revealed by metagenomic analysis. Sci Rep 7:1024. https://doi.org/10.1038/s41598-017-00896-0

Kurth D, Farías ME (2016) Metagenómica de rodopsinas microbianas en lagunas andinas. Paper presented at the XXIII Congreso Latinoamericano de Microbiologia, Rosario, 26–30 Sep 2016

Lynch RC, King A, Farías ME, Sowell P, Vitry C, Schmidt SK (2012) The potential for microbial life in the highest-elevation (>6000 m a.s.l.) mineral soils of the Atacama region. J Geophys Res Biogeosciences 117. https://doi.org/10.1029/2012JG001961

Maldonado MJ, Albarracín VH, Lara JA, Ferrero MA, Farías ME (2018) Culture-dependent and-independent methods reveal dominance of halophilic Euryarchaeota in high-altitude Andean lakes. Aquat Microb Ecol 81(2):171–188

Ordoñez OF, Rasuk MC, Soria MN, Contreras M, Farías ME (2018) Haloarchaea from the Andean Puna: biological role in the energy metabolism of arsenic. Microb Ecol 76:1–11. https://doi.org/10.1007/s00248-018-1159-3.

Rascovan N, Maldonado J, Vazquez MP, Farías ME (2016) Metagenomic study of red biofilms from Diamante Lake reveals ancient arsenic bioenergetics in haloarchaea. ISME J 10:299–309. https://doi.org/10.1038/ismej.2015.109

Rasuk MC, Fernández AB, Kurth D, Contreras M, Novoa F, Poiré D et al (2016a) Bacterial diversity in microbial mats and sediments from the Atacama Desert. Microb Ecol 71:44–56. https://doi.org/10.1007/s00248-015-0649-9

Rasuk MC, Ferrer GM, Moreno JR, Farías ME, Albarracín VH (2016b) The diversity of microbial extremophiles. In: Bruce T, Trindade Silva AE (eds) Molecular diversity of environmental prokaryotes. Boca Raton, CRC, pp 87–126. https://doi.org/10.1201/9781315381909-5

Rasuk MC, Kurth D, Flores MR, Contreras M, Novoa F, Poire D et al (2014) Microbial characterization of microbial ecosystems associated to evaporites domes of gypsum in Salar de Llamara in Atacama Desert. Microb Ecol 68, 483. https://doi.org/10.1007/s00248-014-0431-4

Risacher F, Alonso H, Salazar C (2003) The origin of brines and salts in Chilean salars: a hydrochemical review. Earth-Sci Rev 63:249–293. https://doi.org/10.1016/S0012-8252(03)00037-0

Ruvindy R, White RA, Neilan BA, Burns BP (2016) Unravelling core microbial metabolisms in the hypersaline microbial mats of Shark Bay using high-throughput metagenomics. ISME J 10:183–196. https://doi.org/10.1038/ismej.2015.87

Schröder H (2000) Morfología climática y morfogénesis del Volcán LLullaillaco (Chile/Argentina).

Slotznick SP, Fischer WW (2016) Examining Archean methanotrophy. Earth Planet Sci Lett 441:52–59. https://doi.org/10.1016/j.epsl.2016.02.013

Stal LJ, Défarge C (2005) Structure and dynamics of exopolymers in an intertidal diatom biofilm. Geomicrobiol J 22:341–352. https://doi.org/10.1080/01490450500248721

Stivaletta N, Ĺopez-Garćia P, Boihem L, Millie DF, Barbieri R (2010) Biomarkers of endolithic communities within gypsum crusts (southern Tunisia). Geomicrobiol J 27:101–110. https://doi.org/10.1080/01490450903410431

Stivaletta N, Barbieri R, López-García P, Cevenini F (2011) Physicochemical conditions and microbial diversity associated with the evaporite deposits in the Laguna de la Piedra (Salar de Atacama, Chile). Geomicrobiol. J. 28:83–95

Summers-Engel A et al (2012) Chemotrophy. Encycl. Caves, 125–134

Tang K-H, Tang YJ, Blankenship RE (2011) Carbon metabolic pathways in phototrophic bacteria and their broader evolutionary implications. Front Microbiol 2:165. https://doi.org/10.3389/fmicb.2011.00165

Thiel V, Tank M, Neulinger SC, Gehrmann L, Dorador C, Imhoff JF (2010) Unique communities of anoxygenic phototrophic bacteria in saline lakes of Salar de Atacama (Chile): Evidence for a new phylogenetic lineage of phototrophic Gammaproteobacteria from pufLM gene analyses. FEMS Microbiol. Ecol. 74:510–522

Underwood GJC, Fietz S, Papadimitriou S, Thomas DN, Dieckmann G (2010) Distribution and composition of dissolved extracellular polymeric substances (EPS) in Antarctic sea ice. Mar Ecol Prog Ser 404:1–19. https://doi.org/10.3354/meps08557
Urbani R, Paola S, Pletikapić G, Mišić Radić T, Svetličić V, Žutić, V. (2012). Diatom polysaccharides: extracellular production, isolation and molecular characterization. In: Karunaratne, D. N. (ed). The complex world of polysaccharides. IntechOpen. https://www.intechopen.com/books/the-complex-world-of-polysaccharides/diatom-polysaccharides-extracellular-production-isolation-and-molecular-characterization. https://doi.org/10.5772/2947
Valero-Garcés BL, Arenas C, Delgado-Huertas A (2001) Depositional environments of Quaternary lacustrine travertines and stromatolites from high-altitude Andean lakes, northwestern Argentina. Can J Earth Sci 38(8):1263–1283. https://doi.org/10.1139/e01-014

Chapter 18
Linear Megaplasmids Spreading in the Andean Resistome

Martin Wagenknecht, María Florencia Pérez, and Julián Rafael Dib

18.1 Introduction—Extreme Environments

Extreme environments, such as the Andean Puna, are characterized by conditions that hinder the development of living beings. Among them, the following stand out: exposure to high levels of ultraviolet (UV) radiation, low oxygen content, high thermal amplitude, and high concentrations of arsenic—conditions very similar to those at the origins of life. However, these pristine environments at an elevation of more than 4000 m above sea level comprise wetlands with hypersaline lagoons—places that represent prolific sources of microorganisms with exceptional phenotypic and genotypic characteristics. Microorganisms that inhabit these environments need to develop efficient mechanisms and strategies to thrive under such hostile conditions (Albarracín et al. 2015, 2016; Dib et al. 2009; Farías et al. 2013; Rasuk et al. 2014, 2015; Ordoñez et al. 2009).

M. Wagenknecht
Boehringer Ingelheim RCV GmbH & Co KG, Vienna, Austria

M. F. Pérez
Laboratorio de Investigaciones Microbiológicas de Lagunas Andinas (LIMLA), Planta Piloto de Procesos Industriales Microbiológicos (PROIMI), Centros Científico Tecnológicos (CCT), Consejo Nacional de Investigaciones Científicas y Técnicas (CONICET), Tucumán, Argentina

J. R. Dib (✉)
Laboratorio de Investigaciones Microbiológicas de Lagunas Andinas (LIMLA), Planta Piloto de Procesos Industriales Microbiológicos (PROIMI), Centros Científico Tecnológicos (CCT), Consejo Nacional de Investigaciones Científicas y Técnicas (CONICET), Tucumán, Argentina

Instituto de Microbiología, Facultad de Bioquímica, Química y Farmacia, Universidad Nacional de Tucumán, Tucumán, Argentina
e-mail: jdib@proimi.org.ar

M. E. Farías (ed.), *Microbial Ecosystems in Central Andes Extreme Environments*, https://doi.org/10.1007/978-3-030-36192-1_18

18.2 Actinobacteria

The class Actinobacteria represents one of the largest taxonomic units within the domain Bacteria. It stands out because of the large number of identified species and also because of their wide variety. Actinobacteria comprise Gram-positive bacteria with a high G+C content; it ranges from approximately 50% to more than 70% (Ventura et al. 2007). The latter is particularly found in *Streptomyces* species. Actinobacteria can be divided into two groups: *Streptomyces* and non-*Streptomyces* (also known as rare actinobacteria) (Azman et al. 2015). The important actinobacterial order Actinomycetales, commonly referred to as actinomycetes, is represented by genera, such as *Brevibacterium*, *Micrococcus*, *Mycobacterium*, and *Streptomyces*.

Actinobacteria excel because they have a wide variety of diverse physiological properties and metabolic traits. For example, they produce extracellular enzymes and secondary metabolites (Schrempf 2001). In particular, many of the latter attract attention as potent antibiotics (Demain 1999). In particular, streptomycetes produce many antimicrobial compounds, which account for their substantial pharmaceutical and economic importance (Bérdy 2005).

Actinobacteria also stand out in terms of their morphological diversity. Besides coccoid cells and fragmenting hyphal forms, cells switching between a coccoid and a rod shape exist. Many species produce spores and are characterized by highly differentiated branched mycelial growth (Embley and Stackebrandt 1994). Moreover, actinobacteria are found in both terrestrial and aquatic habitats (Embley and Stackebrandt 1994; Das et al. 2008). A multitude of different lifestyles is realized: actinobacteria exist as plant commensals, pathogens, nitrogen-fixing symbionts, inhabitants of the gastrointestinal tract, and—first and foremost—soil inhabitants (Ventura et al. 2007).

18.3 Actinobacteria from High-Altitude Wetlands and Lakes

In the northwest of Argentina and in some other South American countries, Andean high-altitude wetlands are found. These aquatic environments are located at elevations of more than 4000 m above sea level and are considered pristine. The extreme and hostile nature of these regions is demonstrated by attributes, such as high salinity, oligotrophy, and heavy metals—in particular, arsenic. Moreover, these environments are characterized by high levels of UV radiation, low nutrient concentrations, and high daily air temperature fluctuations (Dib et al. 2008; Flores et al. 2009; Ordoñez et al. 2009). The hostile conditions present in these high-altitude wetlands result in exceptional phenotypic and genotypic adaptations realized by the inhabiting microorganisms. Unexpectedly, a high degree of diversity of the microbial communities from these environments has been revealed by several studies (Demergasso et al. 2004; Ordoñez et al. 2009; Rascovan et al. 2016). Actinobacterial species isolated from such habitats showed resistance to high levels of UV radiation and arsenic (Ordoñez et al. 2009; Dib et al. 2008) as an adaptation to the hostile conditions.

Moreover, multiple forms of resistance to antibiotics—in particular, macrolides—were shown (Dib et al. 2008). However, because of the pristine nature of these habitats, the dissemination and functional relevance of this attribute are still unclear. In addition, actinobacteria were identified to be part of the microbial communities of living stromatolites found in the wetlands (Belluscio 2009, 2010). Many of them harbor extrachromosomal DNA elements ranging between some 10 and more than 100 kb in size (Dib et al. 2010a, b, 2013a, b, c, 2015).

18.4 Plasmids

Plasmids, as self-replicating extrachromosomal genetic elements, are fundamental for genetic exchange between microorganisms (Halary et al. 2010) and contribute significantly to the evolution of prokaryotic genomes (Koonin and Wolf 2008). In addition to coding for the functions required for their replication and mobilization, plasmids also contain accessory genes that contribute to their host's phenotypic diversity (Kav et al. 2012). These genes usually confer an advantage on their hosts within a certain ecological niche (Crossman 2005; Frost et al. 2005; Gil et al. 2006; Heuer et al. 2008; Novick 2003; Van der Meer and Sentchilo 2003). Therefore, it is expected that certain accessory plasmid-borne functions are enriched in a specific ecological niche. Because of this important ecological role, characterization of plasmids and understanding of the function of the encoded attributes are of great importance for characterization and understanding of microbial environments.

18.4.1 Linear Plasmids

Originally, plasmids were considered to be covalently closed circular molecules. However, about four to five decades ago, linear elements were reported to also exist. At first, they were found in the mitochondria of maize (Pring et al. 1977). Since then, a large number of linear plasmids have been discovered in a multitude of organisms, including both prokaryotes and eukaryotes. Nowadays, novel linear elements are identified in the course of frequently performed whole-genome-sequencing studies (Baker et al. 2007; Sekine et al. 2006).

Linear plasmids of prokaryotes are found in both Gram-positive bacteria (Hinnebusch and Tilly 1993; Meinhardt and Klassen 2007) and Gram-negative bacteria (Baker et al. 2007). However, the majority of them have been found in Gram-positive species, particularly in actinomycetes. The first described bacterial linear plasmid is pSLA2, harbored by *Streptomyces rochei* (Hayakawa et al. 1979). Linear plasmids of eukaryotes exist in yeasts, filamentous fungi, and higher plants.

Usually, prokaryotic linear elements are much larger than eukaryotic ones. The latter range in size from about 1 kb to less than approximately 20 kb. Linear bacterial elements often reach sizes of several hundred kilobases. Examples of these replicons (which are often called giant plasmids or megaplasmids) are pSCP1 (356 kb,

Streptomyces coelicolor (Bentley et al. 2004)) and pRHL2 (443 kb, *Rhodococcus jostii* RHA1 (Shimizu et al. 2001)). One of the largest representatives of linear megaplasmids is the 1.8-Mb-spanning pSCL4 (*Streptomyces clavuligerus* ATCC 27064 (Medema et al. 2010)). pSCL1 of *S. clavuligerus* (Keen et al. 1988) has a size of about 12 kb and is considered one of the shortest linear elements in prokaryotes.

Addressing the cellular localization of linear plasmids, it becomes evident that in higher plants and filamentous fungi, they are associated with the mitochondria (Griffiths 1995; Meinhardt et al. 1990; Warren et al. 2016). In bacteria and yeasts, localization in the cytoplasm is realized. Only a few exceptions have been described—for example, the yeast linear plasmids pPK1 of *Pichia kluyveri* and pPH1 of *Pichia heedi*, which are located in the mitochondria (Blaisonneau et al. 1999), and the linear plasmid of the green alga *Chlamydomonas moewusii*, which is chloroplast associated (Turmel et al. 1986).

The linear state of these elements draws attention to their DNA ends—the telomeres. Two types of telomeres have been revealed by means of molecular analyses, exhibiting substantial structural differences. On that basis, linear elements are divided into elements with proteins attached to their DNA 5′ ends (Fig. 18.1A) and the so-called hairpin elements (Fig. 18.1B).

18.4.1.1 Linear Hairpin Elements

At each terminus of a hairpin element, the two DNA single strands are covalently linked, leading to the characteristic hairpin loops (Kikuchi et al. 1985). In addition, terminal inverted repeats (TIRs) are found at the termini of the linear element. The most important bacterial instance harboring plasmids of this type is the genus *Borrelia* (Barbour and Garon 1987; Hinnebusch and Barbour 1991; Hinnebusch et al. 1990). It includes human pathogens, such as *B. hermsii* and *B. burgdorferi*, which cause relapsing fever and Lyme borreliosis, respectively (Burgdorfer et al. 1982; Dworkin et al. 2002; Miller et al. 2013). Some of the pathogenicity determinants are encoded in *Borrelia* linear plasmids (Saint Girons et al. 1994).

Further examples of linear hairpin elements are seen in the genomes of a few bacteriophages that infect Gram-negative bacteria. However, they appear as linear hairpin elements only when the prophages undergo replication (Hertwig 2007).

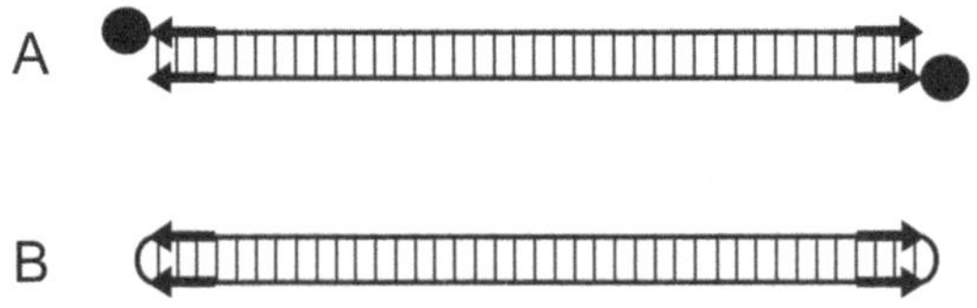

Fig. 18.1 Schematic representation of linear elements with 5′-linked terminal proteins (**A**) and hairpin elements (**B**). The terminal proteins (TPs) and the terminal inverted repeats (TIRs) are represented by *black circles* and *arrows*, respectively

Other examples are the genomes of a few eukaryotic viruses; for example, African swine fever virus, vaccinia virus, and poxvirus possess a linear hairpin structure.

18.4.1.2 Linear Elements with 5′-Attached Proteins

These linear elements are characterized by TIRs and terminal proteins (TPs). The TPs are covalently linked to the 5′ ends of the DNA molecule. Elements possessing this structure are called invertrons (Sakaguchi 1990). These molecules form the largest group of linear elements, and they are found in diverse filamentous fungi, bacteria, yeasts, and plants. Also, the genomes of some viruses and bacteriophages (e.g., *Bacillus* phage Φ29 (Anderson et al. 1966)) may possess TIRs and TPs (for compilations, see Meinhardt et al. (1990) and Meinhardt and Klassen (2007)).

Most of the linear elements found in filamentous fungi show a rather uniform and simple genetic organization (Meinhardt et al. 1997). Because for many of them a correlation with a clear phenotype is not possible, they are regarded as parasitic or selfish elements (Klassen and Meinhardt 2007).

In yeast, plasmids of the invertron type are found in various species, especially in the genera *Debaryomyces* and *Pichia* (Klassen and Meinhardt 2007). Often they occur as systems comprising two or three elements that differ in size.

Linear bacterial plasmids with 5′-attached proteins have been identified in many actinomycetes, first of all in the genus *Streptomyces* (Chater and Kinashi 2007; Chen 2007), but also in *Rhodococcus* (Kalkus et al. 1990; Stecker et al. 2003), *Mycobacterium* (Picardeau and Vincent 1997), and *Arthrobacter* (Overhage et al. 2005; Wagenknecht and Meinhardt 2011a, b). Apart from a few exceptions, a phenotype can be assigned for most of the actinobacterial linear plasmids.

Interestingly, also, the chromosomes of streptomycetes are linear and belong to the invertron type, and genome sequencing studies have shown that the chromosomes of two *Rhodococcus* species are linear too (Takarada et al. 2009; McLeod et al. 2006). Notably, the same bacterial host may harbor more than one linear plasmid, and linear plasmids are sometimes accompanied by circular ones (Dib et al. 2010b).

18.5 Phenotypes Attributed to Actinobacterial Linear Plasmids

Circular bacterial plasmids provide beneficial attributes for their hosts, but also their linear counterparts code for many metabolic and physiological qualities found in actinobacteria. Examples include gene clusters for catabolic traits allowing degradation and metabolism of various organic compounds (Fetzner et al. 2007).

Other linear elements code for the ability to resist numerous antibiotics and heavy metals (e.g., arsenic and mercury). Plasmid-associated resistance determinants have been identified for numerous streptomycetes and some rhodococci.

Linear plasmids of *Streptomyces* not only confer resistance to diverse antibiotics (such as lankacidin, lankamycin, and methylenomycin) but also provide a genetic tool kit for their biosynthesis (Chater and Kinashi 2007; Kinashi 2011).

Moreover, phytopathogenic capacity may be a plasmid-associated quality too, as was shown for the large linear plasmid pFiD188 of the plant pathogen *Rhodococcus fascians* D188 (Francis et al. 2007). Conjugal transfer is a common property for most actinobacterial linear plasmids (Chen 2007; Meinhardt et al. 1997). This enables such elements to disseminate within the same species, and it also allows transfer to closely related organisms. Hence, many actinomycetes possess the potential to share and profit from plasmid-borne attributes.

18.6 Characterization of Novel Linear Megaplasmids of Actinobacteria from High-Altitude Lakes

Although large plasmids have been described in many actinobacteria, rather little is known as to the occurrence and diversity of linear megaplasmids in actinobacteria from extreme environments (Dib et al. 2010a, b).

A great diversity of genera within the so-called rare actinobacteria was isolated from surface water and flamingo fecal samples taken from high-altitude Andean lakes (HAALs). Nine of these strains contained one or more megaplasmids, ranging in size from approximately 75 to 245 kb (Fig. 18.2). *Rhodococcus* sp. Ch13 (194 kb), *Kocuria* sp. N33 (143 kb), and *Micrococcus* sp. V7 (92 kb), A1 (109 kb), A7 (82 kb), and D12 (75 kb) strains contained one plasmid each; *Brevibacterium* sp. Ap13 (89 and 210 kb) and *Micrococcus* sp. H5 (109 and 215 kb) contained two plasmids each; and *Nocardia* sp. A5 even harbored three large plasmids (93, 177, and 244 kb) (Dib et al. 2010a, b, 2015, 2018). It is obvious from these pulsed-field gel electrophoresis (PFGE) data that large plasmids are rather widespread in actinobacterial strains isolated from these HAALs. With use of different PFGE running conditions, no changes in the electrophoretic mobility of plasmids pLMA1, pLMV7, pLMA7, pLMH5, pJD12, and pAP13 were observed, indicating that the plasmids were linear. Besides that, these linear replicons were shown to possess TPs covalently bound to the 5′ ends, which was verified by testing for exonuclease sensitivity.

Moreover, Southern analysis and restriction endonuclease digestion patterns revealed that pLMA1 and pLMH5 were indistinguishable plasmids although they were found in different strains isolated from two distant wetlands—Laguna Azul and Laguna Huaca Huasi (Dib et al. 2010a). This points to the dissemination of these elements by horizontal transfer.

Plasmid nucleotide sequences were determined as a prerequisite for the investigation of the plasmids' potential function in the survival and adaptation of their host to the harsh environments (Table 18.1) (Dib et al. 2013a, b). For example, in the annotation of pAP13 (the 89 kb linear plasmid hosted by *Brevibacterium* sp. Ap13), genes involved in the repair of UV-induced DNA damage (*umu*C and *umu*D)

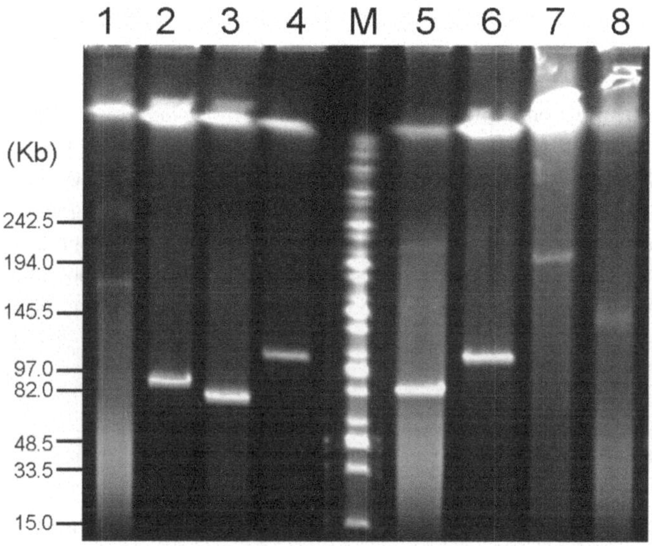

Fig. 18.2 Pulsed-field gel electrophoresis (PFGE) of DNA plugs of actinobacteria isolated from high-altitude Andean wetlands. Lane 1: *Nocardia* sp. A5; lane 2: *Micrococcus* sp. V7; lane 3: *Micrococcus* sp. A7; lane 4: *Micrococcus* sp. A1; lane 5: *Brevibacterium* sp. Ap13; lane 6: *Micrococcus* sp. H5; lane 7: *Rhodococcus* sp. Ch13; lane 8: *Kokuria* sp. N33; lane M: MidRange I PFG Marker (New England BioLabs, Ipswich, MA, USA). The ramping time was 1–25 s for 24 h (6 V/cm at 14 °C). Part of this figure was originally published by Dib et al. (2010a)

Table 18.1 Sequenced linear megaplasmids of actinobacteria isolated from high-altitude Andean lakes

Plasmid	Strain	Isolation source of strain	Plasmid size (kb)	GenBank accession number of plasmid	References
pLMA1	*Micrococcus* sp. A1	Laguna Azul	109	LK056645.1	Dib et al. (2013b, 2018); Wagenknecht et al. (2010)
pLMH5	*Micrococcus* sp. H5	Laguna Huaca Huasi	109	Sequence not determined	Dib et al. (2010b)
pLMV7	*Micrococcus* sp. V7	Laguna Vilama	92	KF577591	Dib et al. (2010b, 2013b)
pLMA7	*Micrococcus* sp. A7	Laguna Azul	82	KJ599675.1	Dib et al. (2018)
pJD12	*Micrococcus* sp. D12	Laguna Diamante	75	KR152226.1	Dib et al. (2015)
pAP13	*Brevibacterium* sp. Ap13	Laguna Aparejos	89	KF577590	Dib et al. (2010b, 2013b)

were found, potentially enhancing cell survival in highly irradiated environments (Dib et al. 2013a). Moreover, a gene cluster related to amino acids and fatty acid metabolism was identified (Dib et al. 2013a).

In addition, the linear plasmid pJD12 from *Micrococcus* sp. D12, which was isolated from the high-altitude volcanic Laguna Diamante in the northwest of Argentina, was completely sequenced and annotated (Dib et al. 2015). The element encodes plasmid-typical genes, including those required for conjugation and replication. Interestingly, it also contains genetic information encoding a glutaredoxin and a putative cobalt–zinc cadmium efflux system. These attributes are potentially involved in coping with oxidative stress and heavy-metal poisoning, and may be favorable for the host's survival and growth in the hostile environment (Dib et al. 2015).

18.7 Plasmidome

Of note, most of the microbial population of these extreme environments is considered noncultivable under laboratory conditions. This fact obviously limits the study of plasmids (linear and circular) because it restricts the available microbiological methods. Even in recent metagenomic studies (Rascovan et al. 2016; Kurth et al. 2017), the impact of plasmids, particularly linear replicons, has been underestimated because it is difficult to discriminate between chromosomal DNA and plasmid DNA, especially when the latter occurs at a low copy number and lower concentrations (Dib et al. 2015). A metagenomic study focusing on mobile extrachromosomal elements and using protocols suitable for the extraction of (linear and circular) plasmids would provide an insight into the plasmid population as a whole and allow better understanding of their ecological impact in these extreme microbial niches.

18.8 Conclusions

The important role of plasmids in the dissemination of genes for antibiotic resistance (Taylor et al. 2004), for metal resistance (Ravel et al. 1998), as determinants of virulence (Stinear et al. 2005), and for production of secondary metabolites and organic compound catabolism has been extensively studied (Top and Springael 2003). However, most of the plasmids that have been studied reside within a few restricted bacterial clades; nearly 70% of the sequences in the Plasmid Genome Database (http://www.genomics.ceh.ac.uk/plasmiddb/) originate from Proteobacteria and Firmicutes.

Among actinobacteria, megaplasmids have been of particular interest and have been described as occurring in various species, such as *Streptomyces* (Ravel et al. 1998, 2000), *Mycobacterium* (Stinear et al. 2005), *Gordonia* (Bröker et al. 2004), *Rhodococcus* (Yang et al. 2007), and *Arthrobacter* (Jerke et al. 2008). However, the knowledge concerning the occurrence and diversity of megaplasmids in actinobacteria from extreme environments, including encoded traits potentially required to tolerate the extreme conditions in such habitats, is rather limited.

Linear megaplasmids, ranging in size between approximately 75 and 245 kb, have been found to be widespread among the actinobacterial strains isolated from HAALs. This suggests that horizontal gene transfer (HGT) is an important driving force for the adaptation of actinobacteria to these extreme and hostile habitats.

The hostile conditions present in the HAALs may act as selection pressure toward gene mutations and HGT, thus driving gene dissemination and forcing widespread phenotypic properties in the population, such as antibiotic and metal resistance (Dib et al. 2008, 2009; Ordoñez et al. 2009). Thereby, in the particular environment of HAALs, HGT may be expected to occur frequently, as it provides the indigenous microbial community with a wide and beneficial gene pool that increases resilience to resist the extreme environmental conditions (Gogarten and Townsend 2005).

Acknowledgements This work was partially supported by the Agencia Nacional de Promoción Científica y Tecnológica (Proyectos de Investigación Científica y Tecnológica (PICT) number 2014-1716) and by the Alexander von Humboldt Foundation.

References

Albarracín VH, Gärtner W, Farías ME (2016) Forged under the Sun: life and art of extremophiles from Andean lakes. Photochem Photobiol 92(1):14–28

Albarracín VH, Kurth D, Ordoñez OF, Belfiore C, Luccini E, Salum GM, Piacentini RD, Farías ME (2015) High-up: a remote reservoir of microbial extremophiles in central Andean wetlands. Front Microbiol 6:1404

Anderson DL, Hickman DD, Reilly BE (1966) Structure of *Bacillus subtilis* bacteriophage φ29 and the length of φ29 deoxyribonucleic acid. J Bacteriol 91(5):2081–2089

Azman AS, Othman I, Velu S, Chan KG, Lee LH (2015) Mangrove rare actinobacteria: taxonomy, natural compound, and discovery of bioactivity. Front Microbiol 6:856

Baker S, Hardy J, Sanderson KE, Quail M, Goodhead I, Kingsley RA, Dougan G (2007) A novel linear plasmid mediates flagellar variation in *Salmonella typhi*. PLoS Pathog 3(5):e59

Barbour AG, Garon CF (1987) Linear plasmids of the bacterium *Borrelia burgdorferi* have covalently closed ends. Science 237(4813):409–411

Belluscio A (2009) A high window in the past. Nat Dig 6:34–36

Belluscio A (2010) Hostile volcanic lake teems with life. Nat News. https://doi.org/10.1038/news.2010.161

Bentley SD, Brown S, Murphy LD, Harris DE, Quail MA, Parkhill J, Yamasaki M (2004) SCP1, a 356023 bp linear plasmid adapted to the ecology and developmental biology of its host, *Streptomyces coelicolor* A3(2). Mol Microbiol 51(6):1615–1628

Bérdy J (2005) Bioactive microbial metabolites. J Antibiot 58(1):1–26

Blaisonneau J, Nosek J, Fukuhara H (1999) Linear DNA plasmid pPK2 of *Pichia kluyveri*: distinction between cytoplasmic and mitochondrial linear plasmids in yeasts. Yeast 15(9):781–791

Bröker D, Arenskötter M, Legatzki A, Nies DH, Steinbüchel A (2004) Characterization of the 101-kilobase-pair megaplasmid pKB1, isolated from the rubber-degrading bacterium *Gordonia westfalica* Kb1. J Bacteriol 186(1):212–225

Burgdorfer W, Barbour AG, Hayes SF, Benach JL, Grunwaldt E, Davis JP (1982) Lyme disease—a tick-borne spirochetosis? Science 216(4552):1317–1319

Chater KF, Kinashi H (2007) *Streptomyces* linear plasmids: their discovery, functions, interactions with other replicons, and evolutinonary significance. In: Meinhardt F, Klassen R (eds) Microbial linear plasmids. Springer, Berlin, pp 1–32

Chen CW (2007) *Streptomyces* linear plasmids: replication and telomeres. In: Meinhardt F, Klassen R (eds) Microbial linear plasmids. Springer, Berlin, pp 33–61

Crossman LC (2005) Plasmid replicons of *Rhizobium*. Biochem Soc Trans 33:157–158

Das S, Ward LR, Burke C (2008) Prospects of using marine actinobacteria as probiotics in aquaculture. Appl Microbiol Biotechnol 81(3):419–429

Demain AL (1999) Pharmaceutically active secondary metabolites of microorganisms. Appl Microbiol Biotechnol 52(4):455–463

Demergasso C, Casamayor EO, Chong G, Galleguillos P, Escudero L, Pedrós-Alió C (2004) Distribution of prokaryotic genetic diversity in athalassohaline lakes of the Atacama Desert, Northern Chile. FEMS Microbiol Ecol 48(1):57–69

Dib JR, Angelov A, Liebl W, Döbber J, Voget S, Schuldes J, Daniel R (2015) Complete genome sequence of the linear plasmid pJD12 hosted by *Micrococcus* sp. D12, isolated from a high-altitude volcanic lake in Argentina. Genome Announc 3(3): e00627-15

Dib JR, Liebl W, Wagenknecht M, Farías ME, Meinhardt F (2013) Extrachromosomal genetic elements in *Micrococcus*. Appl Microbiol Biotechnol 97(1):63–75

Dib JR, Motok J, Fernandez Zenoff V, Ordoñez O, Farías ME (2008) Occurrence of resistance to antibiotics, UV-B and arsenic in bacteria isolated from extreme environments in high altitude (above 4400 m) Andean wetlands. Curr Microbiol 56(5):510–517

Dib JR, Perez MF, Schuldes J, Poehlein A, Wagenknecht M, Farías ME, Daniel R (2018) Complete genome sequences of pLMA1 and pLMA7, two large linear plasmids of *Micrococcus* strains isolated from a high-altitude lake in Argentina. Genome Announc 6(8): e00010-18

Dib JR, Schuldes J, Thürmer A, Farias ME, Daniel R, Meinhardt F (2013a) Complete genome sequence of pAP13, a large linear plasmid of a *Brevibacterium* strain isolated from a saline lake at 4,200 meters above sea level in Argentina. Genome Announc 1(6): e00878-13

Dib JR, Schuldes J, Thürmer A, Farias ME, Daniel R, Meinhardt F (2013b) First complete sequence of a giant linear plasmid from a *Micrococcus* strain isolated from an extremely high-altitude lake. Genome Announc 1(6): e00885-13

Dib JR, Wagenknecht M, Farías ME, Meinhardt F (2015) Strategies and approaches in plasmidome studies—uncovering plasmid diversity disregarding of linear elements? Front Microbiol 6:463

Dib JR, Wagenknecht M, Hill RT, Farías ME, Meinhardt F (2010a) First report of linear megaplasmids in the genus *Micrococcus*. Plasmid 63(1):40–45

Dib JR, Wagenknecht M, Hill RT, Farías ME, Meinhardt F (2010b) Novel linear megaplasmid from *Brevibacterium* sp. isolated from extreme environment. J Basic Microbiol 50(3):280–284

Dib JR, Weiss A, Neumann A, Ordoñez O, Estévez MC, Farías ME (2009) Isolation of bacteria from remote high altitude Andean wetlands able to grow in the presence of antibiotics. Recent Pat Antiinfect Drug Discov 4(1):66–76

Dworkin MS, Schwan TG, Anderson DE (2002) Tick-borne relapsing fever in North America. Med Clin 86(2):417–433

Embley TM, Stackebrandt E (1994) The molecular phylogency and systematics of the actinomycetes. Annu Rev Microbiol 48(1):257–289

Farías ME, Rascovan N, Toneatti DM, Albarracín VH, Flores MR, Poiré DG, Collavino MM, Aguilar OM, Vazquez MP, Polerecky L (2013) The discovery of stromatolites developing at 3570 m above sea level in a high-altitude volcanic lake Socompa, Argentinean Andes. PLoS One 8(1):e53497

Fetzner S, Kolkenbrock S, Parschat K (2007) Catabolic linear plasmids. In: Meinhardt F, Klassen R (eds) Microbial linear plasmids. Springer, Berlin, pp 63–98

Flores MR, Ordoñez OF, Maldonado MJ, Farías ME (2009) Isolation of UV-B resistant bacteria from two high altitude Andean lakes (4,400 m) with saline and non saline conditions. J Gen Appl Microbiol 55(6):447–458

Francis I, Gevers D, Karimi M, Holsters M, Vereecke D (2007) Linear plasmids and phytopathogenicity. In: Meinhardt F, Klassen R (eds) Microbial linear plasmids. Springer, Berlin, pp 99–116

Frost LS, Leplae R, Summers AO, Toussaint A (2005) Mobile genetic elements: the agents of open source evolution. Nat Rev Microbiol 3:722–732

Gil R, Sabater-Muñoz B, Perez-Brocal V, Silva FJ, Latorre A (2006) Plasmids in the aphid endosymbiont *Buchnera aphidicola* with the smallest genomes: a puzzling evolutionary story. Gene 370:17–25

Gogarten JP, Townsend JP (2005) Horizontal gene transfer, genome innovation and evolution. Nat Rev Microbiol 3(9):679

Griffiths AJ (1995) Natural plasmids of filamentous fungi. Microbiol Rev 59(4):673–685

Halary S, Leigh JW, Cheaib B, Lopez P, Bapteste E (2010) Network analyses structure genetic diversity in independent genetic worlds. Proc Natl Acad Sci U S A 107(1):127–132

Hayakawa T, Otake N, Yonehara H, Tanaka T, Sakaguchi K (1979) Isolation and characterization of plasmids from *Streptomyces*. J Antibiot 32(12):1348–1350

Hertwig S (2007) Linear plasmids and prophages in Gram-negative bacteria. In: Meinhardt F, Klassen R (eds) Microbial linear plasmids. Berlin: Springer, pp 141–162

Heuer H, Abdo Z, Smalla K (2008) Patchy distribution of flexible genetic elements in bacterial populations mediates robustness to environmental uncertainty. FEMS Microbiol Ecol 65:361–371

Hinnebusch J, Barbour AG (1991) Linear plasmids of *Borrelia burgdorferi* have a telomeric structure and sequence similar to those of a eukaryotic virus. J Bacteriol 173(22):7233–7239

Hinnebusch J, Bergstrom S, Barbour AG (1990) Cloning and sequence analysis of linear plasmid telomeres of the bacterium *Borrelia burgdorferi*. Mol Microbiol 4(5):811–818

Hinnebusch J, Tilly K (1993) Linear plasmids and chromosomes in bacteria. Mol Microbiol 10(5):917–922

Jerke K, Nakatsu CH, Beasley F, Konopka A (2008) Comparative analysis of eight *Arthrobacter* plasmids. Plasmid 59(2):73–85

Kalkus J, Reh M, Schlegel HG (1990) Hydrogen autotrophy of *Nocardia* opaca strains is encoded by linear megaplasmids. Microbiology 136(6):1145–1151

Kav AB, Sasson G, Jami E, Doron-Faigenboim A, Benhar I, Mizrahi I (2012) Insights into the bovine rumen plasmidome. Proc Natl Acad Sci U S A 109(14):5452–5457

Keen CL, Mendelovitz S, Cohen G, Aharonowitz Y, Roy KL (1988) Isolation and characterization of a linear DNA plasmid from *Streptomyces clavuligerus*. Mol Gen Genet 212(1):172–176

Kikuchi Y, Hirai K, Gunge N, Hishinuma F (1985) Hairpin plasmid—a novel linear DNA of perfect hairpin structure. EMBO J 4(7):1881–1886

Kinashi H (2011) Giant linear plasmids in *Streptomyces*: a treasure trove of antibiotic biosynthetic clusters. J Antibiot 64:19–25

Klassen R, Meinhardt F (2007) Linear protein-primed replicating plasmids in eukaryotic microbes. In: Meinhardt F, Klassen R (eds) Microbial linear plasmids. Berlin: Springer, pp 187–226

Koonin EV, Wolf YI (2008) Genomics of bacteria and archaea: the emerging dynamic view of the prokaryotic world. Nucleic Acids Res 36(21):6688–6719

Kurth D, Amadio A, Ordoñez OF, Albarracín VH, Gärtner W, Farías ME (2017) Arsenic metabolism in high altitude modern stromatolites revealed by metagenomic analysis. Sci Rep 7(1):1024

McLeod MP, Warren RL, Hsiao WW, Araki N, Myhre M, Fernandes C, Dosanjh M (2006) The complete genome of *Rhodococcus* sp. RHA1 provides insights into a catabolic powerhouse. Proc Natl Acad Sci U S A 103(42):15582–15587

Medema MH, Trefzer A, Kovalchuk A, van den Berg M, Müller U, Heijne W, Bovenberg RA (2010) The sequence of a 1.8-Mb bacterial linear plasmid reveals a rich evolutionary reservoir of secondary metabolic pathways. Genome Biol Evol 2:212–224

Meinhardt F, Kempken F, Kämper J, Esser K (1990) Linear plasmids among eukaryotes: fundamentals and application. Curr Genet 17(2):89–95

Meinhardt F, Klassen R (2007) Microbial linear plasmids, vol 7. Springer, Berlin

Meinhardt F, Schaffrath R, Larsen M (1997) Microbial linear plasmids. Appl Microbiol Biotechnol 47:329–336

Miller SC, Porcella SF, Raffel SJ, Schwan TG, Barbour AG (2013) Large linear plasmids of *Borrelia* species that cause relapsing fever. J Bacteriol 195(16):3629–3639

Novick RP (2003) Mobile genetic elements and bacterial toxinoses: the superantigen-encoding pathogenicity islands of *Staphylococcus aureus*. Plasmid 49:93–105

Ordoñez OF, Flores MR, Dib JR, Paz A, Farías ME (2009) Extremophile culture collection from Andean lakes: extreme pristine environments that host a wide diversity of microorganisms with tolerance to UV radiation. Microb Ecol 58(3):461–473

Overhage J, Sielker S, Homburg S, Parschat K, Fetzner S (2005) Identification of large linear plasmids in *Arthrobacter* spp. encoding the degradation of quinaldine to anthranilate. Microbiology 151(2):491–500

Picardeau M, Vincent V (1997) Characterization of large linear plasmids in mycobacteria. J Bacteriol 179(8):2753–2756

Pring DR, Levings CR, Hu WWL, Timothy DH (1977) Unique DNA associated with mitochondria in the "S"-type cytoplasm of male-sterile maize. Proc Natl Acad Sci U S A 74(7):2904–2908

Rascovan N, Maldonado J, Vazquez MP, Farías ME (2016) Metagenomic study of red biofilms from Diamante Lake reveals ancient arsenic bioenergetics in haloarchaea. ISME J 10(2): 299–309

Rasuk MC, Kurth D, Flores MR, Contreras M, Novoa F, Poire D, Farias ME (2014) Microbial characterization of a gypsum endoevaporitic ecosystem in Salar de Llamara Chile. Microb Ecol 68(3):483–494

Rasuk MC, Kurth D, Flores MR, Contreras M, Novoa F, Poire D, Farias ME (2015) Bacterial diversity in microbial mats and sediments from Atacama Desert. Microb Ecol 71(1):44–56

Ravel J, Schrempf H, Hill RT (1998) Mercury resistance is encoded by transferable giant linear plasmids in two Chesapeake Bay *Streptomyces* strains. Appl Environ Microbiol 64:3383–3388

Ravel J, Wellington EM, Hill RT (2000) Interspecific transfer of *Streptomyces* giant linear plasmids in sterile amended soil microcosms. Appl Environ Microbiol 66(2):529–534

Saint Girons I, Old IG, Davidson BE (1994) Molecular biology of the *Borrelia*, bacteria with linear replicons. Microbiology 140(8):1803–1816

Sakaguchi K (1990) Invertrons, a class of structurally and functionally related genetic elements that includes linear DNA plasmids, transposable elements, and genomes of adeno-type viruses. Microbiol Rev 54(1):66–74

Schrempf H (2001) Recognition and degradation of chitin by streptomycetes. Antonie Van Leeuwenhoek 79(3–4):285–289

Sekine M, Tanikawa S, Omata S, Saito M, Fujisawa T, Tsukatani N, Nishiko R (2006) Sequence analysis of three plasmids harboured in *Rhodococcus erythropolis* strain PR4. Environ Microbiol 8(2):334–346

Shimizu S, Kobayashi H, Masai E, Fukuda M (2001) Characterization of the 450-kb linear plasmid in a polychlorinated biphenyl degrader, *Rhodococcus* sp. strain RHA1. Appl Environ Microbiol 67(5):2021–2028

Stecker C, Johann A, Herzberg C, Averhoff B, Gottschalk G (2003) Complete nucleotide sequence and genetic organization of the 210-kilobase linear plasmid of *Rhodococcus erythropolis* BD2. J Bacteriol 185(17):5269–5274

Stinear TP, Hong H, Frigui W, Pryor MJ, Brosch R, Garnier T, Cole ST (2005) Common evolutionary origin for the unstable virulence plasmid pMUM found in geographically diverse strains of *Mycobacterium ulcerans*. J Bacteriol 187(5):1668–1676

Stinear TP, Pryor MJ, Porter JL, Cole ST (2005) Functional analysis and annotation of the virulence plasmid pMUM001 from *Mycobacterium ulcerans*. Microbiology 151(3):683–692

Takarada H, Sekine M, Hosoyama A (2009) *Rhodococcus opacus* B4, complete genome. NCBI reference sequence: NC_012522.1. National Center for Biotechnology Information. https://www.ncbi.nlm.nih.gov/nuccore/NC_012522. Accessed 25 Aug, 2019

Taylor DE, Gibreel A, Tracz DM, Lawley TD (2004) Antibiotic resistance plasmids. In: Plasmid biology. American Society of Microbiology Press, Washington, DC, pp 473–492

Top EM, Springael D (2003) The role of mobile genetic elements in bacterial adaptation to xenobiotic organic compounds. Curr Opin Biotechnol 14(3):262–269

Turmel M, Bellemare G, Lee RW, Lemieux C (1986) A linear DNA molecule of 5.9 kilobase-pairs is highly homologous to the chloroplast DNA in the green alga *Chlamydomonas moewusii*. Plant Mol Biol 6(5):313–319

Van der Meer JR, Sentchilo V (2003) Genomic islands and the evolution of catabolic pathways in bacteria. Curr Opin Biotechnol 14:248–254

Ventura M, Canchaya C, Tauch A, Chandra G, Fitzgerald GF, Chater KF, van Sinderen D (2007) Genomics of actinobacteria: tracing the evolutionary history of an ancient phylum. Microbiol Mol Biol Rev 71(3):495–548

Wagenknecht M, Dib JR, Thürmer A, Daniel R, Farías ME, Meinhardt F (2010) Structural peculiarities of linear megaplasmid, pLMA1, from *Micrococcus luteus* interfere with pyrosequencing reads assembly. Biotechnol Lett 32:1853–1862

Wagenknecht M, Meinhardt F (2011a) Copy number determination, expression analysis of genes potentially involved in replication, and stability assays of pAL1—the linear megaplasmid of *Arthrobacter nitroguajacolicus* Rü61a. Microbiol Res 166:14–26

Wagenknecht M, Meinhardt F (2011b) Replication-involved genes of pAL1, the linear plasmid of *Arthrobacter nitroguajacolicus* Rü61a—phylogenetic and transcriptional analysis. Plasmid 65:176–184

Warren JM, Simmons MP, Wu Z, Sloan DB (2016) Linear plasmids and the rate of sequence evolution in plant mitochondrial genomes. Genome Biol Evol 8(2):364–374

Yang JC, Lessard PA, Sengupta N, Windsor SD, O'Brien XM, Bramucci M, Sinskey AJ (2007) TraA is required for megaplasmid conjugation in *Rhodococcus erythropolis* AN12. Plasmid 57(1):55–70

Chapter 19
Arsenic and Its Biological Role: From Early Earth to Current Andean Microbial Ecosystems

Luis Alberto Saona Acuña, Mariana N. Soria, Patricio G. Villafañe, Tatiana Stepanenko, and María Eugenia Farías

19.1 Introduction

Arsenic (As) is present in the Earth's crust and is widely distributed in the environment. It is frequently a component of sulfidic ores in the form of arsenides of nickel, cobalt, copper, and iron (Mandal and Suzuki 2002). The main sources of As are natural, mostly associated with volcanic areas and hydrothermal vents (Kurth et al. 2017); further, it can also originate as a result of human activities: mining, waste treatment, and industrial activities, among others (Mandal and Suzuki 2002).

Arsenic is a redox-active metalloid and exists in nature in four oxidation states: arsine [As(-III)], elemental [As(0)], arsenate [As(V)], and arsenite [As(III)]. These states vary according to changes in pH and the redox environment (Oremland et al. 2009; Oremland and Stolz 2003). The first two forms are relatively rare; naturally,

L. A. Saona Acuña · M. N. Soria · T. Stepanenko
Laboratorio de Investigaciones Microbiológicas de Lagunas Andinas (LIMLA), Planta Piloto de Procesos Industriales Microbiológicos (PROIMI), Centros Científico Tecnológicos (CCT), Consejo Nacional de Investigaciones Científicas y Técnicas (CONICET), San Miguel de Tucumán, Tucumán, Argentina

P. G. Villafañe
Laboratorio de Investigaciones Microbiológicas de Lagunas Andinas (LIMLA), Planta Piloto de Procesos Industriales Microbiológicos (PROIMI), Centros Científico Tecnológicos (CCT), Consejo Nacional de Investigaciones Científicas y Técnicas (CONICET), San Miguel de Tucumán, Tucumán, Argentina

Instituto Superior de Correlación Geológica (INSUGEO), Universidad Nacional de Tucumán–Consejo Nacional de Investigaciones Científicas y Técnicas (CONICET), Yerba Buena, Tucumán, Argentina

M. E. Farías (✉)
LIMLA-PROIMI-CONICET, Tucuman, Argentina

M. E. Farías (ed.), *Microbial Ecosystems in Central Andes Extreme Environments*, https://doi.org/10.1007/978-3-030-36192-1_19

As is found as As(V) or As(III). As(V) is predominant in oxygenated aqueous environments, while As(III) is found in reduced or anoxic conditions.

Arsenic is a highly dangerous element, as As(III) is 100 times more toxic than As(V) (Andres and Bertin 2016; Rascovan et al. 2016). Its greatest toxicity is due to the fact that it can bind to sulfhydryl groups, affecting the correct functioning of many enzymes and proteins (Belfiore et al. 2013; Kurth et al. 2017; Ordoñez et al. 2015, 2018). As(V), on the other hand, is a chemical analog of phosphate, so it can interact and eventually replace it in early steps in different ways (Tawfik and Viola 2011; Wolfe-Simon et al. 2008). On our planet, there are environments with a high arsenic content.

Our research group has worked in bioprospecting in Andean microbial ecosystems (AMEs) in the Atacama Desert, Bolivian Altiplano, and Argentine Puna (the so-called Puna–High Andes region). In all of these places, there are hypersaline lakes at altitudes higher than 3000 m above sea level (asl) (Albarracín et al. 2015; Farías et al. 2013; Fernandez et al. 2016; Rasuk et al. 2016). These lakes share the common characteristic of high concentrations of arsenic, normally ranging between 12 and 230 mg L^{-1} (Farías et al. 2013; Rascovan et al. 2016). This range of As concentrations is one of the highest ranges reported for hypersaline lakes (Table 19.1).

Despite the high toxicity of As, microorganisms throughout evolution have acquired specific genes that allow them to subsist in environments with high concentrations of As. These microorganisms are known as extremophiles and can be found in high-altitude Andean lakes (HAALs). They cope with arsenic toxicity in different ways: extracellular precipitation, chelation, intracellular sequestration, active extrusion from the cell, or biochemical transformation (by redox processes or methylation). On the other hand, some microorganisms utilize this metalloid as a metabolic energy source (Ordoñez et al. 2018; Rascovan et al. 2016) or they also can involve it in their biogeochemical cycles; for example, microbial mats in Laguna La Brava incorporate this metalloid into exopolysaccharides (EPSs) (Sancho-Tomás et al. 2018).

Arsenic has attracted worldwide attention because of its environmental impact and concerns about its effects on human health. High concentrations of As in

Table 19.1 The main physicochemical parameters of hypersaline lake water (Kurth et al. 2017)

	Ion concentration (mg L^{-1})					
Ions	Laguna Diamante	Socompa Lake	Tebenquiche Lake	Laguna La Brava	Mono Lake	Searles Lake
Arsenic	347[a]	18.5	4.0	15.0	10.0	290.0
Phosphate	25–700	25.0	1.0	4.8	60.0	1, 9000
Sulfate	10,000	31,800	14,671	8, 294	8, 300	91,000
Nitrate	5.0	60.0	1, 100	1, 416	10.0	NA
Sodium	43,800–53,300	37,100	47,110	47,630	28,000	223,800
Magnesium	20.0	4, 090	3, 109	4, 344	40.0	NA

NA not available

[a]Arsenic concentration at the most recent measurement of Laguna Diamante water in 2017. Previous works reported arsenic concentrations of 120–230 mg L^{-1} (Rascovan et al. 2016)

water and soil are the subject of study by many research groups. These studies are grouped into three different categories: (1) the biogeochemical cycle (Visscher et al. 2014), (2) arsenic contamination and bioremediation (Demergasso et al. 2007), and (3) the biological role of arsenic (Demergasso et al. 2007; Rascovan et al. 2016).

In this chapter, we explore the origin of arsenic, discussing mainly its source in the primitive Earth, and the role it could have had in biology before evolution of the first cell, and where arsenic was distributed. A hypothesis about the role of arsenic in the first steps of life is presented, as are its abundance in HAALs and its biological role in the microbial metabolism of AMEs.

19.2 Arsenic in the Origin of Life

19.2.1 Primitive Earth Characteristics

The Archean eon was the time between the appearance of the first material geological evidence (formation of the first rocks) 4 billion years ago and the start of the Proterozoic eon 2.5 billion years ago. During the Archean eon, there were unique and unrepeatable phenomena in the history of our planet, such as the formation of the first crust from an ocean of magma, the formation of the seas and the atmosphere, and the origin of life, among others (Benedetto 2010).

Knowledge of the physicochemical composition of the primitive Earth is essential for astrobiologists. On the basis of these parameters we can propose new hypotheses about the origin and evolution of life. Today there are many chemical elements that play a role in the cellular metabolism of existing organisms; however, there are only six fundamental ones: C, O, N, H, S, and P (Karl 2000). These elements are an essential part of the entire structure of life: DNA, RNA, membranes, proteins, organelles, and endless vital chemical reactions (Kornberg et al. 1999). The big question we still ask ourselves is whether, 3.4 billion years ago, these elements were equally important for the first cells or if they were fed and breathed through alternative routes according to the existing conditions.

The primitive Earth 3.4 billion years ago did not have the same physicochemical composition as today. The atmosphere of that time was formed gradually by the degasification of the mantle and the gases contributed by the great meteorite bombardments. Although its original composition is still a matter of speculation, there is a consensus on it. In the atmospheric composition, gases such as H_2O, CO, CO_2, H_2, HCN, NH_3, and CH_4 initially predominated; later, volcanic activity would have contributed gases such as N_2, CO_2, and H_2O (Holland 1994; Pla-García and Menor-Salván 1990). The first organisms to inhabit the Earth were prokaryotic microorganisms that carried out anoxygenic photosynthesis, which is characterized by not releasing dioxygen as a waste product (Kulp et al. 2008; Shih

2015). Later, Cyanobacteria evolved and developed oxygenic photosynthesis with which they began to release oxygen into the atmosphere (Holland 1997; Kasting 2001; Summons et al. 1999). In this way, it was not until 2.5 billion years ago that the level of oxygen in the atmosphere began to increase in an event known as the great oxygenation event (GOE) (Shih 2015). This event allowed anaerobic life to begin to breathe oxygen and evolve, giving way to the great diversity of aerobic organisms that inhabit the Earth today.

Like oxygen, phosphorus was not present in the same proportions nor in the chemical form it exists in today. Phosphorus is one of the scarcest essential elements for life at present and has one of the slowest cycles. It is even estimated that phosphorus was not available for the metabolic activity of life on the primitive Earth (Goldford et al. 2017). This generates questions about why, when, and how life began to use phosphorus in all of the pathways and biomolecules in which it is now used. The main sources of phosphorus 3.4 billion years ago were minerals, which would require enzymatic activity to be developed for it to be possible to metabolize it, and the presence of phosphorylated biomolecules in the early oceans was not favored thermodynamically, so it is unlikely that this element was an essential component of primitive cells (Pasek 2019). By means of computer simulations, the possible reactions that could have been present on the primitive Earth have been analyzed, and it was concluded that life could have originated without the need for phosphate (the pentavalent oxidation state of phosphorus) (Goldford et al. 2017).

In this metabolic context—historically speaking—the role of other elements that could have been key for the development of life when oxygen and phosphate (among others) were not present has been widely studied.

Of the elements that have been studied, arsenic is the focus of this chapter. It is a natural constituent of the Earth's crust, forming part of the chemical structure of more than 200 minerals. It is considered a chalcophile element because of its tendency to concentrate in sulfides (Bundschuh et al. 2008). It is assumed that the main sources of arsenic on the Earth were meteorites, which, in their base composition, contain 0.1% arsenic (Hamaguchi et al. 1969; Olsen et al. 1999; Oremland et al. 2009). During the Archean eon, the highest concentrations of As were found in rock bodies called "green rock belts" (metamorphized into green amphibolite shale facies with intrusive granite bodies) because they contained important and massive sulfide deposits, which concentrated the As (Benedetto 2010; Bundschuh et al. 2008; Lahaye et al. 2001).

The water of the early Earth—where life originated—came from different sources: from magmatic degasification and from inside meteorites that hit during the great bombardment and formed parts of comets. The Archean ocean was more acidic, and its main sources of minerals were submarine vents, which contributed elements such as Na, K, Ni, and Fe. The contribution of As to the hydrosphere would have been produced mainly by the dissolution of mineral phases (Smedley and Kinniburgh 2002).

19.2.2 Arsenic in the First Steps of Life

The reducing environment of the primitive Earth, with an absence of both oxygen and bioavailable phosphorus, presented favorable conditions for metabolic reactions and for establishment and evolution of protocells. In this context, could As have been a key element in the evolution of life more than 3.4 billion years ago? It may seem unlikely that an element with the toxic characteristics of As would have been important at the origin of life; however, various antecedents would say it was.

Today we know that the ability of microbes to metabolize As may have arisen more than 3.4 billion years ago. Some of the modern environments in which arsenic metabolism occurs are prominently anoxic like the Precambrian oceans. In an article published in 2014 in *Nature Geoscience*, it was demonstrated, by means of Raman spectroscopy and x-ray fluorescence, that arsenic is associated with precipitation of carbonates by biological activity in fossil stromatolites (Sforna et al. 2014). The authors postulated that (1) these stromatolites were formed from photosynthetic microbial mats that grew in an anoxic lake saturated with salt, (2) the metabolism of arsenic was centered on redox reactions in which the oxidation of As(III) occurred by anoxigenic photosynthesis and/or by denitrification, and (3) the consequent accumulation of As(V) in hypersaline lakes could have created niches in which anaerobic chemoautotrophic or heterotrophic microorganisms could breathe As(V). These results are found not only in fossil stromatolites but also in extant extreme environments such as Laguna Diamante (see Chap. 8 of this book) (Ordoñez et al. 2018; Rascovan et al. 2016) and Laguna La Brava. In Laguna La Brava, ecosystems of microbial mats have been reported where there would be a widely distributed arsenic cycle. Arsenic has been found to be part of the EPS of the microorganisms of the microbial mats and has been found in organic globules in which there is no presence of other metals. In addition, the abundant presence of arsenic and sulfur in the water column, added to the heterogeneous distribution between arsenate and arsenite, suggests the presence of an arsenic–sulfur cycle (Sancho-Tomás et al. 2018). This background raises the possibility that there was an As cycle in the Archean eon, similar to what is observed in extant habitats today (Kulp et al. 2008; Oremland and Stolz 2003).

19.2.3 Bioenergetic Role of Arsenic in Andean Microbial Ecosystems

The waters where the first cells originated were fed by volcanic activity. They presented high salinity due to the constant evaporation of the waters where salt basins were formed, similar to those present on Mars. In addition, there was high ultraviolet radiation—due to the absence of an ozone-rich atmosphere—and a very low oxygen concentration. While it is impossible to find such intact environmental conditions on Earth today, there is the possibility of finding natural laboratories that have conditions as similar as possible to that primitive Earth. The Argentinean Puna is one of these places, since it has hypersaline lakes, evaporitic basins, stromatolites, and microbial mats, with high ultraviolet radiation, low oxygen pressure, and high concentrations of metals and of As, among other features.

In particular, As is a common chemical component for most of the HAALs prospected to date. In these lakes, the concentration of this metalloid can range from 12 to 230 mg L^{-1} (Farías et al. 2013, 2014; Rascovan et al. 2016; Zannier et al. 2019). Several microorganisms have evolved resistance mechanisms to survive and detoxify the cell. These strategies are carried out by genes related to As resistance, usually encoded in the *ars* operon, with genes to express different proteins such as repressor ArsR, an arsenate reductase ArsC, and arsenite efflux pumps ArsB and Acr3 (Páez-Espino et al. 2009) (Fig. 19.1). The *acr3* gene was found and described for the first time in the genome of *Exiguobacterium* sp. S17, isolated from Socompa Lake stromatolites (Ordoñez et al. 2013). This isolated species had high tolerance of arsenite (10 mM), higher than that of *Exiguobacterium* sp. N139, which was isolated from water in Laguna Negra, where the *acr3* gene was not found (Ordoñez et al. 2015). Therefore, Ordoñez et al. (2015) postulated that the Acr3 efflux pump was the main mechanism of the resistance of *Exiguobacterium* sp. S17 to As. Similar findings of Acr3 abundance was reported in Diamante and Socompa lake biofilm and microbialite metagenomes, respectively (Rascovan et al. 2016; Kurth et al. 2017). On the other hand, in water and sediments from Andean salt deposits in Northern Chile, the

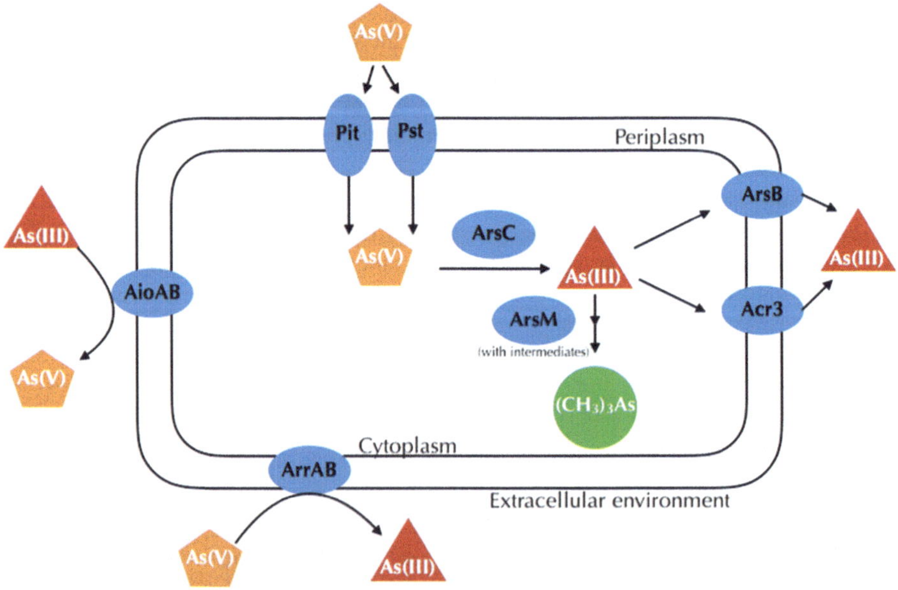

Fig. 19.1 Scheme of the arsenic metabolic cycle. Enzymes responsible for the electrochemical transformation of extracellular arsenic: AioAB (arsenite oxidase) and ArrAB (arsenate respiratory reduction); arsenite efflux pumps: ArsB and Acr3; ArsC (arsenate reductase) and ArsM (arsenite methyltransferase). On the other hand, phosphate transports are also observed through which arsenate enters the cell: Pit (inorganic phosphate transporter) and Pst (phosphate specific transport)

presence of *arsC*, *arrA*, *aioA*, and *acr3* genes was reported (Escudero et al. 2013). That work showed that *arsC* experienced high diversification and that bacterial lineages inherited variants of the gene, resulting in different ArsC families providing different degrees of protection against arsenic (Escudero et al. 2013). In addition, the presence of a likely new kind of *arsC* was shown in a *Fusibacter* sp. 3D3 genome isolated from the Salar de Ascotán in Northern Chile (Serrano et al. 2017).

Not only have most microorganisms developed arsenic resistance mechanisms (e.g., the *ars* operon), but also some of them are able to use it in their metabolism (Páez-Espino et al. 2009). In work published by Rascovan et al. (2016), the presence of the genes *aioA* and *aioB* was reported in Laguna Diamante archeal biofilms located in the crater of the Galán volcano. These genes encode the heterodimeric enzyme arsenite oxidase, which participates in detoxification of As (Fig. 19.1) in heterotrophic bacteria (Muller et al. 2003), as well as generation of energy in chemoheterotrophic bacteria (Vanden Hoven and Santini 2004) and chemolithoautotrophy at the expense of As(III), which is used as a source of electrons (Oremland et al. 2002). The presence of the ArrA enzyme was also found; it is responsible for the process of the dissimilatory reduction of As(V) (Fig. 19.1), which is used as the final acceptor of an electron chain (anaerobic respiration of arsenic). The same was confirmed in work with microorganisms isolated from biofilms in Laguna Diamante, such as *Halorubrum* sp. DM2, which was shown to grow better in the presence of As (Ordoñez et al. 2018). The evolution of these enzymes, as well as a more complete description of arsenic metabolism in Laguna Diamante, are discussed in Chap. 9 of this book.

Microbial metabolism mediated by As was also described by Kurth et al. (2017). Stromatolite microbial communities in Socompa Lake were analyzed by metagenomic DNA sequencing and compared with the microbial communities present in Shark Bay stromatolites (Kurth et al. 2017). The Socompa stromatolites are inhabited mainly by Proteobacteria, Bacteroidetes, and Firmicutes, and they showed a complete set of arsenic metabolism genes, including *arsM*, coding for a methyltransferase that is responsible for As(III) methylation (Fig. 19.1). In addition, *aioA*, *arrA*, and *arxA* genes were found in that metagenomic analysis. This novel *arxA* gene, coding for ArxA oxidases, is related to ArrA proteins, but rather than reducing arsenate, ArxA proteins act as an arsenite oxidase (Kurth et al. 2017). In this sense, the microbial community present in stromatolites in Socompa Lake have a complete arsenic cycle, even using it as an energy source. Besides, through analysis of binning of the assembled contigs, it was demonstrated that arsenic resistance metabolism is a key factor for growth of hypothetical individual strains, since all of them carry resistance genes in their genomes.

Another microbial influence over the As biogeochemical cycle was observed in microbial populations from evaporitic basins at the Salar de Ascotán and Salar de Atacama. With the goal of knowing which ecotypes are involved in the arsenic cycle, the authors used different kinds of samples to prepare culture enrichments. They observed that all enrichments presented the As electrochemical transformation process, as well as high arsenic tolerance levels (Lara et al. 2012). Additionally, bacterial arsenic sulfide precipitation in the biogeochemical cycle was demonstrated (Demergasso et al. 2007). The authors concluded that the microbial populations have a fully developed arsenic cycle, since (1) they could obtain energy from As sources, (2) they could accumulate arsenic, (3) they had the necessary mechanisms to produce electrons as potential sources for arsenic reduction, and (4) they could tolerate high arsenic levels.

It is not only Laguna Diamante (the lake with the highest arsenic concentration (230 mg L^{-1})) that has genes related to arsenic metabolism; they are distributed in most of the HAALs. Knowing the biological role of arsenic in the AMEs of each lake is a fundamental part of our research. Waiting for future findings, we can confirm again that the HAALs represent a unique environment on the Earth, where we can go back to the origin of life and provide knowledge to elucidate one of the great unknowns that still remain in science.

References

Albarracín VH, Kurth D, Ordoñez OF, Belfiore C, Luccini E, Salum GM et al (2015) High-up: a remote reservoir of microbial extremophiles in central Andean Wetlands. Front Microbiol 6:1404. https://doi.org/10.3389/fmicb.2015.01404

Andres J, Bertin PN (2016) The microbial genomics of arsenic. FEMS Microbiol Rev 40:299–322. https://doi.org/10.1093/femsre/fuv050

Belfiore C, Ordoñez OF, Farías ME (2013) Proteomic approach of adaptive response to arsenic stress in *Exiguobacterium* sp. S17, an extremophile strain isolated from a high-altitude Andean lake stromatolite. Extremophiles 17:421–431. https://doi.org/10.1007/s00792-013-0523-y

Benedetto JL (2010) El continente Gondwana a través del tiempo. Una introducción a la geología histórica. Academia Nacional de Ciencias. http://www.scielo.org.ar/scielo.php?script=sci_arttext&pid=S0002-70142010000200010. Accessed 11 Mar 2019

Bundschuh J, Gimene-Forcada E, Guerequiz E, Pérez Carrera A, García ME Mello J et al (2008) Fuentes geogénicas de arsénico y su liberación al medio ambiente. CYTED, Ciencia y Tecnología para el Desarrollo. https://eprints.usq.edu.au/26283/. Accessed 11 Mar 2019

Demergasso CS, Guillermo CD, Lorena EG, Mur JJP, Pedrós-Alió C (2007) Microbial precipitation of arsenic sulfides in Andean salt flats. Geomicrobiol J 24:111–123. https://doi.org/10.1080/01490450701266605

Escudero LV, Casamayor EO, Chong G, Pedrós-Alió C, Demergasso C (2013) Distribution of microbial arsenic reduction, oxidation and extrusion genes along a wide range of environmental arsenic concentrations. PLoS One 8:e78890. https://doi.org/10.1371/journal.pone.0078890

Farías ME, Contreras M, Rasuk MC, Kurth D, Flores MR, Poiré DG et al (2014) Characterization of bacterial diversity associated with microbial mats, gypsum evaporites and carbonate microbialites in thalassic wetlands: Tebenquiche and La Brava, Salar de Atacama, Chile. Extremophiles 18:311–329. https://doi.org/10.1007/s00792-013-0617-6

Farías ME, Rascovan N, Toneatti DM, Albarracín VH, Flores MR, Poiré DG et al (2013) The discovery of stromatolites developing at 3570 m above sea level in a high-altitude volcanic lake Socompa, Argentinean Andes. PLoS One 8:e53497. https://doi.org/10.1371/journal.pone.0053497

Fernandez AB, Rasuk MC, Visscher PT, Contreras M, Novoa F, Poire DG et al (2016) Microbial diversity in sediment ecosystems (evaporites domes, microbial mats, and crusts) of hypersaline Laguna Tebenquiche, Salar de Atacama, Chile. Front Microbiol 7:1284. https://doi.org/10.3389/fmicb.2016.01284

Goldford JE, Hartman H, Smith TF, Segrè D (2017) Remnants of an ancient metabolism without phosphate. Cell 168:1126–1134.e9. https://doi.org/10.1016/j.cell.2017.02.001

Hamaguchi H, Onuma N, Hirao Y, Yokoyama H, Bando S, Furukawa M (1969) The abundances of arsenic, tin and antimony in chondritic meteorites. Geochim Cosmochim Acta 33:507–518. https://doi.org/10.1016/0016-7037(69)90130-6

Holland HD (1994) Early Proterozoic atmospheric change. In: Bengston S (ed) Early life on Earth. Columbia University Press, New York, pp 237–244

Holland HD (1997) Evidence for life on Earth more than 3850 million years ago. Science 275:38–39. https://doi.org/10.1126/SCIENCE.275.5296.38

Karl DM (2000) Aquatic ecology: phosphorus, the staff of life. Nature 406:31–33. https://doi.org/10.1038/35017683.

Kasting JF (2001) Earth history: the rise of atmospheric oxygen. Science 293:819–820. https://doi.org/10.1126/science.1063811

Kornberg A, Rao NN, Ault-Riché D (1999) Inorganic polyphosphate: a molecule of many functions. Annu Rev Biochem 68:89–125. https://doi.org/10.1146/annurev.biochem.68.1.89

Kulp TR, Hoeft SE, Asao M, Madigan MT, Hollibaugh JT, Fisher JC et al (2008) Arsenic(III) fuels anoxygenic photosynthesis in hot spring biofilms from Mono Lake, California. Science. 321:967–970

Kurth D, Amadio A, Ordoñez OF, Albarracín VH, Gärtner W, Farías ME (2017) Arsenic metabolism in high altitude modern stromatolites revealed by metagenomic analysis. Sci Rep 7:1024. https://doi.org/10.1038/s41598-017-00896-0

Lahaye Y, Barnes S-J, Frick LR, Lambert DD (2001) Re-Os isotopic study of komatiitic volcanism and magmatic sulfide formation in the Southern Abitibi Greenstone Belt, Ontario, Canada. https://constellation.uqac.ca/2075/1/Yann_Lahaye.pdf. Accessed 11 Mar 2019

Lara J, Escudero González L, Ferrero M, Chong Díaz G, Pedrós-Alió C, Demergasso C (2012) Enrichment of arsenic transforming and resistant heterotrophic bacteria from sediments of two salt lakes in Northern Chile. Extremophiles 16:523–538. https://doi.org/10.1007/s00792-012-0452-1

Mandal BK, Suzuki KT (2002) Arsenic round the world: a review. Talanta 58:201–235. http://www.ncbi.nlm.nih.gov/pubmed/18968746. Accessed 8 Mar 2019

Muller D, Lievremont D, Simeonova DD, Hubert J-C, Lett M-C, Coppee J-Y et al (2003) Arsenite oxidase aox genes from a metal-resistant—proteobacterium. J Bacteriol 185:135–141. https://doi.org/10.1128/JB.185.1.135-141.2003

Olsen EJ, Kracher A, Davis AM, Steele IM, Hutcheon ID, Bunch TE (1999) The phosphates of IIIAB iron meteorites. Meteorit Planet Sci 34:285–300. https://doi.org/10.1111/j.1945-5100.1999.tb01752.x

Ordoñez OF, Lanzarotti E, Kurth D, Cortez N, Farías ME, Turjanski AG (2015) Genome comparison of two *Exiguobacterium* strains from high altitude Andean lakes with different arsenic resistance: identification and 3D modeling of the Acr3 efflux pump. Front Environ Sci 3:50. https://doi.org/10.3389/fenvs.2015.00050

Ordoñez OF, Lanzarotti E, Kurth D, Gorriti MF, Revale S, Cortez N et al (2013) Draft genome sequence of the polyextremophilic *Exiguobacterium* sp. strain S17, isolated from hyperarsenic lakes in the Argentinian Puna. Genome Announc 1:e00480–e00413. https://doi.org/10.1128/genomeA.00480-13

Ordoñez OF, Rasuk MC, Soria MN, Contreras M, Farías ME (2018) Haloarchaea from the Andean Puna: biological role in the energy metabolism of arsenic. Microb Ecol 76(3):695–705. https://doi.org/10.1007/s00248-018-1159-3

Oremland RS, Hoeft SE, Santini JM, Bano N, Hollibaugh RA, Hollibaugh JT (2002) Anaerobic oxidation of arsenite in Mono Lake water and by a facultative, arsenite-oxidizing chemoautotroph, strain MLHE-1. Appl Environ Microbiol 68:4795–4802. https://doi.org/10.1128/AEM.68.10.4795-4802.2002

Oremland RS, Saltikov CW, Wolfe-Simon F, Stolz JF (2009) Arsenic in the evolution of Earth and extraterrestrial ecosystems. Geomicrobiol. J. 26:522–536. https://doi.org/10.1080/01490450903102525

Oremland RS, Stolz JF (2003) The ecology of arsenic. Science 300:939–944. https://doi.org/10.1126/science.1081903

Páez-Espino D, Tamames J, de Lorenzo V, Cánovas D (2009) Microbial responses to environmental arsenic. Biometals 22:117–130. https://doi.org/10.1007/s10534-008-9195-y

Pasek M (2019) A role for phosphorus redox in emerging and modern biochemistry. Curr Opin Chem Biol 49:53–58. https://doi.org/10.1016/J.CBPA.2018.09.018

Pla-García J, Menor-Salván C (1990) La composición química de la atmósfera primitiva del planeta Tierra. Real Sociedad Española de Química. http://analesdequimica.es/index.php/AnalesQuimica/article/view/940. Accessed 11 Mar 2019

Rascovan N, Maldonado J, Vazquez MP, Eugenia Farías M (2016) Metagenomic study of red biofilms from Diamante Lake reveals ancient arsenic bioenergetics in haloarchaea. ISME J 10:299–309. https://doi.org/10.1038/ismej.2015.109

Rasuk MC, Fernández AB, Kurth D, Contreras M, Novoa F, Poiré D et al (2016) Bacterial diversity in microbial mats and sediments from the Atacama Desert. Microb Ecol 71:44–56. https://doi.org/10.1007/s00248-015-0649-9

Sancho-Tomás M, Somogyi A, Medjoubi K, Bergamaschi A, Visscher PT, Van Driessche AES et al (2018) Distribution, redox state and (bio)geochemical implications of arsenic in present day microbialites of Laguna Brava, Salar de Atacama. Chem Geol 490:13–21. https://doi.org/10.1016/j.chemgeo.2018.04.029

Serrano AE, Escudero LV, Tebes-Cayo C, Acosta M, Encalada O, Fernández-Moroso S et al (2017) First draft genome sequence of a strain from the genus *Fusibacter* isolated from Salar de Ascotán in Northern Chile. Stand Genomic Sci 12:43. https://doi.org/10.1186/s40793-017-0252-4

Sforna MC, Philippot P, Somogyi A, van Zuilen MA, Medjoubi K, Schoepp-Cothenet B et al (2014) Evidence for arsenic metabolism and cycling by microorganisms 2.7 billion years ago. Nat Geosci 7:811–815. https://doi.org/10.1038/ngeo2276

Shih PM (2015) Photosynthesis and early Earth. Curr Biol 25:R855–R859. https://doi.org/10.1016/J.CUB.2015.04.046

Smedley P, Kinniburgh D (2002) A review of the source, behaviour and distribution of arsenic in natural waters. Appl Geochem 17:517–568. https://doi.org/10.1016/S0883-2927(02)00018-5

Summons RE, Jahnke LL, Hope JM, Logan GA (1999) 2-Methylhopanoids as biomarkers for cyanobacterial oxygenic photosynthesis. Nature 400:554–557. https://doi.org/10.1038/23005

Tawfik DS, Viola RE (2011) Arsenate replacing phosphate: alternative life chemistries and ion promiscuity. Biochemistry 50:1128–1134. https://doi.org/10.1021/bi200002a

Vanden Hoven RN, Santini JM (2004) Arsenite oxidation by the heterotroph *Hydrogenophaga* sp. str. NT-14: the arsenite oxidase and its physiological electron acceptor. Biochim Biophys Acta Bioenerg 1656:148–155. https://doi.org/10.1016/J.BBABIO.2004.03.001

Visscher P, Farias ME, Contreras M, Novoa-Cortez F, Rasuk MC, Patterson MM et al. (2014) The role of arsenic cycling in carbonate precipitation in microbialites through time. Paper presented at the 2014 Geological Society of America Annual Meeting, Vancouver, 19–22 Oct 2014

Wolfe-Simon F, Davies PCW, Anbar A (2008) Did nature also choose arsenic? Nat Prec. https://doi.org/10.1038/NPRE.2008.1482.1

Zannier F, Portero LR, Ordoñez OF, Martinez LJ, Farías ME, Albarracin VH (2019) Polyextremophilic bacteria from high altitude Andean lakes: arsenic resistance profiles and biofilm production. Biomed Res Int 2019:1–11. https://doi.org/10.1155/2019/1231975

Index

M. E. Farías (ed.), *Microbial Ecosystems in Central Andes Extreme Environments*, https://doi.org/10.1007/978-3-030-36192-1

The manufacturer's authorised representative in the EU is Springer Nature Customer Service Centre GmbH, Europaplatz 3, 69115 Heidelberg, Germany. If you have any concerns regarding our products, please contact ProductSafety@springernature.com

Printed and bound by CPI Group (UK) Ltd, Croydon, CR0 4YY
05/10/2026
02228124-0009